H. Takeshita   B.K. Siesjö   J.D. Miller (Eds.)

Advances in Brain Resuscitation

With 99 Illustrations

Springer-Verlag
Tokyo Berlin Heidelberg
New York London Paris
Hong Kong Barcelona

Hiroshi Takeshita, M.D.
Emeritus Professor, Yamaguchi University School of Medicine, Department of Anesthesiology and Critical Care Medicine, 1144 Kogushi, Ube, Yamaguchi, 755 Japan

Bo K. Siesjö, M.D.
Director, Laboratory for Experimental Brain Research, Head Research Departments,
Lund Hospital, S-221 85 Lund, Sweden

James Douglas Miller, M.D.
Professor for Surgical Neurology and Chairman, Department of Clinical Neurosciences,
University of Edinburgh, Western General Hospital, Edinburgh EH4 2XU, Scotland

ISBN-13: 978-4-431-68540-1   e-ISBN-13: 978-4-431-68538-8
DOI: 10.1007/978-4-431-68538-8

Library of Congress Cataloging-in-Publication Data
Advances in brain resuscitation / H. Takeshita, B.K. Siesjö, J.D. Miller (eds.). p.   cm. Papers
from the International Symposium on Brain Resuscitation, held in Ube, Yamaguchi, Japan, Oct. 31–
Nov. 2, 1988 and sponsored by Yamaguchi University and the Japanese Ministry of Education.
Includes bibliographical references. Includes index. ISBN-13: 978-4-431-68540-1
                 1. Brain resuscitation—Congresses.   2. Cerebral ischemia—Congresses.
I. Takeshita, Hiroshi, 1928– . II. Siesjö, Bo K.   III. Miller, J. Douglas (James Douglas),
1937– . IV. Yamaguchi Daigaku.   V. Japan. Monbushō.   VI. International Symposium on Brain
Resuscitation (1988: Ube-shi, Japan). [DNLM: 1. Brain—physiopathology—congresses. 2.
Cerebral   Anoxia—complications—congresses.   3.   Cerebral   Ischemia—complications—
congresses. 4. Resuscitation—congresses. WL 354 A2445] RC388.5.A346 1991. 616.8—dc20
DNLM/DLC for Library of Congress. 91-4600

Typesetting: Asco Trade Typesetting Ltd., Hong Kong

Preface

Brain resuscitation is the therapeutic intervention for critically ill patients with severe brain damage, particularly the types caused by ischemia and hypoxia. The The objective of the International Symposium on Brain Resuscitation held in Ube, Yamaguchi Japan October 31 to November 2 1988, and sponsored by Yamaguchi University and the Japanese Ministry of Education, was to review our recent progress in brain resuscitation and to discuss controversies both basic and clinical. To my knowledge, this symposium was the first held in Japan. Our understanding of neuronal dysfunction due to ischemic/hypoxic insults at organ, cellular, and molecular levels has advanced significantly in the last two decades. We had therefore intended that this international symposium should broadly cover the topics which are of interest to both basic researchers and clinicians.

Three hundred and twenty-five attendants, including twenty scientists from eight different countries, actively participated in discussion and exchange of new ideas and thoughts concerning brain resuscitation. This book comprises the reports presented during the symposium which consisted of two main parts: basic and clinical. Although one single meeting can never be expected to solve any problems, meetings often highlight areas of ignorance and problems which are ripe for solving. It has been hard to review all the papers because of the multiplicity of the discussed topics, but the overview on brain resuscitation by Professor Bo K. Siesjö and the summary by Professor J. Douglas Miller highlight the main subjects of the symposium and suggest future directions for brain resuscitation.

I anticipated that the symposium would serve as a mere stepping stone for the development of interest in patho-physiology and patho-chemistry and for the better management of patients with severe brain damage. However, I feel certain that we have been greatly encouraged by the recent advances in techniques and by the fact that knowledge obtained through basic research can be applied to the clinical arena, although it seems that our clinical impression of these applications is still far from our goal. In view of the quality of papers and number of attendants at this symposium, more than expected was achieved, and in this sense this book will prove a valuable resource for interdisciplinary investigation.

Finally, I am pleased to say that a large number of original papers submitted to the symposium are herein being published after peer review by Professors Siesjö and Miller. In addition, Drs. Sakabe and Murakawa assisted the editors. I am sure that this book would never have been completed without their painstaking efforts and patience. Again, I would like to take this last opportunity to express my many thanks to those who contributed to this symposium, particularly those who took the time for this symposium despite their busy schedules.

Hiroshi Takeshita

Contents

Contents

Contents IX

List of Contributors

Akimura, T. 141
Aoki, H. 141
Buchan, A.M. 59
Bullock, R. 233
Dearden, N.M. 221
Drummond, J.C. 45
Fujiwara, N. 115
Fukuda, S. 115
Furuse, M. 283
Hallmayer, J. 99
Harada, K. 141
Hashimoto, H. 173
Hirakawa, M. 131
Hoshiyama, M. 283
Hossmann, K.A. 99
Ichihara, K. 283
Ikeda, T. 173
Inao, S. 283
Ishikawa, T. 123, 165
Ishimatsu, S. 151
Ito, H. 141
Kamiryo, T. 141
Kaneoke, Y. 283
Kirino, T. 183
Kohama, A. 151
Kondo, M. 173
Kosaka, F. 173

Maekawa, T. 165
Maenosone, A. 151
Manaka, S. 183
Miller, J.D. 221, 315
Miyauchi, Y. 123
Motegi, Y. 283
Nakajima, K. 267
Nakayama, H. 123
Newberg Milde, L. 195
Nishizaki, T. 141
Ohkawa, M. 173
Oka, S. 173
Okamoto, T. 173
Ono, S. 275
Orita, T. 141
Paschen, W. 99
Pulsinelli, W.A. 59
Röhn, G. 99
Ropper, A.H. 301
Sakabe, T. 123, 165
Sano, K. 131
Sano, T. 123
Sasaki, M. 183
Saso, K. 283
Sato, M. 173
Shimizu, H. 275
Shimoda, Y. 173

Shimoji, K. 115
Shiogai, T. 247
Siesjö, B.K. 3
Smith, M.-L. 3
Steen, P.A. 211
Suzuki, A. 267
Suzuki, K. 151
Taga, K. 115
Takahata, Y. 115
Takasu, N. 151
Takeda, Y. 173
Takeshita, H. 123, 165
Takeuchi, K. 247
Tamura, A. 131
Tateishi, A. 123
Teasdale, G.M. 233
Tsubokawa, T. 289
Ueda, T. 165
Watanabe, I. 275
Wieloch, T. 21
Xie, Y. 99
Yaida, Y. 173
Yamamoto, T. 289
Yanagihara, T. 77
Yasui, N. 267
Yatsuzuka, H. 173

The page numbers refer to chapter opening pages of the contributors.

I
Basic Aspects

1

Brain Resuscitation: Yesterday, Today and Tomorrow

Bo K. Siesjö and Maj-Lis Smith[1]

The Past: General Principles

It is common knowledge that brain cells, notably neurons, are very sensitive to oxygen deprivation. Several decades ago, the prevailing dogma was that brain cells could tolerate anoxia for only 3–4 min. This concept was based on both clinical experience and on several experimental studies in which the circulation to the brain was temporarily arrested. The experience of clinicians attempting to resuscitate patients after cardiac arrest is exemplified by the title of an article published more than 30 years ago, which stated that brain damage was incurred following cardiac arrest of more than 4 min duration [1]. Some previous experimental studies [2], were in agreement with these clinical findings revealing clear evidence of brain damage in animals with transient ischemia lasting longer than 4–5 min. However, experimental work conducted in the period around 1960 by Hirsch, Schneider, and their collaborators indicated the likelihood that such short revival times reflected the susceptibility of the heart to anoxia, rather than the brain [3,4]. The revival times were then defined as the longest ischemic periods which the brain could tolerate without functional or structural damage. Thus, when Hirsch et al. took precautions to protect the heart against anoxic damage, the revival times rose to 8–10 min at temperatures of around 37°C. They established that resuscitation of the brain was optimal only if the perfusion pressure of the brain could be promptly restored at the termination of the ischemia. This finding seemed to provide an explanation for the short revival periods following cardiac arrest, a condition in which post-ischemic blood pressure is often suboptimal. As a result, the findings inspired both clinical and experimental workers to raise postischemic perfusion pressure by the administration of pressure agents, such as catecholamines. When these principles were adopted, revival times for recovery of neurological function in experimental

---

[1] Laboratory of Experimental Brain Research, Forskningsavd 4, Lund University Hospital, S-221 85 Lund, Sweden

3

ischemia were prolonged, sometimes to 10–12 min [5,6], and in some conditions, even longer. However, it is possible that accidental hypothermia could have contributed to the most dramatic results.

Another interesting finding was reported by Hirsch et al., using their long-term recovery model in rabbits. They showed ischemic brain damage to be very sensitive to hypothermia. In fact, Hirsch and Müller [4] found that differences in brain temperatures of only 1–2°C changed the revival times after ischemia. We will discuss this interesting finding later. For now, we will only emphasize the fact that although the findings of a protective effect of hypothermia have been extensively exploited in thoracic and neurosurgery, they seemed to provide no help in the resuscitation of brain function *after* ischemia.

Subsequent work centered on the reasons for brain cells tolerating anoxia better in vitro than in vivo. Thus, even with what seemed to be adequate restoration of cerebral perfusion pressure, brain cells appeared inappropriately sensitive to anoxia. For example, while brain damage seemed inevitable after ischemic periods lasting longer than 8–10 min, Ames and Gurian reported that retinal tissue in vitro tolerated 20 min of a combined oxygen and glucose deprivation [7]. The hypothesis was advanced that resuscitation of the brain was limited by vascular factors: following ischemia of a certain duration, an adequate reflow is hindered by the increased viscosity of stagnating blood, and by the swelling of both parenchymal and endothelial cells, with intravascular coagulation probably playing a contributory role. This was the so-called no-reflow phenomenon theory of Ames et al. [8]. These workers probably exaggerated the phenomenon since they assessed reflow by perfusing the tissue with a carbon black solution at a low perfusion pressure, and with a non-pulsatile flow. Nonetheless, results from our own laboratory [9] demonstrated that following *complete* ischemia of more than 10 min duration, central parts of the cerebral hemispheres were not adequately reperfused, in spite of an adequate perfusion pressure (Fig. 1). After ischemic periods of 30 min, the perfusion defects were extensive, and could last for 30–60 min. It was remarkable, however, that such reperfusion defects did not appear after severe *incomplete* ischemia, i.e. a condition in which a trickle of flow persisted during the ischemia [10]. These findings suggest that with a complete arrest of capillary blood flow, there is a critical opening pressure which must be exceeded for adequate reflow to occur.

What therapeutic inferences can be drawn from such studies? In this field, the early work by Hossmann, Kleihues and their collaborators is important [6,11,12]. These authors made painstaking efforts to raise postischemic perfusion pressure above normal ("vascular flush"), to decrease postischemic swelling by infusion of hyperosmotic solutions, and to combat intravascular coagulation. The results were striking since a remarkable, short-term recovery of neurophysiology and metabolic brain functions was obtained after complete ischemia of up to 60 min duration. Using similar principles, Safar and colleagues [13] could achieve good long-term recovery after such a severe insult as 12 min of cardiac arrest. The combination treatment instituted at the time of resuscitation was pressure check agents, heparinization, and hemodilution.

Ischemia probably leads to brain damage because it disrupts brain energy

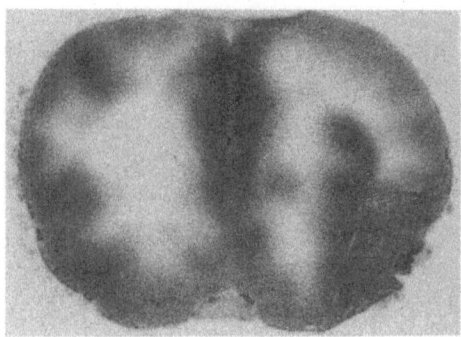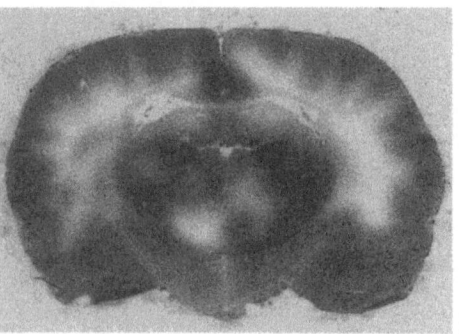

a
b

Fig. 1. Autoradiographic pictures from coronal sections of the caudoputamen/ sensorimotor cortex region **a** and the thalamus/parietal cortex region **b**, showing no-reflow phenomenon following 15 min of complete compression ischemia and 5 min of recirculation

metabolism. As Fig. 2 shows, complete ischemia leads to a very rapid break-down of phosphocreatine and to a somewhat slower decrease in the adenylate energy charge. The graph demonstrates that little or no energy is available after 5–7 min of anoxia. In the closed system created by complete ischemia, all gly-cogen and glucose are quickly converted to lactate. As will be discussed below, failure of ATP synthesis leads to secondary changes such as loss of ion homeo-stasis and to lipolysis with accumulation of free fatty acids, as shown in the figure. Under optimal conditions, these changes are reversed, but this does not mean that cell damage is absent. Typically, a selective loss of neurons will occur.

It has been known since the days of Spielmayer (1925) [14], and of the Vogts (1937) [15], that some neurons are selectively sensitive to insults [16]. The largest vulnerability to ischemia is found among pyramidal cells in the CA1 and CA4 sectors of the hippocampus, and in layers 3–6 in the neocortex, among middle-sized and small neurons in the dorsolateral part of the caudoputamen and in some thalamic nuclei [17,18]. Purkinje cells in the cerebellum are also quite sensitive to ischemia [19]. A selective neuronal vulnerability is also a strik-ing feature of hypoglycemic and epileptic brain damage, although neither the distribution nor the time course of the neuronal necrosis is identical [20]. This often raised the question of what metabolic and/or functional characteristics de-termine the vulnerability of some neurons to various insults and the resistance of others?

It soon became clear that short-time recovery after brain ischemia was not synonymous with long-term recovery. Thus, following the original observation from Dr. Klatzo's laboratory, published by Ito et al. in 1975 [21], Kirino (1982) [22] and Pulsinelli and collaborators [17] described in detail that following tran-sient ischemia, death of some neuronal populations could be delayed by 1–2 days, or even more. It was subsequently shown that neuronal activity [23] and energy metabolism [24] returned before the cells lost function, suffered a secon-dary loss of energy homeostasis, and developed light and electron microscopial

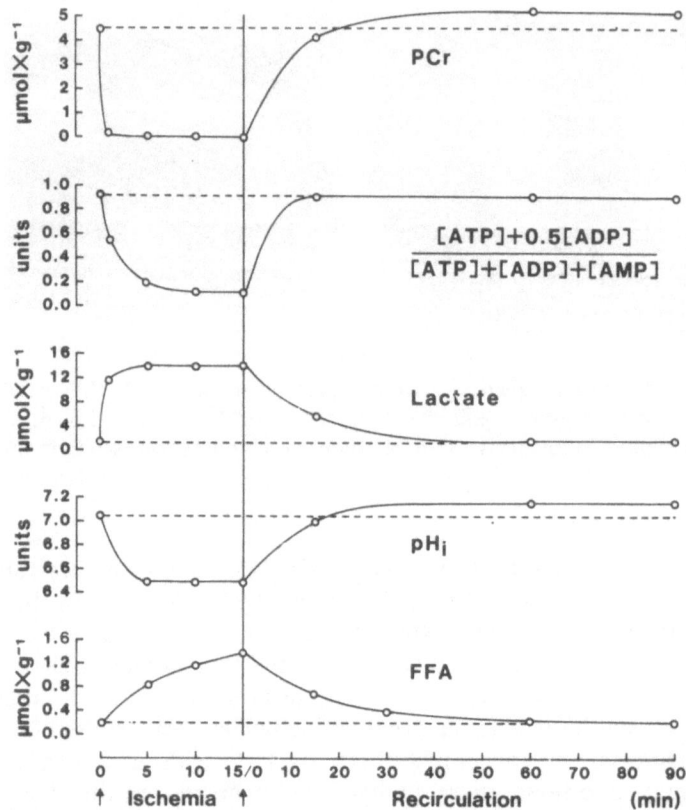

Fig. 2. Changes in cerebral cortical concentrations of phosphocreatine (PCr), lactate, and free fatty acids (FFA), as well as in the adenylate energy charge and intracellular pH (pH$_i$), during and following transient complete ischemia. (Reproduced by permission from [101])

signs of irreversible damage. The rate of maturation following ischemia seemed to vary between regions, with neurons in the caudoputamen showing a faster developing cell necrosis than those in the neocortex and, particularly, in the CA1 region of the hippocampus. These results, which underscored the necessity to use long-term recovery models for assessing neuronal damage following transient ischemia, suggested that pharmacological intervention may be possible in the free interval between the primary insult and the secondary damage. However, Klatzo's suggestion of an inverse relation between the duration of the ischemia and the length of the maturation period has been substantiated. Thus, with a long ischemic period, any therapeutic window that is present must by necessity be brief.

All this has been known for a long time and belongs to the discoveries of yesterday. During recent years we have seen the evolution of some major concepts, and the formulation of more specific hypotheses which have been proposed to explain why brain damage is incurred in ischemia. One major concept relates to differences in pathogenesis between selective neuronal vulnerability and pan-necrosis, or infarction. Two hypotheses predict that loss of calcium homeostasis and excessive acidosis, respectively, are responsible for these two types of brain damage.

The Past: Evolution of Concepts, With Major Hypotheses

Selective Neuronal Vulnerability and Loss of Cellular Calcium Homeostasis

The calcium hypothesis of ischemic/anoxic cell damage was originally proposed as an explanation for cell necrosis in the ischemic heart and the dystrophic muscle [25]. One possible mechanism for such damage is massive calcium loading of cells upon reperfusion with subsequent mitochondrial failure. Many years ago, we proposed that loss of calcium homeostasis could explain selective neuronal necrosis in ischemia, hypoglycemia and status epilepticus [26]. One reason for accepting this was that burst-firing cells with high calcium conductances in their apical dendrites, known to be present in several brain regions, seemed to be identical to the selectively vulnerable neurons. The second reason was the demonstration by Nicholson et al. [27] showing that anoxia is accompanied by cellular efflux of K^+, and influx of Ca^{2+} [25,28]. However, it is not likely that reperfusion following *brain* ischemia is followed by sudden calcium overload, simply because the blood-brain barrier is only slightly permeable to calcium. As a result, the initial load to which brain cells are exposed is the calcium contained in the extracellular fluid of the tissue. With time, however, calcium can be transported from the blood to the brain and accumulate in cells which have lost their ability to regulate the free, cytosolic calcium concentration (Ca^{2+}_i) [29,30]. Since the total tissue calcium concentration increases with time in cells which are destined to die after a transient insult [31,32], the question arises whether calcium accumulation is the cause of cell death or merely a result of it. Our own results suggest that calcium accumulation in the CA1 sector of the hippocampus precedes cell death [32–34]. Consequently, we have proposed that the initial insult leads to membrane damage, which results in an increased calcium cycling and a rise in calcium concentration sufficient to gradually overload the mitochondria [32].

In general, it seems more likely that the molecular damage caused by calcium overload is a result of excessive stimulation of lipases and proteases, or of a change in the phosphorylation state of receptor and channel proteins [35,36] (Fig. 3). Considerable attention has been recently directed towards the effects of enhanced protease activity since this, among other factors, could cleave the proteins which anchor the plasma membrane to the cytoskeleton, causing bleb formation and pathologically enhanced membrane permeability to calcium [37–39].

It now seems likely that calcium can trigger reactions which give rise to reperfusion damage by the formation of free radicals. According to the mechanism

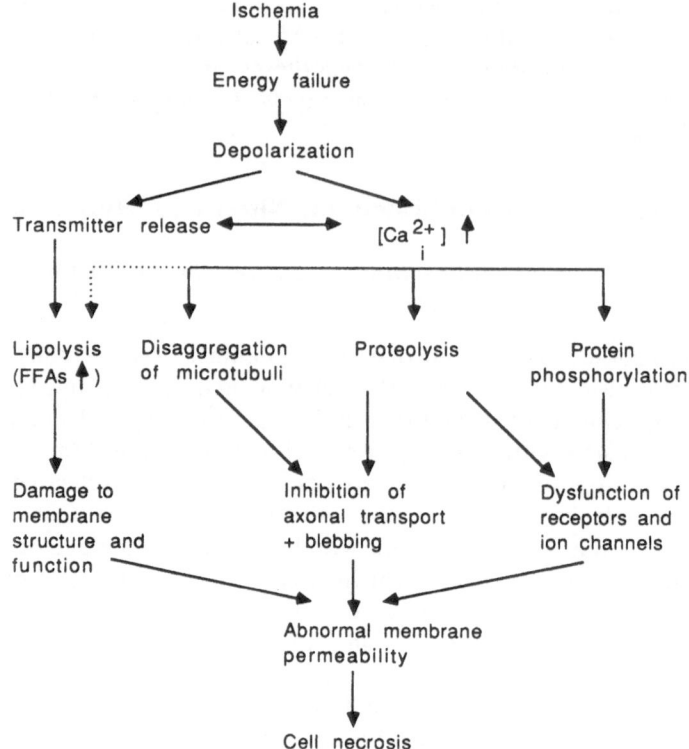

Fig. 3. Schematic diagram illustrating the coupling between ischemia, depolarization, loss of calcium homeostasis, and presumed calcium-related pathological events. (Reproduced by permission from [102])

originally proposed by McCord [40], calcium activates a protease which converts xanthine dehydrogenase to xanthine oxidase. The latter can then oxidize the hypoxanthine which had accumulated during the ischemia, forming uric acid and superoxide anions which, by virtue of their toxicity, represent a potential threat to the tissue. Free radicals may also arise when arachidonic acid, formed during the ischemia, is oxidized by cyclo-oxygenase and lipoxygenase in the recovery period [26]. Since phospholipid degradation is enhanced by calcium, this production of free radicals is also a secondary result of the loss of cellular calcium homeostasis.

Given all these facts, one can define three major factors which constitute possible causes of reperfusion or secondary ischemic damage and which are potentially amenable to treatment: (1) a low perfusion pressure and/or a hindered microvascular perfusion, (2) membrane damage with an increased calcium cycling across damaged membranes, and (3) free radical-induced damage. Clinicians know how to combat the first of these disorders. Unfortunately, the effect of calcium antagonists given *after* the insult is moderate, to say the least

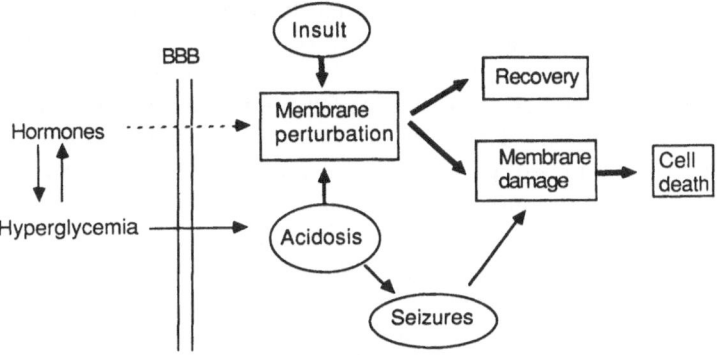

Fig. 4. Schematic diagram illustrating the aggravating effect of hyperglycemia on a primary ischemic insult

[25], and, up to very recently, the free radical hypothesis has not resulted in therapeutic measures which can be instituted in clinics.

Acidosis and Brain Damage

The question about the causes of brain damage is even more complicated since there are actually two types of ischemic brain damage: selective neuronal necrosis and pan-necrosis, or infarction [41]. One can deduce from a number of studies that one factor which can convert selective neuronal vulnerability to infarction is the duration of ischemia, longer ischemic periods being prone to cause infarction. For example, following 30 min of ischemia in fed animals, and 90 min of recovery, the tissue damage is characterized by ischemic neuronal changes plus extensive edema due to swelling of dendrites and glial cells [42,43]. The end stage of this process is infarction.

Studies performed more than 10 years ago in several laboratories [43–48], showed that feeding, or glucose infusion, aggravated the ischemic damage, converting selective neuronal vulnerability to infarction [49,50]. Thus, in hyperglycemic subjects even brief periods of ischemia can give rise to a syndrome with postischemic seizures, edema developement, and subsequent infarction. Even this type of damage shows a delay, the length of which varies inversely with the duration of the ischemia [51]. This is evidenced by the delayed edema and seizures which develop after about 24 h in hyperglycemic animals subjected to 10 min of ischemia.

The above results can be summarized by stating that an ischemic insult causes a membrane perturbation which can either be followed by overt membrane damage and cell death or by recovery (Fig. 4). Acidosis clearly enhances the membrane damage and somehow leads to seizures. These factors and the initial exaggeration of membrane perturbation are obviously what increases the brain damage.

This exaggeration of ischemic lesions caused by preischemic hyperglycemia or by a continuous glucose supply is not easily combatted in the intensive care setting where the insult has already occurred. However, since this type of pan-necrotic lesion occurs after a delay, there is hopefully a therapeutic window.

It is clear that the research of yesterday has not left us with much which can be used to treat the patients. However, today's research will probably determine the therapeutic measures of tomorrow. Thus, intensive attempts are being made to define the cellular and molecular mechanisms of ischemic damage and to find pharmacological means of ameliorating the damage. Three mediators of cell damage are in the focus of interest: calcium overload, acidosis, and free radical formation, i.e., the same mediators which we have already considered. We will discuss each of these mediators in turn, and speculate about the possibilities to ameliorate the ischemic damage. It will be obvious from this discussion that what the research of today has brought are new insights into the pathophysiology involved. In a way, previous investigators seem to have been working at the right problem from the wrong angle.

The Present: Cellular and Molecular Mechanisms Involved

Calcium-related Mechanisms of Brain Damage

Current results strongly indicate that enhanced calcium influx into cells is a major factor in the pathogenesis of ischemic, hypoglycemic, and epileptogenic damage [52–54]. However, whereas the original calcium hypothesis of neuronal necrosis assumed the influx of Ca^{2+} through voltage-sensitive calcium channels (VSCCs) [26], it is now believed that the pathogenerically important route of calcium entry is via agonist-operated calcium channels (AOCCs), notably those gated by glutamate and related excitatory amino acid (EAAs). In fact, it now appears likely that the toxicity of such EAAs is mediated via enhanced calcium influx [53–55]. We have to assume, therefore, that cellular domains which con-tain a high density of receptors gating such calcium-permeable channels (e.g. dendritic spines and apical dendrites) are subjected to such high calcium concen-trations that the series of events depicted in Fig. 3 is elicited.

In order to discuss calcium-related damage it is necessary to take a closer look at calcium homeostasis at the cellular level [25,41,56–59]. As Fig. 5 shows, cal-cium is known to enter cells by VSCCs and AOCCs. Furthermore, since activation of receptors linked to phospholipase C leads to the release of calcium from the endoplasmic reticulum (ER), and since the ER may be refilld from the outside via a separate "channel" [60], there are three major routes of entry for calcium. Export of calcium occurs by an ATP-driven translocase and by $3Na^+/Ca^{2+}$ exchange. When calcium enters cells, it is buffered by proteins and many other compounds and is sequestered by the ER by an ATP-dependent uptake mechanism. Calcium uptake by the mitochondria and calcium cycling across their inner membranes may primarily serve to regulate intramitochondrial de-hydrogenases and, thereby, respiratory functions of the mitochondria [61,62]. However, with large cellular calcium loads, the mitochondria have the capacity to sequester appreciable amounts of calcium at the expense of ATP production

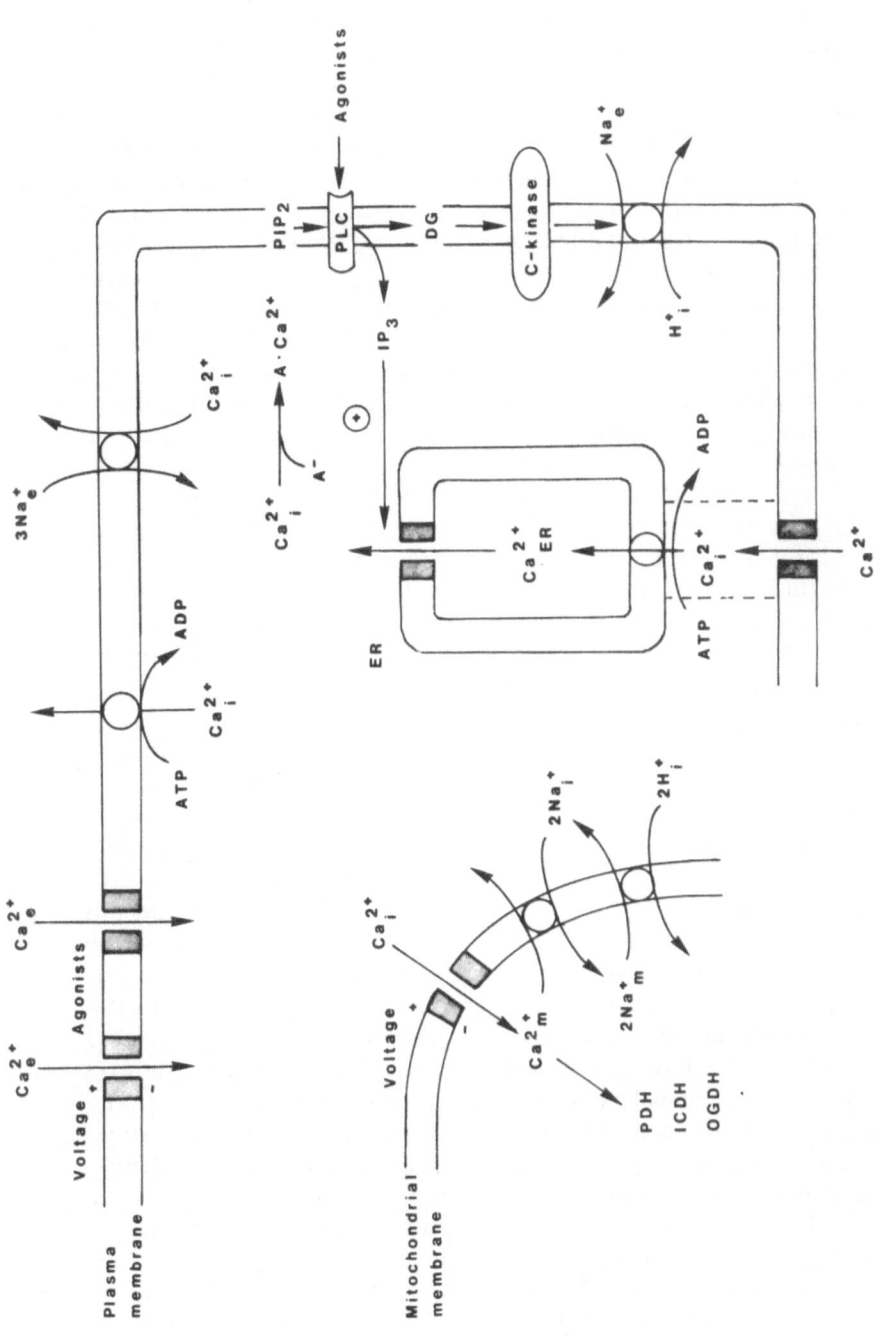

Fig. 5. Schematic diagram illustrating voltage-sensitive channels in pre- and post-synaptic structures, and post-synaptic agonist-operated channels. The voltage-sensitive channels have been assumed to be of the L, T, and N types, the L-type being sensitive and the other types insensitive to calcium antagonists of the dihydropyridine class. Agonist-operated channels, which are activated by either glutamate or a related amino acid, have been assumed to be gated by kainate/quisqualate (K/A) or by NMDA (N) receptors. Normally, only the latter is supposed to open a conductance for Ca^{2+}. (Reproduced by permission from [41])

[58,63]. We know that sequestration of calcium by the ER and mitochondria requires ATP and/or oxygen. Thus, anoxia will seriously curtail calcium sequestration. It will probably also decrease calcium buffering because the excess H^+ produced during anaerobic glycolysis can successfully compete Ca^{2+} for common binding sites [25,64,65].

Results obtained on cultured neurons in vitro demonstrate that many neurons contain VSCCs blocked by the dihydropyridine (DHP)-type of calcium antagonists [66,67]. However, it seems likely that the pathologically most important calcium channels are the AOCCs, notably those linked to glutamate receptors [68–71]. These are usually considered to be of two types. One is the receptor selectively activated by kainate/quisqualate and believed to gate a channel permeable to all monovalent cations. The other, selectively activated by NMDA, is the channel which is permeable to calcium as well. The latter is blocked by Mg^{2+} but the Mg^{2+} block is voltage-dependent, meaning that depolarisation of the membrane relieves the Mg^{2+} block, allowing Ca^{2+} to enter. Since glutamate is a mixed agonist which activates both channels, it causes downhill fluxes of all cations, i.e., K^+, Na^+ and Ca^{2+}, and most definitely H^+.

With this background, we can discuss ion fluxes in ischemia/anoxia. Following the induction of anoxia, the extracellular K^+ concentration (K^+_e) slowly increases to 10–15 mM. At that point, K^+ efflux accelerates while Na^+ and Ca^{2+} suddenly enter the cells. Since Cl^- is also taken up by cells (probably secondary to Na^+ influx) there is flux of water from the extra- to the intracellular fluids, leading to cell swelling.

It is tempting to assume that the "shock" opening of ionic gates after 1–2 min of anoxia is due to the sudden release of glutamate, with activation of both K/Q and NMDA receptors. This assumption received support from the results reported by Benveniste et al. [72], showing that deafferentation of CA1 pyramidal cells prevents influx of Ca^{2+} into the dendrites of such cells during a 10 min period of ischemia. Interestingly, although local injection of a competitive NMDA antagonist delayed and attenuated the decrease in Ca^{2+}_e, it did not prevent it altogether. However, it has been difficult to explain why systemically administered non-competitive NMDA antagonists do not prevent, delay, or curtail Ca^{2+} influx in dense ischemia or in hypoglycemia, especially since such antagonists block electrically induced spreading depression (SD), and the associated calcium influx [73,74]. We have proposed that SD, hypoglycemia, and anoxia lead to the activation of a non-selective cation conductance mechanism [73]. The tentative opinion is that the activation is due to a rise in Ca^{2+}_i to a threshold value. The conductance mechanism could either be a non-selective calcium-activated cation channel of the type described for many other cells [75], or the transformation of the glutamate-activated receptor complex to a substate in which the K/Q receptor also gates a calcium-permeable channel [76]. However, it is possible to explain the failure of NMDA antagonists to block calcium influx by invoking alternative routes of entry, notably VSCCs and a reversal of the $3Na^+/Ca^{2+}$ exchange [74].

The discrepant results reported in the literature can be most readily explained if one assumes that the initial calcium influx is through channels linked to the

NMDA receptor, and that a rise in $Ca^{2+}{}_i$ (or another molecular event) activates a high-conductance calcium-permeable channel. Thus, since an NMDA antagonist blocks the major route of calcium entry, an SD cannot easily be elicited. Furthermore, the salutary effects of NMDA antagonists in moderately severe ischemia (and in hypoglycemia, [77,78]) are of similar origin. Since a major route of calcium entry is blocked, the meager stores of ATP suffice to keep pace with the ionic leaks. In contrast, since complete ischemia opens a major calcium conductance, the stimulation of phospholipase C by glutamate activation of metabolotropic (α) receptors may cause IP_3-triggered calcium release, and the influx of calcium may also occur by reversal of the $3Na^+/Ca^{2+}$ antiporter, NMDA antagonists cannot be expected to ameliorate brain damage under this condition. The situation is different in moderately severe ischemia, such as that in the penumbra zone of a stroke lesion. Clearly, NMDA antagonists could also be expected to ameliorate any ischemic damage which is incurred in the recovery period following transient ischemia, whether that is due to postischemic hypotension, to increased calcium cycling across leaky membranes, or to epileptogenic activity.

Molecular Mechanisms of Acidosis-related Damage

The question now arises as to whether there are factors other than the loss of calcium homeostasis involved in the final damage. We can answer to this by referring to the experiments of Siemkowiez and Hansen [79] who found that preischemic hyperglycemia shortened the duration of the calcium transient (defined as the period of calcium uptake) during 10 min of ischemia; yet, hyperglycemia *aggravates* ischemic damage (see above). This suggests that acidosis per se is harmful.

Exaggerated tissue acidosis has been linked to postischemic seizures, edema formation, destruction of glial cells, and infarct development [25,80–83]. It is not likely that acidosis acts by enhancing the rise in $Ca^{2+}{}_i$. Although the molecular mechanisms have not been defined, it seems more likely that acidosis acts primarily by accelerating the formation of harmful free radicals, thereby contributing to the postischemic membrane damage. Part of this damage probably affects the microvasculature which, by virtue of its high content of xanthine dehydrogenase, has the potential of forming superoxide anions ($\cdot O_2{}^-$) during reoxygenation [84]. Thus, promising results have been obtained with the administration of superoxide dismutase in conjugated form or trapped in liposomes [85]. Furthermore, brain damage and edema formation during moderate ischemia of long duration have been ameliorated by the administration of a free radical scavenger such as dimethylthiourea [84,86].

Many now believe that acidosis wrecks the mechanism by releasing pro-oxidant iron from proteins, particularly of the transferrin type, and that iron, chelated to low molecular weight compounds, triggers the formation of toxic free radicals, such as $\cdot OH$ [80,87,88]. Circumstantial evidence for this was obtained in experiments in which acidosis was shown to markedly enhance the formation of free radicals in vitro [89,90]. The participation of iron in this process was evident from the fact that the iron chelator desferral blocked the reaction [90]. Under normal circumstances, the superoxide radicals and hydrogen peroxide formed in the cell do not react to yield hydroxide radicals. However, in the presence of pro-oxidant iron the reaction proceeds as described [91,92].

In theory, acidosis could act by converting the superoxide anion to its acid form, the species HO_2 being more lipid soluble than $\cdot O_2^-$, as well as a stronger oxidant [93]. However, it seems even more likely that acidosis acts by releasing iron from transferrin. Thus, iron is bound to the transferrin via a bicarbonate bridge which is attacked by protons [89,90,94].

A dramatic support for the free radical hypothesis of ischemic damage was recently obtained when it was shown that a novel group of free radical scavengers (21-aminosteroids, or "lazaroids") seem to markedly ameliorate ischemic damage, at least that incurred by reversible focal ischemia [95,96].

The Future: Prevention of Ischemic Damage

Finally, with a view to future developments, we should like to recall that novel results have reconfirmed the protective effects of hypothermia, and have shown that a lowering of brain temperature by only 2–4°C can have pronounced effects on selective neuronal vulnerability [97,98]. Even more interestingly, amelioration of ischemic damage seems possible with *post*-ischemic brain cooling [99,100]. In one of these series [100] two animals with dense forebrain ischemia of 20 min duration and post-ischemic cooling to 27°C for 1 h suffered virtually no brain damage. In contrast, all animals subjected to ischemia under identical conditions, but with normothermia maintained in the postischemic period, had extensive brain damage. These results emphasize that ischemic brain damage is, in reality, incurred in the postischemic period, and suggest that effective therapy will become eventually available. One can foresee that pharmacological treatment during brain resuscitation will include a combination of at least two types of calcium antagonists (directed against VSCCs and AOCCs), and of free radical scavengers/iron chelators, perhaps supplemented with moderate hypothermia.

References

1. Cole SL, Corday E (1956) Four-minute limit for cardiac resuscitation. J Am Med Assoc 161:1454–1458
2. Weinberger LM, Gibbon MH, Gibbon JH (1940) Temporary arrest of the circulation to the central nervous system. Arch Neurol (Chicago) 43:961–986
3. Hirsch H, Euler KH, Schneider M (1957) Über die Erholung und Wiederbelebung des Gehirns nach Ischämie bei Normothermie. Pflugers Arch 265:281–313
4. Hirsch H, Müller HA (1962) Funktionelle und Histologische Veränderungen des Kaninchengehirns nach kompletter Gehirnischämie. Pflugers Arch 275:277–291
5. Siesjö BK (1978) Brain Energy Metabolism. Wiley, New York, pp 380–397
6. Hossmann K-A (1985) Post-ischemic resuscitation of the brain: Selective vulnerability versus global resistance. In: Kogure K, Hossman K-A, Siesjö BK, Welsh FA (eds) Progress in brain research, vol 63. Elsevier, Amsterdam, pp 3–17

7. Ames A III, Gurian BS (1963) Effects of glucose and oxygen deprivation on function of isolated mammalian retina. J Neurophysiol 26:617–634

8. Ames A III, Wright RL, Kowada M, Thurston JM, Majno G (1968) Cerebral ischemia: II. The no-reflow phenomenon. Am J Pathol 52:437–453

9. Kågström E, Smith M-L, Siesjö BK (1983) Local cerebral blood flow in the recovery period following complete cerebral ischemia in the rat. J Cereb Blood Flow Metab 3:170–182

10. Kågström E, Smith M-L, Siesjö BK (1983) Recirculation in the rat brain following incomplete ischemia. J Cereb Blood Flow Metab 3:183–192

11. Hossmann K-A, Kleihues P (1973) Reversibility of ischemic brain damage. Arch Neurol 29:375–384

12. Hossmann K-A (1982) Treatment of experimental cerebral ischemia. J Cereb Blood Flow Metab 2:275–297

13. Safar P, Stezoski W, Nemoto EM (1976) Amelioration of brain damage after 12 minutes' cardiac arrest in dogs. Arch Neurol (Chicago) 33:91–95

14. Spielmeyer W (1925) Zur Pathogenese örtlich elektiver Gehirnveränderungen. Z Ges Neurol Psychiatr 99:756–776

15. Vogt C, Vogt O (1937) Sitz und Wesen der Krankheiten im Lichte der topistischen Hirnforschung und des Variierens der Tiere. J Psychol Neurol (Leipzig) 47:237–457

16. Brierley JB, Graham DI (1984) Hypoxia and vascular disorders of the central nervous system. In: Adams JH, Corsellis JAN, Duchen LW (eds) Greenfield's Neuropathology, 4th edn. Edward, London pp 125–207

17. Pulsinelli WA, Brierley JB, Plum F (1982) Temporal profile of neuronal damage in a model of transient forebrain ischemia. Ann Neurol 11:491–498

18. Smith M-L, Auer RN, Siesjö BK (1984) The density and distribution of ischemic brain injury in the rat following two to ten minutes of forebrain ischemia. Acta Neuropathol (Berl) 64:319–332

19. Diemer NH, Siemkowicz E (1981) Regional neurone damage after cerebral ischaemia in the normo- and hyperglycaemic rat. Neuropathol Appl Neurobiol 7:217–227

20. Auer RN, Siesjö BK (1988) Biological differences between ischemia, hypoglycemia and epilepsy. Ann Neurol 24:699–707

21. Ito U, Spatz M, Walker JT, Klatzo I (1975) Experimental cerebral ischemia in Mongolian gerbils: 1. Light microscope observations. Acta Neuropathol (Berl) 32:209–223

22. Kirino T (1982) Delayed neuronal death in the gerbil hippocampus following ischemia. Brain Res 239:57–69

23. Suzuki R, Yamaguchi T, Li C-L, Klatzo I (1983) The effects of 5 min ischemia in Mongolian gerbils: II. Changes of spontaneous neuronal activity in cerebral cortex and CA2 sector of hippocampus. Acta Neuropathol (Berl) 60:217–222

24. Pulsinelli WA, Duffy TE (1983) Regional energy balance in rat brain after transient forebrain ischemia. J Neurochem 40:1500–1515

25. Siesjö BK (1988) Historical overview. Calcium, ischemia, and death of brain cells. Ann NY Acad Sci 522:638–661

26. Siesjö BK (1981) Cell damage in the brain: A speculative synthesis. J Cereb Blood Flow Metab 1:155–185

27. Nicholson C, Bruggencate GT, Steinberg R, Stöckle H (1977) Calcium modulation in brain extracellular microenvironment demonstrated with ion-selective micropipette. Proc Natl Acad Sci USA 74:1287–1290

28. Hansen AJ (1985) Effects of anoxia on ion distribution in the brain. Physiol Rev 65(1):101–148
29. Yanagihara T, McCall J (1982) Ionic shifts in cerebral ischemia. Life Sci 30:1921–1925
30. Hossmann K-A, Paschen W, Csiba L (1983) Relationship between calcium accumulation and recovery of cat brain after prolonged cerebral ischemia. J Cereb Blood Flow Metab 3:346–353
31. Dienel GA (1984) Regional accumulation of calcium in postischemic rat brain. J Neurochem 43:913–925
32. Deshpande JK, Siesjö BK, Wieloch T (1987) Calcium accumulation and neuronal damage in the rat hippocampus following cerebral ischemia. J Cereb Blood Flow Metab 7:89–95
33. Martins E, Inamura K, Themner K, Malmqvist KG, Siesjö BK (1988) Accumulation of calcium and loss of potassium in the hippocampus following transient cerebral ischemia: A proton microprobe study. J Cereb Blood Flow Metab 8:531–538
34. Dux E, Mies G, Hossmann K-A, Siklós L (1987) Calcium in the mitochondria following brief ischemia of gerbil brain. Neurosci Lett 78:295–300
35. Siesjö BK, Wieloch T (1985) Cerebral metabolism in ischemia: Neurochemical basis for therapy. Br J Anaesth 57:47–62
36. Siesjö BK, Wieloch T (1986) Epileptic brain damage: Pathophysiology and neurochemical pathology. In: Delgado-Escueta AV, Ward AA Jr, Woodbury DM, Porter RJ (eds) Advances in Neurology, vol 44. Raven, New York, pp 813–847
37. Jennings RB, Reimer KA, Steenbergen C (1986) Myocardial ischemia revisited. The osmolar load, membrane damage, and reperfusion. J Mol Cell Cardiol 18:769–780
38. Nicotera P, Harzell P, Davis G, Orrenius S (1986) The formation of plasma membrane blebs in hepatocytes exposed to agents that increase cytosolic Ca^{2+} is mediated by the activation of a non-lysosomal proteolytic system. FEBS Lett 209:139–144
39. Orrenius S, McConkey DJ, Jones DP, Nicotera P (1988) Ca^{2+}-activated mechanisms in toxicity and programmed cell death. ISI atlas of science. Pharmacology 2:319–324
40. McCord JM (1985) Oxygen-derived free radicals in postischemic tissue injury. N Engl J Med 312(3):159–163
41. Siesjö BK (1989) Calcium and cell death. Magnesium 8:223–237
42. Nordström C-H, Rehncrona S, Siesjö BK (1978) Effects of phenobarbital in cerebral ischemia. Part two: Restitution of cerebral energy state, as well as of glycolytic metabolites, citric acid cycle intermediates and associated amino acids after pronounced, incomplete ischemia. Stroke 9:335–343
43. Rehncrona S, Rosén I, Siesjö BK (1981) Brain lactic acidosis and ischemic cell damage: 1. Biochemistry and neurophysiology. J Cereb Blood Flow Metab 1:297–311
44. Myers RE, Yamaguchi M (1977) Nervous system effects of cardiac arrest in monkeys. Arch Neurol 34:65–74
45. Siemkowicz E, Hansen J (1978) Clinical restitution following cerebral ischemia in hypo-, normo-, and hyperglycemic rats. Acta Neurol Scand 58:1–8
46. Myers RE (1979) Lactic acid accumulation as cause of brain edema and cerebral necrosis resulting from oxygen deprivation. In: Korobkin R, Guilleminault G (eds) Advances in perinatal neurology. Spectrum, New York pp 85–114
47. Myers RE (1979) A unitary theory of causation of anoxic and hypoxic brain patholo-

gy. In: Fahn S, Davis JN, Rowland LP (eds) Cerebral hypoxia and its consequences Adv Neurol 26, Raven, New York, pp 195–213

48. Rehncrona S, Rosén I, Siesjö BK (1980) Excessive cellular acidosis: An important mechanism of neuronal damage in the brain. Acta Physiol Scand 110:435–437

49. Welsh FA, Ginsberg MD, Rieder W, Budd WW (1980) Deleterious effect of glucose pretreatment on recovery from diffuse cerebral ischemia in the cat: II. Regional metabolite levels. Stroke 11(4):355–363

50. Pulsinelli WA, Waldman S, Rawlinson D, Plum F (1982) Moderate hyperglycemia augments ischemic brain damage: A neuropathologic study in the rat. Neurolog 32:1239–1246

51. Warner DS, Smith ML, Siesjö BK (1987) Ischemia in normo- and hyperglycemic rats: Effects on brain water content and electrolytes. Stroke 18:464–471

52. Choi DW (1985) Glutamate neurotoxicity in cortical cell culture is calcium dependent. Neurosci Lett 58:293–297

53. Choi DW (1988) Calcium-mediated neurotoxicity: relationship to specific channel types and role in ischemic damage. Trends Neurosci 11:465–469

54. Rothman SM, Olney JW (1986) Glutamate and the pathophysiology of hypoxic-ischemic brain damage. Ann Neurol 19:105–111

55. Olney J (1978) Neurotoxicity of excitatory amino acids. In: McGeer EG, Olney JW, McGeer PL (eds) Kainic as a tool in neurobiology. Raven, New York pp 95–121

56. Berridge MJ (1979) Modulation of nervous activity by cyclic nucleotides and calcium. In: Schmitt FO, Woden FG (eds) The neurosciences: 4th study program. MIT Press, Cambridge, Mass, pp 873–889

57. Berridge MJ (1984) Inositol triphosphate and diacylglycerol as second messengers. Biochem J 221.345–360

58. Rasmussen H, Waisman DM (1983) Modulation of cell function in the calcium messenger system. Rev Physiol Biochem Pharmacol 95:111–148

59. Carafoli E (1987) Intracellular calcium homeostasis. Annu Rev Biochem 56:395–433

60. Putney JW Jr (1986) A model for receptor-regulated calcium entry. Cell Calcium 7:1–12

61. Denton RM, McCormack JG (1985) Ca^{2+} transport by mammalian mitochondria and its role in hormone action. Am J Physiol 249:E543–E554

62. Hansford RG (1985) Relation between mitochondrial calcium transport and control of energy metabolism. Rev Physiol Biochem Pharmacol 102:1–72

63. Nicholls DG (1986) Intracellular calcium homeostasis. Br Med Bull 42(4):353–358

64. Busa WB, Nuccitelli R (1984) Metabolic regulation via intracellular pH. Am J Physiol 246:R409–R438

65. Abercombie RF, Hart CE (1986) Calcium and proton buffering and diffusion in isolated cytoplasm from Myxicola axons. Am J Physiol 250 (J Cell Physiol 19): C391–C405

66. Kudo Y, Oguro A (1986) Glutamate-induced increase in intracellular Ca^{2+} concentration in isolated hippocampal neurones. Br J Pharmacol 89:191–198

67. Thayer SA, Murphy SN, Miller RJ (1986) Widespread distribution of dihydropyridine-sensitive calcium channels in the central nervous system. Mol Pharmacol 30:505–509

68. Collingridge GJ (1985) Long-term potentiation in the hippocampus: mechanisms of initiation and modulation by neurotransmitters. Trends Pharmacol Sci 6:407–411

69. Foster AC, Fagg GE (1987) Taking apart NMDA receptors. Nature 329:395–396

70. Mayer ML, Westbrook GL (1987) Cellular mechanisms underlying excitotoxicity. Trends Neurosci 10:59–61
71. Miller RJ (1987) Multiple calcium channels and neuronal function. Science 235:46–52
72. Benveniste H, Jørgensen MB, Diemer NH, Hansen AJ (1988) Calcium accumulation by glutamate receptor activation is involved in hippocampal cell damage after ischemia. Acta Neurol Scand 78:529–536
73. Siesjö BK, Bengtsson F (1989) Calcium fluxes, calcium antagonists, and calcium-related pathology in brain ischemia, hypoglycemia, and spreading depression: A unifying hypothesis. J Cereb Blood Flow Metab 9:127–140
74. Siesjö BK, Bengtsson F, Grampp W, Theander S (1989) Calcium, excitotoxins and neuronal death in the brain. Ann NY Acad Sci 568:234–251
75. Partridge LD, Swandulla D (1988) Calcium-activated non-specific cation channels. Trends Neurosci 11(2):69–72
76. Mayer M (1987) Two channels reduced to one. Nature 325:480–481
77. Wieloch T (1985) Hypoglycemia-induced neuronal damage is prevented by a N-methyl-D-aspartate receptor antagonist. Science 230:681–683
78. Westerberg E, Kehr J, Ungerstedt U, Wieloch T (1988) The NMDA-antagonist MK-801 reduces extracellular amino acid levels during hypoglycemia and prevents striatal damage. Neurosci Res 3(3):151–158
79. Siemkowicz E, Hansen J (1981) Brain extracellular ion composition and EEG activity following 10 minutes ischemia in normo- and hyperglycemic rats. Stroke 12:236–240
80. Siesjö BK (1985) Acid-base homeostasis in the brain: Physiology, chemistry, and neurochemical pathology. In: Kogure K, Hossmann K-A, Siesjö BK, Welsh FA (eds) Progress in brain research, vol 63. Elsevier Science, Amsterdam, pp 121–154
81. Siesjö BK, Smith M-L, Warner DS (1987) Acidosis and ischemic brain damage. In: Raichle ME, Powers WJ (eds) Cerebrovascular diseases. Raven, New York, pp 83–95
82. Kraig RP, Pulsinelli WA, Plum F (1986) Carbonic acid buffer changes during complete brain ischemia. Am J Physiol 150:R348–R357
83. Kraig RP, Petito CK, Plum F, Pulsinelli WA (1987) Hydrogen ions kill brain at concentrations reached in ischemia. J Cereb Blood Flow Metab 7:379–386
84. Betz AL (1985) Identification of hypoxanthine transport and xanthine oxidase in brain capillaries. J Neurochem 44:574–579
85. Michelson AM, Jadot G, Puget K (1988) Treatment of brain trauma with liposomal superoxide dismutase. Free Rad Res Comm 4:209–224
86. Patt A, Harken AH, Burton LK, Rodell TC, Piermattei D, Schorr WJ, Parker NB, Berger EM, Horesh IR, Terada LS, Linas SL, Cheronis JC, Repine JE (1988) Xanthine oxidase-derived hydrogen peroxide contributes to ischemia reperfusion-induced edema in gerbil brains. J Clin Invest 81:1556–1562
87. White BC, Krause GS, Aust SD, Eyster GE (1985) Postischemic tissue injury by iron-mediated free radical lipid peroxidation. Ann Emerg Med 14:804–809
88. Krause GS, White BG, Aust SD, Nayini NR, Kumar K (1988) Brain cell death following ischemia and reperfusion: A proposed biochemical sequence. Crit Care Med 16:714–721
89. Siesjö BK, Bendek G, Koide T, Westerberg E, Wieloch T (1985) Influence of acidosis on lipid peroxidation in brain tissues in vitro. J Cereb Blood Flow Metab 5:253–258
90. Rehncrona S, Nielsen Hauge H, Siesjö BK (1989) Enhancement of iron-catalyzed

free radical formation by acidosis in brain homogenates: Difference in effect by lactic acid and CO_2. J Cereb Blood Flow Metab 9:65–70

91. Halliwell B, Gutteridge JMC (1984) Oxygen toxicity, oxygen radicals, transition metals and disease. Biochem J 219:1–14

92. Halliwell B, Gutteridge JMC (1985) Oxygen radicals and the nervous system. Trends Neurosci 8:22–26

93. Gebicki JM, Bielski BH (1981) Comparison of the capacities of the perhydroxyl and the superoxide radicals to initiate chain oxidation of linoleic acid. J Am Chem Soc 103:7020–7022

94. Aisen P (1979) Some physiochemical aspects of iron metabolism. In: CIBA Foundation Symposium, No. 51: Iron Metabolism. Elsevier, Amsterdam pp 1–17

95. Silvia RC, Piercey MF, Hoffmann WE, Chase RL, Braughler JM, Tang AH (1987) U-74006F, an inhibitor of lipid peroxidation, protects against lesion development following experimental stroke in the cat: Histological and metabolic analysis (abstract). Soc Neurosci 13:1495

96. Hall ED, Pazara KE, Braughler MJ (1988) 21-aminosteroid lipid peroxidation inhibitor U74006 protects against cerebral ischemia in gerbils. Stroke 19:997–1002

97. Busto R, Dietrich WD, Globus MY-T, Valdés I, Scheinberg P, Ginsberg MD (1987) Small differences in intraischemic brain temperature critically determine the extent of ischemic neuronal injury. J Cereb Blood Flow Metab 7:729–738

98. Minamisawa H, Nordström C-H, Smith M-L, Siesjö BK (1990) The influence of mild body and brain hypothermia on ischemic brain damage. J Cereb Blood Flow Metab 10:365–374

99. Busto R, Dietrich WD, Globus My-T (1989) Postischemic moderate hypothermia inhibits CA1 hippocampal injury. Neurosci Lett 101:299–304

100. Boris-Möller F, Smith M-L, Siesjö BK (1989) Effect of hypothermia on ischemic brain damage: A comparison between preischemic and postischemic cooling. Neurosci Res Comm 5:87–94

101. Siesjö BK (1984) Cerebral circulation and metabolism. J. Neurosurg 60:883–908

102. Siesjö BK (1988) Mechanisms of ischemic brain damage. Critical Care Med 16: 954–963

2

Glutamate Neurotoxicity and Ischemic Neuronal Damage

Tadeusz Wieloch[1]

Introduction

The first observation of glutamate neurotoxicity was reported in 1957 when Lucas and Newhouse found that glutamate given intraperitoneally to newborn mice induced neurodegeneration in the retina [1]. Later, in a similar experimental paradigm, Olney observed neuronal degeneration in the hypothalamus of infant mice brain [2], and found that glutamate causes extensive swelling of dendrites and while some axons are spared. Based on a series of additional observations, he formulated the excitotoxic hypothesis of cell damage, stating that glutamate causes a continuous depolarization of the plasma membrane by interacting with receptors on the vulnerable neurons. The depolarization causes energy exhaustion due to a disturbed ion homeostasis, and triggers adverse intracellular reactions causing cell damage [3]. Two mechanisms have been proposed to explain glutamate neurotoxicity, both based on experiments performed on dissociated neurons in culture. One suggests that glutamate-induced membrane depolarization enhances sodium ion influx with accompanying chloride ions and water causing massive cellular swelling and acute osmolysis of the cells due to mechanical disruption. Chloride and sodium are essential for this acute glutamate neurotoxicity while calcium ions are not [4]. Osmotic cytolysis was also observed when dissociated neuronal cultures were exposed to anoxia [5]. The other mechanism favors the calcium ion as the decisive factor in glutamate-induced delayed neuronal death [6]. Delayed neuronal necrosis of dissociated neurons in culture exposed to anoxia also requires calcium ions [7]. However, the connection between glutamate and hypoxia dates back to the late 1950s, when van Harreveld discovered that glutamate could induce cortical spreading depression [8] and cause dendritic swelling, similar to that observed during

[1]Laboratory for Experimental Brain Research, Forskningsavd 4, Lund University Hospital, S-221 85 Lund, Sweden

21

hypoxia [9]. Later, in 1982 Jørgensen and coworkers proposed that the correlation between glutamate uptake sites and selective neuronal damage in the hippocampus indicate that glutamate may be involved in selective neuronal damage to the hippocampus, and that glutamate antagonists may be cerebroprotective [10]. Since these initial observations, a vast number of investigations on glutamate toxicity in ischemia have been published. Although many reports favor glutamate as a major pathogenetic factor in ischemia and support the notion that glutamate antagonists—notably NMDA antagonists—are cerebroprotective agents following ischemia and may be ready for clinical use [11], several recent investigations reached the opposite conclusion [12,13]. This review discusses glutamate neurotoxicity in cerebral ischemia and its relation to the mechanisms of ischemic brain damage.

Glutamate and Glutamate Receptors

Glutamate

Glutamate is an amino acid found as a structural element in proteins and as an important intermediary compound in cell metabolism. Glutamate is a precursor for neurotransmitters such as GABA and aspartate, and is the major excitatory neurotransmitter in the brain [14]. The total concentration of glutamate in brain tissue is approximately 10 mM, while the extracellular level is in the micromolar range. It is indeed a paradox that the brain accommodates such huge stores of a potent neurotoxin such as glutamate. Consequently, the brain is equipped with mechanisms that control the release and uptake of glutamate, and that do not permit the extracellular levels of glutamate to reach toxic concentrations. For example, in order to induce neuronal necrosis by intracerebroventricular injections of glutamate, glutamate solutions of 0.5 M are required [15]. Additionally, if glutamate uptake sites are removed [16] or uptake is inhibited [17], intracerebral injections of a small amount of glutamate is sufficient to cause neuronal necrosis.

Figure 1 shows a schematic drawing of the glutamatergic synapse. During neurotransmission, the intracellular level of calcium in the nerve endings is transiently elevated by the opening of voltage-operated channels (VOC) in plasma membranes, causing a calcium flux down its concentration gradient [18], and stimulating the release of glutamate [19]. Protein kinase C (PKC) enhances the release of glutamate. It has been shown that inhibitors of PKC decrease and activators of PKC increase evoked transmitter release [20–22]. The released glutamate in the synaptic cleft activates the glutamate receptor on the postsynaptic neuron. Its action is terminated by sequestration into glia cells and presynaptic elements via a sodium-dependent uptake mechanism [14] that is highly dependent on the sodium gradient across plasma membranes and thus dependent on ATP formation. If intracellular levels of sodium are elevated, the symporter may function in the reversed mode and thus increase the glutamate levels in the synaptic cleft [23].

Several situations can be envisaged where synaptic glutamate levels may reach

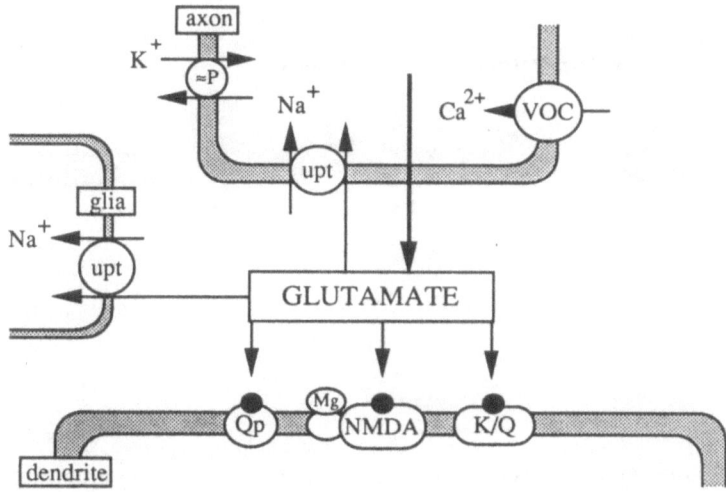

Fig. 1. A schematic picture of a glutamatergic synapse. The different glutamate receptors and ion channels are defined on the postsynaptic (dendrite) neuron: N-methyl-D-aspartate, NMDA; kainate, K; quisqualate, Q; metabotropic quisqualate receptor, Qp; magnesium ions, Mg. The sodium dependent glutamate uptake sites (upt), on glia and axon are indicated, as is the voltage operated calcium channel (VOC), and energy requiring ion pumps ($\approx P$)

toxic concentrations. This could occur, for example, during excessive release of glutamate from the presynaptic nerve endings overriding the efficiency of the uptake systems, found during neuronal hyperactivity (epilepsy). Severe energy deprivation, such as in cerebral ischemia or hypoglycemia, may elevate extracellular glutamate levels both via the activation of the calcium-dependent exocytotic mechanism and by the reversal of the sodium dependent uptake. Apart from changes in the extracellular levels of glutamate due to an imbalance between release and uptake, changes in the postsynaptic characteristics of glutamate receptors (increased number of receptors or increased affinity of the receptors for glutamate) may also be a contributory factor to glutamate neurotoxicity.

Glutamate Receptors

Four major types of glutamate receptors have been identified (Fig. 2) Three are ionotropic receptors, i.e., receptors coupled to ion channels with multiple conductance states, classified by their affinity for specific glutamate analogues, kainate (K), quisqualate (Q), and NMDA [24]. A metabotropic receptor, Qp, that is activated by quisqualate has also been described [25–27]. This receptor stimulates the hydrolysis of polyphosphoinositol-phosphoglycerides via a G-protein-mediated activation of phospholipase C (PLC), mobilizing diglycerides and intracellular calcium [28]. The role of the Qp receptor in the brain is still elusive. Of importance for the present discussion is the calcium mobilization and diglyceride and arachidonate formation associated with activation of the Qp re-

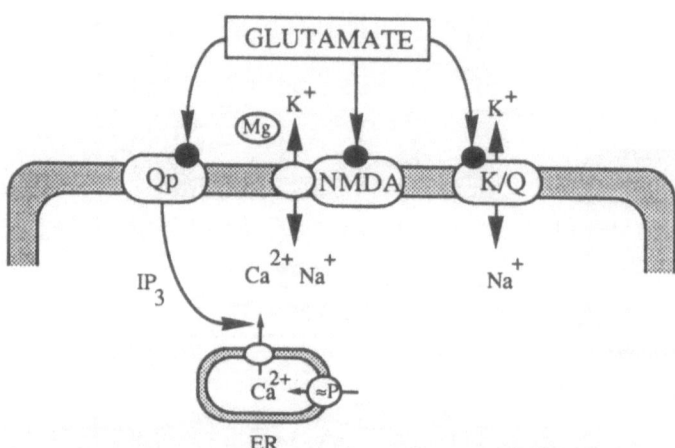

Fig. 2. The four major types of known glutamate receptors in the brain: the ionotropic Kainate, K; Quisqualate Q receptors, that are permeable to potassium and sodium; the N-methyl-D-aspartate, NMDA receptor that in addition to sodium and potassium, is also permeable to calcium ions when the magnesium ion (Mg) block is released. The metabotropic quisqualate activated receptor (Qp), stimulates inositoltrisphosphate (IP$_3$) formation mobilizing intracellular calcium from intracellular stores (ER)

ceptor (see below). Extensive electrophysiological studies of the ionotropic receptors have been conducted [29]. The Q/K receptor-operated ion channels (Q/K-ROC) have intermediate permeabilities for sodium and potassium ions, and participate in fast excitatory transmission evoked by low frequency stimulation [30]; they can be blocked by specific antagonists such as CNQX [31]. The NMDA receptor-operated channel (NMDA-ROC) is a high conductance channel [32–34], that, in addition to sodium and potassium, may also be permeable to calcium ions [35–36]. Under normal conditions, the contribution of NMDA receptors to fast neurotransmission is small [30]. During high frequency stimulation glutamate receptors are activated and the accompanied depolarization of the plasma membrane releases the magnesium block which normally inhibits the NMDA-ROC, rendering the NMDA-ROC permeable to calcium ions [34–37].

The calcium ion has been implicated as an important factor in the pathogenesis of ischemic and hypoglycemic neuronal damage [38]. Thus the NMDA receptor with its high calcium permeability is of particular interest. The receptor has not yet been characterized biochemically, but several regulatory and modulatory sites that affect its electrophysiological properties have been identified (Fig. 3). Glutamate binds to the receptor molecule and can be competitively displaced by antagonists (competitive antagonists) such as AP7 or CPP [39]. As mentioned above, the calcium ion which fluxes through the NMDA-ROC are blocked by magnesium ions in a voltage-dependent manner. The channel can also be blocked by zinc ions [40] and by non-competitive receptor antagonists such as MK-801, dextrorphane, or phencyclidine derivatives [41]. These antagonists require an open channel to exert their inhibitory effects, and are thus use-dependent

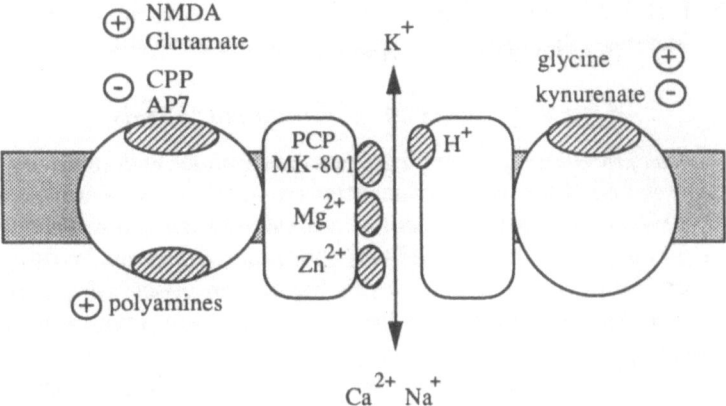

Fig. 3. A hypothetical picture of the NMDA receptor, based on available experimental data. The binding sites (hatched areas) for activators (+) and inhibitors (−) are indicated

[42]. Glycine potentiates the glutamate-induced activation of the NMDA receptor [43], and kynurenate is an antagonist at this site [44]. Recently, it was shown that polyamines bind to the NMDA receptor, and that the protective effects of ifenprodil against ischemic damage could occur at this site [45–47]. The receptor requires ATP for normal function and may be regulated by phosphorylation [48]. Acidosis decreases the activity of the NMDA-ROC in dissociated cell cultures [49]. At a pH of 6.1 (a pH value readily attained during ischemia) the NMDA-ROC is not activated by glutamate receptor agonists.

The activation of the NMDA receptor causes a massive influx of calcium ions that may activate a vast number of intracellular reactions. In isolated neurons and brain homogenates, NMDA receptor agonists increase the levels of arachidonic acid and its oxidative products, probably via activation of a calcium-dependent phospholipase A_2 [50].

Effects of Ischemia on Glutamate Receptors

The binding of agonists to the glutamate receptors does not increase during or following ischemia. Immediately following reperfusion, a transient decrease in AMPA binding to quisqualate receptors is noted [51]. In contrast to the kainate and quisqualate receptors, the NMDA receptor complex is remarkably resistant to neuronal degeneration, and retains its glutamate binding properties during the reperfusion phase. For example, one week following a transient ischemic insult, only a 20% decrease in the NMDA-sensitive glutamate binding is seen, although extensive neuronal degeneration is evident [52,53]. Similar effects can be seen in hypoglycemia [54], and suggest that if it is functional, the NMDA receptor may exert a deleterious effect in the reperfusion phase. However, the functional state of the receptors during and following ischemia in vivo has not been studied. It is feasible that during ischemia, when intracellular pH and the ATP levels decrease, the NMDA receptor is inhibited [48,49].

Effects of Ischemia on Glutamate Levels

Since the balance between glutamate release and uptake is dependent upon the sodium gradient and the calcium concentration in the nerve endings, ischemia sufficient to elevate intracellular sodium and calcium concentrations will cause an increase in the extracellular levels of glutamate. In many investigations, a mani fold increase in the glutamate levels have been observed during and immediately following ischemia [55,56]. Before elaborating on the cause and possible implications of the elevated extracellular glutamate concentrations in ischemia, we will summarize some of the relevant pathophysiology of ischemia.

Pathophysiology of Cerebral Ischemia

The Ischemic Thresholds

During *incomplete ischemia*, when cerebral blood flow (CBF) decreases from normal values (60–100 ml/100 g·min) to around 15–20 ml/(mg·min), brain electrical activity (measured as evoked electrical potentials) ceases [57,58], probably due to increased potassium conductance in presynaptic elements that hyperpolarizes the membrane [59], and/or to an increase in extracellular pH [60]. The phosphocreatine levels decrease [61,62], suggesting that energy shortage may prevail at the plasma membrane [63,64], although the overall brain ATP levels are essentially normal. A progressive increase in the extracellular glutamate levels occurs already prior to complete cessation of electrical activity [56], i.e., at around 20 ml/(100 g·min). The significance of this early release of glutamate for brain damage is not known, but the elevated glutamate levels may activate Q/K- and NMDA-ROC causing an increase in ion cycling across the plasma membrane thus increasing glucose consumption [65], and activating intracellular second messenger systems. Since oxygen supply is limited, a further deterioration of the cellular energy state caused by an increased glucose consumption can push the cell over the threshold of terminal membrane depolarization. This occurs spontaneously when *complete ischemia* is instituted, i.e., when blood flow decreases to below 5–10 ml/(100 g·min). The oxygen supply becomes severely curtailed, inhibiting ATP production [66] and causing membrane ion pump failure, leading to an elevation of intracellular sodium and calcium ion concentrations and extracellular potassium ion concentrations [67,68] (Fig. 4). These ionic shifts may occur both at pre- and postsynaptic sites. Presynaptically, the increased calcium and sodium levels in the nerve terminals may enhance the release of neurotransmitters, including glutamate, by a reversal of the sodium uptake system, and the stimulation of the calcium- and PKC-dependent transmitter release. Postsynaptically, VOCs and ROCs will be activated, including the Q/K- and NMDA-ROC. In addition, the receptor-triggered elevation of second messenger concentrations such as calcium ions, cAMP, and diglycerides,

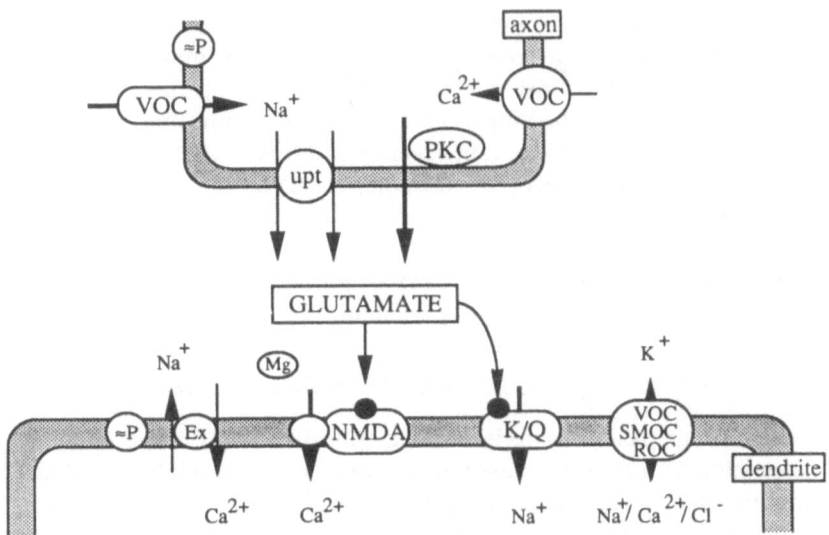

Fig. 4. The possible events taking place at a glutamatergic synapse during ischemia, causing energy depletion severe enough to inhibit the ATPases ($\approx$P). Presynaptically, sodium and calcium will enter through voltage-operated channels (VOC), stimulating the release of glutamate and other neurotransmitters. Protein kinase C (PKC) may also stimulate the release. Postsynaptically, the plasma membrane will be depolarized by activation of receptor operated ion channels (ROC) including the glutamate receptors (NMDA, K, and Q), voltage-operated (VOC) and second-messenger-operated (SMOC) ion channels. The elevated sodium concentration will enhance the influx of calcium ions via the sodium/calcium antiporter (Ex)

may activate second messenger-operated channels (SMOCs). The increased intracellular sodium concentration will drive the sodium-calcium antiporter to exchange intracellular sodium ions for calcium ions, increasing the intracellular calcium concentration [69].

Ischemic Neuronal Damage

If cerebral blood flow is decreased for a prolonged period of time to levels causing membrane depolarization, irreversible damage to brain cells will be incurred [70]. If the ischemic insult is complete with no reperfusion, neurons will degenerate due to the activation of cellular degradative systems such as the lysozomes. If the ischemic insult is transient or incomplete, several factors affecting the severity and distribution of damage can be defined: the ischemic period, the density of ischemia (residual perfusion during the insult), the local environment and biochemistry of the affected neurons, and influence of blood borne compounds (hormones). Thus brain damage studied using in vivo models of cerebral ischemia will vary depending on the procedure of induction of ischemia and the characteristics of the experimental animals. These factors will determine the degree of activation of a variety of adverse inter- and intracellular interactions and reactions, and thereby also determine the efficiency of pharmacological interventions, including the glutamate receptor blockade. For example, a pharmacological treatment directed against one particular adverse mechanism dominating the cause of cell damage following a certain ischemic insult, may not be effective following an insult that is different in severity and that induces neuronal damage that is dependent on other adverse mechanisms. Although it is not the scope of the present article to compare different ischemia models, a short summary of the major features of the experimental models currently used in ischemia research is necessary before discussing the efficiency of glutamate antagonists as cerebroprotective agents during and following ischemia.

Experimental Models of Cerebral Ischemia

Due to the intimate relationship between cerebral blood flow and ischemic damage it is possible to group the vast number of ischemia models described in the literature into two major categories: (1) models of transient global ischemia, and (2) models of focal ischemia, where one or several major arteries in the brain are permanently or transiently occluded.

The 4-vessel occlusion (4-VO) ischemia type (occlusion of the common carotid and vertebral arteries) [71], the 2-vessel occlusion (2-VO) type (occlusion of common carotid arteries combined with hypotension) in the rat [72], the 2-VO, type (occlusion of both common carotid arteries) in the gerbil [73], and the cardiac arrest types (transient cardiac arrest) are models of global complete ischemia (CBF < 5ml/(100 g.min)). These models probably resemble the clinical situation of cardiac arrest with subsequent resuscitation, as well as the severely ischemic areas during stroke that are reperfused following, for example, treatment with anticoagulants. In these models, plasma membrane depolarization commences 1–2 min after the onset of ischemia, preceded by a slow increase in K_e (67,68) (Fig. 5). When blood is reintroduced after a short period of ischemia (10–20min) energy metabolism and ion gradients are normalized. Within 3–72 h of reperfusion selective neuronal necrosis develop in specific vulnerable regions of the brain [71,73,74]. Postischemic pharmacological interventions may decrease neuronal damage [75,76], demonstrating that events in the reperfusion phase, such as neuronal hyperactivity [77] or transient membrane depolarizations [78], may contribute to the development of neuronal damage.

The other category of experimental models resemble stroke, and include models where one major artery, such as the middle cerebral artery (MCA), is occluded [79], or clots or microspheres sufficiently large to occlude vessels are injected intravenously [80]. An area of mixed density of ischemia is formed in the territory supplied by the occluded arteries via collateral vessels. If the vessels are permanently occluded, total tissue necrosis (infarct) develops in the ischemic center, while in the tissue surrounding the ischemic focus (the "penumbra"), the residual blood flow is sufficiently low to induce inhibition of neurotransmission but not membrane depolarization [81]. In this area, energy production is severely curtailed but sufficient ATP production remains [61,62] to preserve some activity of the $Na^+ - K^+ - ATPase$, and to transiently normalize ionic gradients across

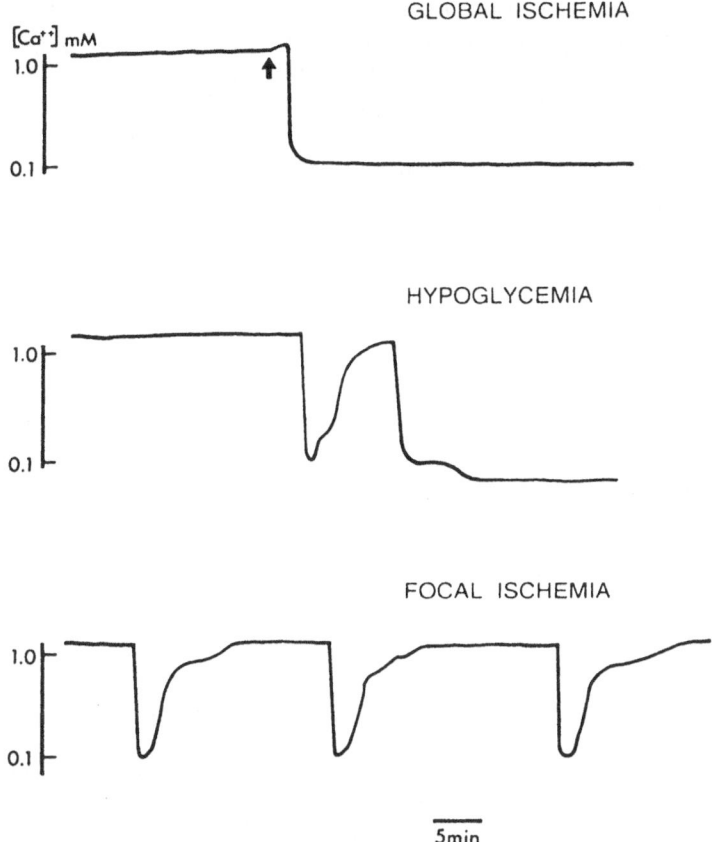

Fig. 5. Changes in the extracellular levels of calcium ions as measured by calcium selective microelectrodes: during global ischemia induced by i.v. injection of potassium chloride, during insulin-induced hypoglycemia (*middle*), and during focal ischemia induced by occlusion of the middle cerebral artery (*bottom*). (From [101])

the plasma membrane [65] (Fig. 5). The "penumbra" area is apparently an area at risk which may succumb if ischemia is allowed to progress without intervention, but can potentially be salvaged by protective agents.

The division into these two categories is not clear-cut. For example, in the rat 2-VO model, occlusion of common carotid arteries will induce complete ischemia to the forebrain but not to the caudal parts of the brain, and thus a region of intermediate flow can be expected [72]. Likewise, in the focal ischemia model, the infarct size varies between animal species [82]. If collateral blood flow in the penumbra zone is inefficient [83], such as in hypertensive subjects, a negligible penumbra zone can be envisaged.

Hypoglycemia is a condition where brain energy stores are partially depleted, and is similar in some pathophysiological aspects to the so-called "penumbra" zone. When blood glucose levels decrease below 1 umol/g for a prolonged period of time, brain ATP production decreases by approximately 50% [84], resulting in persistent membrane depolarization, preceded by one or more transient membrane depolarizations [85] (Fig. 5). A hypoglycemic insult (30–60 min) in the rat causes selective neuronal damage in cortical and subcortical regions. In the cortex, damage develops acutely during the hypoglycemic insult, while damage in the caudate nucleus matures several hours after restitution of glucose levels [86].

A third class of ischemia models are the in vitro systems, such as the brain slice preparations [59] and the dissociated cell cultures subjected only to anoxia [5,7] or in combination with hypoglycemia [87]. These models are valuable when the objective is to study particular cellular mechanisms while the surrounding milieu is controlled. Much important information concerning the mechanism of neuronal damage has been obtained using these models, but the applicability of these results in the in vivo situations has to be considered critically. Of particular importance may be the relatively small extracellular space in the brain compared to the cell culture dish, and the influence of blood components and integrated neuronal circuits present in the intact animal.

Glutamate Antagonists and Ischemic Neuronal Damage

Several possible preventive measures against glutamate toxicity during ischemia can be envisaged: (1) prevention of glutamate release during ischemia, (2) stimulation of glutamate uptake, and (3) pharmacological blockade of the glutamate receptors.

Lesions of the glutamatergic fibres innervating a brain region vulnerable to ischemia prevent the release of glutamate in the projection area. It has been demonstrated that lesions of the glutamatergic innervation to the hippocampus decrease CA1 damage following global ischemia in rats [88,89] and gerbils [90], demonstrating the involvement of glutamatergic neurotransmission in ischemic damage. Similarly, lesions of the glutamatergic cortico-striatal pathways from neocortex to the striatum decrease hypoglycemic damage in the striatum [91]. Selective inhibitors of glutamate release are not yet available, but may be a target for pharmaceutical interventions.

Glutamate uptake is dependent upon the preservation of the sodium gradient across the plasma membrane and may be enhanced by preserving and stimulating the energy production. This can be accomplished by enhancing the reperfusion of the brain following complete ischemia, or by enhancing perfusion in the perifocal area during focal ischemia. Some calcium antagonists may act in this manner [92].

Blockade of postsynaptic glutamate receptors, the NMDA receptor in particular, has been a major focus of ischemia research during recent years. As mentioned in the introduction, evidence supporting the involvement of glutamate and related compounds in ischemia originated both from experiments performed in vivo and in vitro. In dissociated neuronal cultures, hypoxic and hypoglycemic damage can efficiently be blocked by NMDA receptor antagonists [7,87]. On the other hand, results obtained with in vivo models are ambiguous. Protective

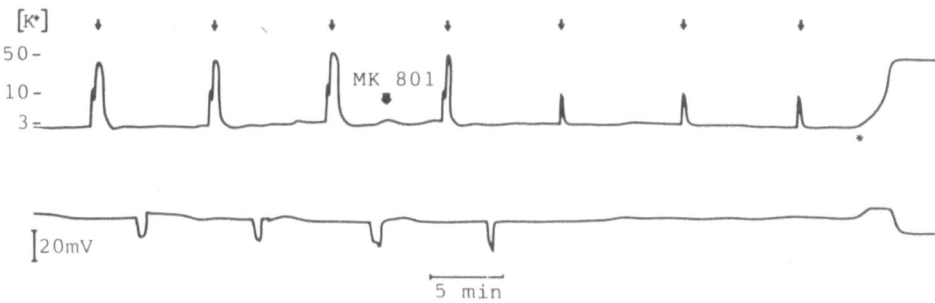

Fig. 6. Changes in the extracellular potassium concentrations following repeated cortical stimulations which elicit spreading depressions. Following intravenous injections of the non-competitive glutamate antagonist MK-801 (10mg/ml) (*arrow*), spreading depression could not be elicited. However, following potassium chloride injections (*asterisk*) a persistent increase in potassium levels was observed signifying membrane depolarization. (From [100])

effects of competitive and non-competitive glutamate antagonist have been reported in models of global ischemia [76,93,94]. However, other reports do not corroborate these findings. In two studies in rats [12,13], competitive or non-competitive antagonists given pre- or postischemia failed to show any protective effect. Similar findings have been reported from experiments using global ischemia models in monkeys and dogs [95,96].

How can these discrepancies be explained? If, indeed, membrane depolarization triggers the reactions that cause neuronal damage and if glutamate antagonists are protective, can glutamate antagonists prevent membrane depolarization during ischemia? Spreading depression of Leao (SD) is a transient reversible membrane depolarization induced by stimulation of the cortical surface, and where the initial extracellular ionic changes are similar to those observed during anoxic membrane depolarization [67]. During SD the energy charge of the tissue is preserved [97]. Spreading depression does not cause neuronal damage [98] and can be completely inhibited by competitive and non-competitive antagonists [99–101]. Fig. 6 shows the transient membrane depolarizations induced by repeated cortical spreading depressions. When MK-801 or competitive antagonists such as CGS-19755 are administered, SD is completely inhibited. However, following induction of ischemia by cardiac arrest in the same animal, terminal membrane depolarization transpires despite a blockade of the NMDA-ROC (Fig. 6). Ion fluxes across the plasma membrane apparently take place through other channels such as VOCs, ROCs, and SMOCs, and/or by the reversal of sodium/calcium exchangers, (Fig. 4). If membrane depolarization triggers the events causing neuronal damage, then one would not expect the NMDA antagonists to be protective in areas with complete ischemia. However, in situations similar to that of spreading depression, i.e., where energy production is partially preserved such as in the "penumbra" area during focal ischemia (see below), or in the reperfusion phase following transient global ischemia, a partial protective

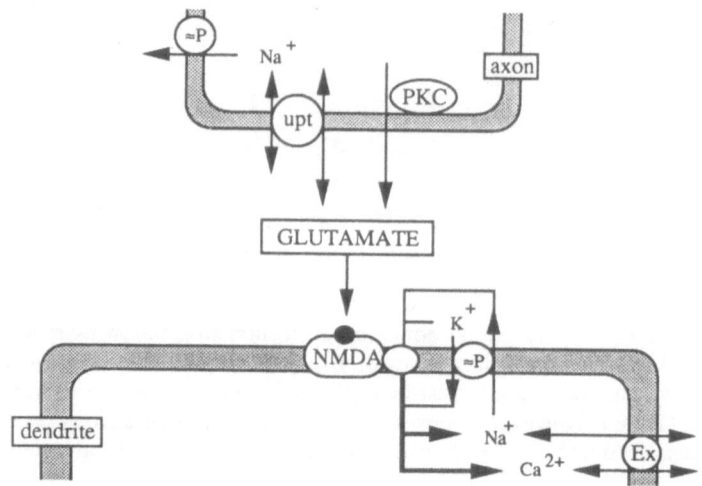

Fig. 7. A hypothetical picture of the events taking place at the plasma membrane; e.g., in the ischemic "penumbra" zone. The transient membrane depolarizations induced by glutamate causes a release of the magnesium block of the NMDA-ROC, increasing ion cycling and energy consumption, and causing an elevation of the intracellular calcium levels. A blockade of the NMDA receptor will preserve energy that can be utilized for glutamate uptake and termination of second messenger action

effect by NMDA receptor blockade can be envisaged. For example, a number of competitive and non-competitive antagonists are effective in the 2-VO model when given immediately following the ischemic insult [76]. In the gerbil 2-VO ischemia model, both competitive and noncompetitive antagonists given in multiple doses postischemia (up to 24 h postischemia) are protective [93,94]. Thus, events taking place during the reperfusion phase may contribute to the final damage. Some of these processes may be affected by NMDA-antagonists, such as postischemic recovery of the membrane potential, postischemic neuronal hyperactivity [102], or transient NMDA receptor-induced membrane depolarizations [78]. Since the NMDA antagonist when given in the reperfusion phase is not always protective [12,13], other non-NMDA receptor-mediated processes must also contribute [75].

In models of focal cerebral ischemia, it has generally been found that the NMDA antagonists decrease infarct size by approximately 50% [80, 103–105]. In these situations, the NMDA receptors may contribute to the spread of membrane depolarization and to the growth of the infarct size. In the "penumbra" zone, where ion cycling probably occurs through the NMDA-ROC, NMDA receptor antagonists may be cerebroprotective by blocking this energy-consuming and calcium-elevating process (Fig. 7). In spontaneously hypertensive rats, occlusion of the MCA causes a larger infarct size than in normotensive rats [82], suggesting that a more complete ischemia is induced and that the penumbra zone is smaller or insignificant. Here NMDA antagonists are not as effective cere-broprotectants [106].

During severe hypoglycemia, local injections of AP7 into the striatum and hippocampus [107,108] and systemic administration of MK-801 decrease neuronal necrosis [109]. Since a significant amount of ATP is produced during hypoglycemia, and the damage in striatum develops in the recovery period when energy metabolism is essentially normalized, the NMDA antagonists may exert a protective effect in a similar manner as in the "penumbra" area. For example, the observed increase in extracellular levels of glutamate during hypoglycemia is curtailed by MK-801, suggesting that a postsynaptic blockade of the NMDA receptors preserves energy that can be utilized for, among others, glutamate uptake [109].

Thus in brain regions where a significant energy production remains, such as in the ischemic "penumbra" zone during reperfusion following transient cere-bral ischemia or hypoglycemia, the energy charge of the tissue may be sufficient to transiently maintain the integrity of the plasma membrane. The activation of ion channels will increase energy consumption, and elevate intracellular messengers that may initiate detrimental processes eventually causing cell death. The NMDA-ROC is particularly sensitive to membrane depolarization, and may thus significantly contribute to these processes. Blockade of the NMDA-ROC will interrupt cycling of ions across the plasma membrane, including calcium influx, thereby preserving energy, preventing activation of detrimental in-tracellular processes, and enhancing glutamate uptake.

Mechanisms of Glutamate Neurotoxicity

The intracellular mechanism underlying neuronal damage following ischemia is unknown. Events associated with the membrane depolarization and the activa-tion of receptors probably trigger the intracellular events eventually leading to neuronal damage, such as excessive influx of calcium ions and enhanced mem-brane lipid turnover, which activate the second messenger systems.

Importance of Intracellular Calcium

In neuronal cell cultures, a clear calcium dependency of ischemic neuronal death has been shown [6,7], and increased cytosolic calcium activities have been de-monstrated during hypoxia and hypoglycemia [110,111]. The importance of cal-cium ions in the reperfusion phase is less clear. It seems reasonable that delete-rious calcium-activated processes may continue in the early reperfusion phase when plasma membranes repolarize, and that the NMDA-ROC may contribute to this calcium influx. Increased intracellular calcium levels may activate harmful reactions [38] such as calpain-stimulated degradation of cytoskeletal proteins [112], and phospholipase and lipase-induced degradation of cell membranes [113]. The relative importance of these reactions for ischemic neuronal damage, has not been clarified.

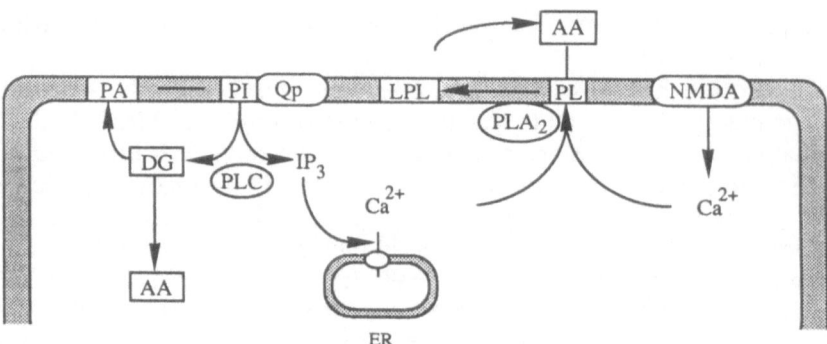

Fig. 8. A schematic picture of the phospholipid turnover cycles possibly activated by glutamate receptors in the brain. Ligand receptor interactions may cause activation of a phospholipase C via a G-protein. The polyphosphoinositolphospholipids (PI) are degraded to diglycerides (DG) and inositoltrisphosphate (IP_3), that can mobilize calcium from intracellular stores (ER). In the presence of ATP, the diglycerides can reform PI via phosphatidic acid (PA). Some of the DG can be further hydrolyzed to arachidonic acid. Phospholipids in general may be subjected to the action of a phospholipase A_2 activated by an elevated intracellular calcium ion concentration caused by a release from intracellular stores (ER) or via entry through the NMDA receptor operated channel. The degradation of PL leads to the formation of lysophospholipids and free fatty acids, including arachidonic acid. The PL can be resynthesized by an ATP requiring process

Importance of Phospholipid Breakdown

Receptor activation and an increase in intracellular calcium ion concentration stimulate the turnover of phospholipids (Fig. 8). During cerebral ischemia when ATP levels are low, an enhanced degradation of phospholipids occurs leading to increased levels of diglycerides and free fatty acids [114]. In ischemia, hypoglycemia, and spreading depression there is a marked increase in the levels of arachidonic acid (AA) [113,115,116,117]. The contribution of glutamate receptor activation to the ischemia-induced phospholipid degradation is not known. Activation of the Qp receptor may elevate the levels of diglycerides and IP_3 [26,27]. Since reacylation and phosphorylation processes are mitigated due to ATP shortage, the diglycerides are further broken down to free fatty acids including AA. The NMDA receptor-mediated calcium influx may activate PLA_2 that lead to hydrolysis of phospholipids and the release of AA [50]. There is a correlation between the release of arachidonic acid and neuronal damage. During ischemia, the levels of arachidonic acid increase more in the ischemia sensitive CAl region than in the less vulnerable CA3 region [118]. Cortical lesions leading to depletion of the glutamatergic input to the striatum mitigate the early increase in arachidonic acid during hypoglycemia and protect striatal neurons against hypoglycemia-induced damage [119].

Two effects of arachidonic acid pertinent to glutamate toxicity can be envisaged: inhibition of glutamate uptake, and stimulation of PKC. Arachidonic acid levels increase immediately following complete ischemia, prior to membrane

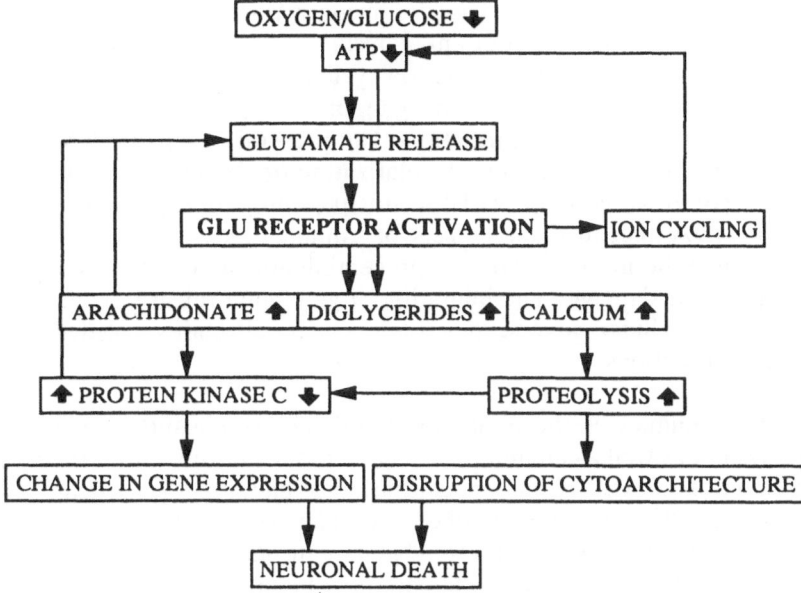

Fig. 9. Some of the cellular events taking place during ischemia, that may cause neuronal damage

depolarization, and may influence early synaptic events, such as glutamate uptake and release. In brain slices, glutamate uptake is inhibited by arachidonic acid or its oxidative products, either by inhibiting the sodium glutamate symporter and/or by inhibiting the $Na^+ - K^+$ ATPase [120,121]. Arachidonic acid and diglycerides are activators of PKC [122,123]. PKC is translocated from the cytosol to cell membranes during and following ischemia [124,125], and downregulated in the postischemic phase [125,126]. The action of PKC is complex and may adversely affect cell function both when excessively activated or when inhibited and downregulated. The translocated PKC may stimulate the release of transmitters and activate calcium ion channels [127]. Downregulation and/or inhibition of PKC may have adverse effects, since some growth factors and survival promoting responses are PKC-dependent [128]. Covalent modifications of proteins incited by calcium-activated processes or excessive receptor activation may ultimately be events causing cell death. Such reactions most probably include proteolysis, protein phosphorylation, and protein ubiquitination [129], and may cause cell damage by changing structural cell components and detrimentally modifying gene expression (Fig. 9).

Conclusions

Glutamate may contribute to ischemic neuronal damage by aggravating and expanding the damage in the penumbra zone, and by imposing an additional insult to neurons in the reperfusion phase following a transient period of complete ischemia. The glutamate receptor antagonists, particularly the NMDA antagonists, may be protective by inhibiting ion cycling across the plasma membrane, thereby preserving energy enhancing glutamate uptake, and terminating intracellular second messenger formation. The presently available NMDA antagonists do not prevent membrane depolarization or decrease neuronal damage induced by complete ischemia (CBF<5ml/100g/min). It is possible that selective non-NMDA receptor antagonists or combinations of different receptor antagonists may be more effective in preventing anoxic membrane depolarization and ischemic damage. The inhibition of glutamate release or the blockade of specific intracellular adverse reactions may be alternative routes for therapeutical interventions.

Summary. Glutamate is the major excitatory neurotransmitter in the central nervous system as well as an amino acid essential for normal brain function and development. The extracellular concentration of glutamate is meticulously regulated by controlled release and effective uptake systems. If the synaptic levels of glutamate are persistently elevated, glutamate may induce neurodegeneration. Such neurotoxic levels of glutamate may be attained under conditions of enhanced neuronal activity or severe energy deprivation, such as during cerebral ischemia and hypoglycemia. Glutamate toxicity is mediated via receptors on glutamate sensitive neurons. The N-methyl-D-aspartate (NMDA) receptor-operated ion channel is permeable to calcium ions when the plasma membrane is depolarized, and mediates glutamate toxicity in several brain regions. Cerebro-protection by NMDA antagonists following cerebral ischemia is ambiguous. In some investigations, NMDA receptor antagonists decrease infarct size and mitigate neuronal damage in selective vulnerable brain regions, while in other investigations NMDA receptor antagonists fail to decrease brain damage. The NMDA antagonists inhibit plasma membrane depolarization and cellular calcium influx during experimental cortical spreading depression, but not following complete cerebral ischemia, indicating that NMDA antagonists may be neuroprotective in tissue where some energy formation remains. This could occur in the area surrounding the ischemic center during focal ischemia by preventing expansion of damage or in areas recovering from a transient ischemic insult, by preventing postischemic complications (such as neuronal hyperactivity) or by enhancing membrane repolarization. The intracellular events leading to glutamate-induced cell death during ischemia could include initial ion cycling across the plasma membrane that causes energy depletion concurrent with detrimental activation of second messenger systems. This could trigger adverse reactions, such as covalent modification of proteins including proteolysis, protein phosphorylation and protein ubiquitination.

Acknowledgements. This work has been supported by the Swedish Medical Research Council (4X-08644) and United States Public Health Services (NS-25302).

References

1. Lucas DR, Newhouse JP (1957) The toxic effect of sodium L-glutamate on the inner layers of the retina. Arch Ophthalmol 58:193–201
2. Olney JW, Ho OL, Rhee V (1971) Cytotoxic effects of acidic and sulphur containing amino acids on the infant mouse central nervous system. Exp Brain Res 14:61–76
3. Olney JW (1978) Neurotoxicity of excitatory amino acids. In: McGeer EG, Olney JW, McGeer PL (eds) Kainic acid as a tool in neurobiology. Raven, New York pp 95–121
4. Rothman SM (1985) The neurotoxicity of excitatory amino acids is produced by passive chloride influx. J Neurosci 5:1483–1489
5. Rothman SM (1984) Synaptic release of excitatory amino acid neurotransmitters mediates anoxic neuronal death. J Neurosci 4:1884–1891
6. Choi DW (1987) Ionic dependence of glutamate neurotoxicity. J Neurosci 7:369–379
7. Goldberg MP, Weiss JH, Pham PC, Choi DW (1987) N-methyl-D-aspartate receptors mediate hypoxic neuronal injury in cortical culture. J Pharmacol Exp Ther 243:784
8. Van Harreveld A (1959) Compounds in brain extracts causing spreading depression of cerebral cortical activity and contraction of crustacean muscle. J Neurochem. 3:300-315
9. Van Harreveld A, Fifkova E (1971) Light- and electron-microscopic changes in central nervous tissue after electrophoretic injection of glutamate. Exp Mol Pathol 15:61–81
10. Jørgensen MB, Johansen FF, Diemer NH (1987) Removal of the entorhinal cortex protects hippocampal CA1 neurons from ischemic damage. Acta Neuropathol (Berl) 73:189–194
11. Albers GW, Goldbergs, Choi DW (1989) N-methyl-D-aspartate antagonists: Ready for clinical trial in brain ischemia? Ann Neurol 25:398–403
12. Block GA, Pulsinelli WA (1987) Excitatory amino acid receptor antagonists: Failure to prevent ischemic neuronal damage. J Cereb Blood Flow Metab (suppl) 1 (7):149
13. Wieloch T, Gustafson I, Westerberg E (1988) Effects of the non-competitive NMDA receptor antagonist MK-801 on ischemic and hypoglycemic brain damage. In: Turski L, Lehmann J, Cavalheiro EA (eds) Frontiers in excitatory amino acid research, Liss, New York pp 715–722
14. Fonnum F (1984) Glutamate: A neurotransmitter in the mammalian brain. J Neurochem 42:1–11
15. Sloviter RS, Dempster DW (1985) "Epileptic" brain damage is replicated qualitatively in the rat hippocampus by central injection of glutamate or aspartate but not by GABA or acetylcholine. Brain Res Bull 15:39–60
16. Köhler C, Schwarcz R (1981) Monosodium glutamate: Increased neurotoxicity after removal of neuronal re-uptake sites. Brain Res 211:485–491
17. McBean GJ, Roberts PJ (1985) Neurotoxicity of L-glutamate and DL-threo-3-hydroxyaspartate in the rat striatum. J Neurochem 44:247–254
18. Carafoli E (1987) Intracellular calcium homeostasis. Annu Rev Biochem 56:395–433
19. Nicholls DG (1989) Release of glutamate, aspartate and γ-aminobutyric acid from isolated nerve terminals. J Neurochem 52:331–341

20. Chandler LJ, Leslie SW (1989) Protein kinase C activation enhances K^+-stimulated endogenous dopamine release from rat striatal synaptosomes in the absence of an increase in cytosolic Ca^{2+}. J Neurochem 52:1905–1912
21. Allgaier C, Hertting G, Huang HY, Jackisch R (1987) Protein kinase C activation and α_2-autoreceptor modulated release of noradrenaline. Br J Pharmacol 92:161–172
22. Diaz-Guerra MJM, Sanchez-Pietro J, Bosca L, Pocock J, Barrie A, Nicholls D (1988) Phorbol ester translocation of protein kinase C in guineapig synaptosomes and the potentiation of calcium-dependent glutamate release. Biochem Biophys Acta 970:157–165
23. Sánchez-Prieto J, Gonzáles P (1988) Occurence of a large Ca^{2+}-independent release of glutamate during anoxia in isolated nerve terminals (synaptosomes). J Neurochem 50:1322–1324
24. Watkins JC, Evans RH (1981) Excitatory amino acid transmitters. Annu Rev Pharmacol Toxicol 21:165–204
25. Sugiyama H, Ito I, Hirono C (1987) A new type of glutamate receptor linked to inositol phospholipid metabolism. Nature 325:531–533
26. Nicoletti F, Iadarola MJ, Wroblewski JT, Costa E (1986) Excitatory amino acid recognition sites coupled with inositol phospholipid metabolism: Developmental changes and interaction with αl-adrenoceptors. Proc Natl Acad Sci USA 83:1931–1935
27. Sladeczek F, Pin J-P, Récasens M, Bockaert J, Weiss S (1985) Glutamate stimulates inositol phosphate formation in striatal neurones. Nature 317:717–719
28. Berridge MJ (1984) Inositol triphosphate and diacylglycerol as second messengers. Biochem J 220:345–360
29. Dingledine R, Boland LM, Chamberlain NL, Kawasaki K, Klickner NW, Traynelis SF, Verdoorn TA (1988) Amino acid receptor and uptake systems in the mammalian central nervous system. CRC Critical Rev J Neurobiol 4:1–96
30. Davies SN, Collingridge GL (1989) Role of excitatory amino acid receptors in synaptic transmission in area CAl of rat hippocampus. Proc R Soc Lond 236:373–384
31. Honore T, Davies SN, Drejer J, Fletcher EJ, Jacobsen P, Lodge D, Nielsen E (1988) Quinoxalinediones: Potent competitive non-N-methyl-D-aspartate glutamate receptor antagonists. Science 241:701–703
32. Cull-Candy SG, Usowicz MM (1987) Multiple-conductance channels activated by excitatory amino acids in cerebellar neurons. Nature 325:525–527
33. Jahr CE, Stevens CF (1987) Glutamate activates multiple single channel conductances in hippocampal neurons. Nature 325:522–525
34. Nowak L, Bregestovski P, Ascher P, Herbet A, Prochiantz A (1984) Magnesium gates glutamate-activated channels in mouse central neurones. Nature 307:462–465
35. Mayer ML, Westbrook GL (1985) The action of N-methyl-D-aspartic acid on mouse spinal neurones in culture. J Physiol (Lond) 361:65–90
36. Asher P, Nowak L (1987) Electrophysiological studies of NMDA receptors. Trends Neurosci 10:284–288
37. MacDermott AM, Dale N (1987) Receptors, ion channels and synaptic potentials underlying the integrative actions of excitatory amino acids. Trends Neurosci 10:280–284
38. Siesjö BK (1981) Cell damage in the brain: A speculative synthesis. J Cereb Blood Flow Metab 1:155–185.
39. Watkins JC, Olverman HJ (1987) Agonists and antagonists for excitatory amino acid receptors. Trends Neurosci 10:265–272

40. Westbrook GL, Mayer ML (1987) Micromolar concentrations of Zn^{2+} antagonize NMDA and GABA responses of hippocampal neurons. Nature 328:640–643
41. Kemp JA, Foster AC, Wong EHF (1987) Non-competitive antagonists of excitatory amino acid receptors. Trends Neurosci 19:294–299
42. Huettner JE, Bean BP (1988) Block of N-methyl-D-aspartate-activated current by the anticonvulsant MK-8091: Selective binding to open channels. Proc Natl Acad Sci USA 85:1307–1311
43. Johnson JW, Ascher P (1987) Glycine potentiates the NMDA response in cultured mouse brain neurons. Nature 325:529–531
44. Johnson KM, Snell LD, Jones SM, Qi H (1988) Glycine antagonist activity of simple glycine analogues and Nmethyl-D-aspartate receptor antagonists. In: Cavalheiro EA, Lehmann J, Turski L (eds) Frontiers in excitatory amino acid research. Alan Liss, New York, pp 551–558
45. Schoemaker H, Allen J, Langer SZ (1990) Binding of ^{3}H-ifenprodil, a novel NMDA antagouist, to a polyamine-sensitive site in the rat cerebral cortex. Eur J Pharmacol 176:249–250
46. Carter C, Benavides J, Legendre P, Vincent DJ, Noel F, Thuret F, Lloyd KG, Arbilla S, Zivkovic B, MacKenzie ET, Scatton B, Langer SZ (1988) Ifenprodil and SL 82.0715 as cerebral anti-ischemic agents: II. Evidence for N-methyl-D-aspartate receptor antagonist properties. J Pharmacol Exp Ther 247:1222–1232
47. Gotti B, Duverger D, Bertin J, Carter C, Dupont R, Frost J, Gaudilliere B, MacKenzie ET, Rousseau J, Scatton B, Wick A (1988) Ifenprodil and SL 82.0715 as cerebral anti-ischemic agents: I. Evidence for efficacy in models of focal cerebral ischemia. J Pharmacol Exp Ther 247:1211–1221
48. MacDonald JF, Salter MW, Mody I (1988) Intracellular regulation of the NMDA channel. In: Cavalheiro EA, Lehmann J, Turski L (eds) Frontiers in Excitatory amino acid research. Alan Liss, New York pp 159–166
49. Giffard RG, Monyer H, Christine CW, Choi DW (1990) Acidosis reduces NMDA receptor activation, glutamate neurotoxicity and oxygen-glucose deprivation neuronal injury in cortical cultures. Brain Res 506:339–342
50. Dumuis A, Sebben M, Haynes L, Pin J-P, Bockaert J (1988) NMDA receptors activate the arachidonic cascade system in striatal neurons. Nature 336:68–70
51. Westerberg E, Monaghan DT, Cotman CW, Wieloch T (1987) Excitatory amino acid receptors and ischemic brain damage in the rat. Neurosci Lett 73:119–124
52. Westerberg E, Monaghan DT, Kalimo H, Cotman CW, Wieloch T (1989) Dynamic changes of excitatory amino acid receptors in the rat hippocampus following transient cerebral ischemia. J Neurosci 9:798–805
53. Dewar D, Wallace MC, Kurumaji A, McCulloch J (1989) Alterations in the N-methyl-D-aspartate receptor complex following focal cerebral ischemia. J Cereb Blood Flow Metabol 9:709–712
54. Westerberg E, Wieloch T (1989) Changes in excitatory amino acid receptor binding in the intact and decorticated rat neostriatum following insulin-induced hypoglycernia. J Neurochem 52:1340–1347
55. Benveniste H, Drejer J, Schousboe A, Diemer NH (1984) Elevation of the extracellular concentrations of glutamate and aspartate in rat hippocampus during transient cerebral ischemia monitored by intracerebral microdialysis. J Neurochem 43:1369–1374
56. Shimada N, Graf R, Rosner G, Wakayama A, George CP, Heiss W-D (1989) Ischemic flow thresholds for extracellular glutamate increase in cerebral cortex. J Cereb Blood Flow Metabol 9:603–606
57. Branston NM, Symon L, Crockard HA, Pasztor E (1974) Relationship between the

cortical evoked potential and local cortical blood flow following acute middle cerebral artery occlusion in the baboon. Exp Neurol 45:195–208

58. Astrup J, Symon L, Branson NM, Lassen NA (1977) Cortical evoked potential and extracellular K+ and H+ at critical levels of brain ischemia. Stroke 8:51–57

59. Hansen AJ, Hounsgaard J, Jahnsen H (1982) Anoxia increases potassium conductance in hippocampal nerve cells. Acta Physiol Scand 115:301–310.

60. Harris RJ, Richards PG, Symon L, Habib AHA, Rosenstein J (1987) pH, K+, and PO$_2$ of the extracellular space during ischemia of primate cerebral cortex. J Cereb Blood Flow Metabol 7:599–604

61. Obrenovitch TP, Garofalo O, Harris RJ, Bordi L, Ono M, Momma F, Bachelard HS, Symon L (1988) Brain tissue concentrations of ATP, phosphocreatine, lactate, and tissue pH in relation to reduced cerebral blood flow following experimental acute middle cerebral artery occlusion. J Cereb Blood Flow Metabol 8:866–874

62. Naritomi H, Sasaki M, Kanashiro M, Kitani M, Sawada T (1988) Flow thresholds for cerebral energy disturbance and Na+ pump failure as studied by in vivo ^{31}P and ^{23}Na nuclear magnetic resonance spectroscopoy. J Cereb Blood Flow Metabol 8:16–23

63. Meyer RA, Sweeney HL, Kushmerick MJ (1984) A simple analysis of the "phosphocreatine shuttle". Am J Physiol 246:C365–277

64. Bessman SP, Carpenter CL (1985) The creatine-creatine phosphate energy shuttle. Annu Rev Biochem 54:831–862

65. Nedergaard M, Astrup J (1986) Infarct rim: Effect of hyperglycemia on direct current potential and [14C]2-deoxyglucose phosphorylation. J Cereb Blood Flow Metabol 6:607–615

66. Siesjö BK (1978) Brain energy metabolism. Wiley, New York, pp 453–526

67. Hansen AJ, Zeuthen T (1981) Extracellular ion concentrations during spreading depression and ischemia in the rat brain cortex. Acta Physiol Scand 113:437–445

68. Harris RJ, Symon L (1984) Extracellular pH, potassium and calcium activities in progressive ischemia of rat cortex. J Cereb Blood Flow Metabol 4:178–186

69. Choi DW (1988) Calcium-mediated neurotoxicity: Relationship to specific channel types and role in ischemic damage. Trends Neurosci 11:465–469

70. Morawetz RB, DeGirolami U, Ojemann RG, Marcoux FW, Crowell RM (1978) Cerebral blood flow determined by hydrogen clearance during middle cerebral artery occlusion in unanesthetized monkeys. Stroke 9:143–149

71. Pulsinelli WA, Brierley JB, Plum F (1982) Temporal profile of neuronal damage in a model of transient forebrain ischemia. Ann Neurol 11:491–499

72. Smith M-L, G. Bendek, N. Dahlgren, I. Rosén, T. Wieloch, Siesjö BK (1984) Models for studying long-term recovery following forebrain ischemia in the rat: 2. A 2-vessel occlusion model. Acta Neurol Scand 69:385–401

73. Kirino T (1982) Delayed neuronal death in the gerbil hippocampus following transient cerebral ischemia. Brain Res 239:57–59

74. Smith M-L, Auer RN, Siesjö BK (1984) The density and distribution of ischemic brain injury in the rat following 2-10 min of forebrain ischemia. Acta Neuropathol (Berl) 64:319–332

75. Gustafson I, Miyauchi Y, Wieloch T (1989) Postischemic administration of Idazoxan, an α-2 adrenergic receptor antagonist, decreases neuronal damage in the rat brain. J Cereb Blood Flow Metab 9:171–174

76. Swan JH, Meldrun BS (1990) Protection by NMDA antagonists against selective cell loss following transient ischemia. J Cereb Blood Flow Metab 10:343–351

77. Suzuki R, Yamaguchi T, Li C-L, Klatzo I (1983). The effects of 5 min ischemia in

Mongolian gerbils: III.Changes of spontaneous neuronal activity in cerebral cortex and CAl sector of hippocampus. Acta Neuropathol (Berl) 60:217–222

78. Andiné P, Jacobson I, Hagberg H (1988) Calcium uptake evoked by electrical stimulation is enhanced postischemically and precedes delayed neuronal death in CA1 of rat hippocampus: Involvement of N-methyl-D-aspartate receptors. J Cereb Blood Flow Metab 8:799–807

79. Tamura A, Graham DI, McCulloch J, Teasdale GM (1981) Focal cerebral ischaemia in the rat: Description of technique and early neuropathological consequences following middle cerebral artery occlusion. J Cereb Blood Flow Metab 1:53–60

80. Kochar A, Zivin J, Lyden P, Mazzarella V (1988) Glutamate antagonist therapy reduces neurologic deficits produced by focal central nervous system ischemia. Arch Neurol 45:148–153

81. Astrup J, Siesjö BK, Symon L (1981) Thresholds in cerebral ischemia–The ischemic penumbra. Stroke 12:723–725

82. Duverger D, MacKenzie ET (1988) The quantification of cerebral infarction following focal ischemia in the rat: influence of strain, arterial pressure, blood glucose concentration and age. J Cereb Blood Flow Metabol 8:449–461

83. Coyle P, Heistad DD (1986) Blood flow through cerebral collateral vessels in hypertensive and normotensive rats. Hypertension 8 (Suppl 2):67–71

84. Agardh CD, Folbergrova J, Siesjö BK (1978) Cerebral metabolic changes in profound insulin-induced hypoglycemia and in the recovery period following glucose administration. J Neurochem 31:1135–1142

85. Harris RT, Wieloch T, Symon L, Siesjö BK (1984) Cerebral extracellular calcium activity in severe hypoglycemia: Relation to extracellular potassium activity and energy state. J Cereb Blood Flow Metab 4:187–193

86. Auer RN, Olsson Y, Siesjö BK (1984) Hypoglycemic brain injury in the rat: Correlation of density of brain damage with the EEG isoelectric time. A quantitative study. Diabetes 33:1090–1098

87. Monyer H, Goldberg MP, Choi DW (1989) Glucose deprivation neuronal injury in cortical culture. Brain Res 483:347–354

88. Wieloch T, Lindvall O, Blomqvist P, Gage F (1985) Evidence for amelioration of ischemic neuronal damage in the hippocampal formation by lesions of the perforant path. Neurol Res 7:24–26

89. Jørgensen MB, Johansen FF, Diemer NH (1987) Removal of the entorhinal cortex protects hippocampal CAl neurons from ischemic damage. Acta Neuropathol (Berl) 73:189–194

90. Onodera H, Sato G, Kogure K (1986) Lesions of the Schaffer collaterals prevent ischemic death of CA1 pyramidal cells. Neurosci Lett 68:169–174

91. Wieloch T, Engelsen B, Westerberg E, Auer RN (1985) Lesions of the glutamatergic cortico-striatal projections in the rat ameliorate hypoglycemic brain damage in the striatum. Neurosci Lett 58:25–30

92. Gotoh O, Mohamed AA, McCulloch J, Graham DI, Harper AM, Teasdale GM (1986) Nimodipine and the haemodynamic and histopathological consequences of middle cerebral artery occlusion in the rat. J Cereb Blood Flow Metabol 6:321–331

93. Gill R, Foster AC, Woodruff GN (1987) Systemic administration of MK-801 protects against ischaemia-induced hippocampal neurodegeneration in the gerbil. J Neurosci 7:3343–3349

94. Boast CA, Gerhardt SC, Pastor G, Lehmann J, Etienne PE, Liebman JM (1988)

The NMDA antagonists CGS 19755 and CPP reduce ischemic brain damage in gerbils. Brain Res 442:345–348

95. Sterz F, Leonov Y, Safar P, Shearman GT, Stezoski SW, Perch H (1989) Effect of excitatory amino acid receptor blocker MK-801 on overall and neurological outcome after prolonged cardiac arrest in dogs. Anesthesiology 71:907–918

96. Michenfelder JD, Lanier WL, Scheithauer BW, Perkins WJ, Shearman GT, Milde JH (1989) Evaluation of glutamate antagonist dizocilpine maleate (MK-801) on neurological outcome in a canine model of complete cerebral ischemia: correlation with hippocampal histopathology. Brain Res 481:228–234

97. Mies G, Paschen W (1984) Regional changes in blood flow, glucose, and ATP content determined on brain sections during a single passage of spreading depression in rat brain cortex. Exp Neurol 84:249–258

98. Nedergaard M, Hansen AJ (1988) Spreading depression is not associated with neuronal injury in the normal brain. Brain Res 449:395–398

99. Goroleva NA, Koroleva VO, Amemori T, Pavlik V, Bures J (1987) Ketamine blockade of cortical spreading depression in rats. Electroencephalogr Clin Neurophysiol 66:440–447

100. Hansen AJ, Lauritzen M, Wieloch T (1988) NMDA antagonists inhibit cortical spreading depression but not anoxic depolarization. In: Cavalheiro EA, Lehmann J, Turski L (eds) Frontiers in excitatory amino acid research. Liss, New York, pp 661–666

101. Zhang E, Lauritzen M, Wieloch T, Hansen AJ (1989) Calcium movements in brain during failure of energy metabolism. In: Hartmann A, Kuschinsky W (eds) Cerebral ischemia and calcium. Springer, Berlin pp 162–168

102. Suzuki R, Yamaguchi T, Li C-L, Klatzo I (1983) The effects of 5 min ischemia in Mongolian gerbils: II. Changes of spontaneous neuronal activity in cerebral cortex and CA2 sector of hippocampus. Acta Neuropathol (Berl) 60:217–222

103. Steinberg GK, George CP, DeLaPaz R, Shibata DK, Gross T (1988) Dextromethorphan protects against cerebral injury following transient focal ischemia in rabbits. Stroke 19:1112–1118

104. Park CK, Nehls DG, Graham DI, Teasdale GM, McCulloch J (1988) The glutamate antagonist MK-801 reduces focal ischemic brain damage in the rat. Ann Neurol 24:543–551

105. Germano IM, Pitts LH, Meldrum BS, Bartkowski HM, Simon RP (1987) Kynurenate inhibition of cell excitation decreases stroke size and deficits. Ann Neurol 22:730–734

106. Roussel S, Pinard E, Peres M, Seylaz J (1989) Kynurenate and R-PIA do not improve the histopathological consequences of MCA-occlusion in spontaneously hypertensive rats. In: Vol 1. Seylaz J, MacKenzie ET (eds) Neurotransmission and cerebrovascular function. Elsevier, Amsterdam, pp 453–456

107. Wieloch T (1985) Hypoglycemia-induced neuronal damage prevented by an N-methyl-D-aspartate antagonist. Science 230:681–683

108. Simon RP, Schmidley JW, Meldrum BS, Swan JH, Chapman AG (1986) Excitotoxic mechanisms in hypoglycaemic hippocampal injury. Neuropath Appl Neurobiol 12:567–576.

109. Westerberg E, Kehr J, Ungerstedt U, Wieloch T (1988) The NMDA-antagonist MK-801 reduces extracellular amino acid levels during hypoglycemia and prevents striatal damage. Neurosci Res Comm 3:151–158

110. Uematsu D, Greenberg JH, Reivich M, Karp A (1988) In vivo measurement of cytosolic free calcium during cerebral ischemia and reperfusion. Ann Neurol 24:420–428

111. Uematsu D, Greenberg JH, Reivich M, Karp A (1989) Cytosolic free calcium, NAD/NADH redox state and hemodynamic changes in the cat cortex during severe hypoglycemia. J Cereb Blood Flow Metab 9:149–155
112. Siman R, Noszek JC (1988) Excitatory amino acids activate calpain I and induce structural protein breakdown in vivo. Neuron 1:279–287
113. Avaldano MI, Bazan NG (1975) Rapid production of diacylglycerols enriched in arachidonate and stearate during early brain ischemia. J Neurochem 25:919–920
114. Wieloch T, Siesjö BK (1982) Ischemic brain injury: The importance of calcium, lipolytic activities, and free fatty acids. Path Biol 30:269–277
115. Agardh C-D, Chapman AG, Nilsson B, Siesjö BK (1981) Endogenous substrates utilized by rat brain in severe insulin-induced hypoglycemia. J Neurochem 36:490–500
116. Yoshida S, Inoh S, Asano T (1980) Effect of transient ischemia on free fatty acids and phospholipids in the gerbil brain. Lipid peroxidation as a possible cause of postischemic injury. J Neurosurg 53:323–333
117. Lauritzen M, Hansen AJ, Wieloch T (1987) Metabolic changes with spreading depression in rat cortex. J Cereb Blood Flow Metab 7 (Suppl 1): S125
118. Westerberg E, Deshpande JK, Wieloch T (1987) Regional differences in arachidonic acid release in rat hippocampal CA1 and CA3 regions during cerebral ischemia. J Cereb Blood Flow Metab 7:189–192
119. Westerberg E, Wieloch T (1986) Lesions to the cortico-striatal pathways ameliorate hypoglycemia-induced arachidonic acid release. J Neurochem 47:1507–1511
120. Chan PH, Kerlan R, Fischman RA (1983) Reductions of γ-aminobutyric acid and glutamate uptake and (Na+-K+)- ATPase activity in brain slices and synaptosomes by arachidonic acid. J Neurochem 40:309–316
121. Chan PH, Fishman RA (1985) Free fatty acids, oxygen free radicals and membrane alterations in brain ischemia and injury. In: Plum F, Pulsinelli WA (eds) Cerebrovascular diseases. Raven, New York pp 201–205
122. Kraft AS, Anderson WB (1983) Phorbol esters increase the amount of Ca^{2+}/ phospholipid dependent protein kinase associatecl with the plasma membrane. Nature (London) 301:621–623
123. Shearman MS, Naor Z, Sekiguchi K, Kishimoto A, Nishizuka Y (1989) Selective activation of the φ-subspecies of protein kinase C from bovine cerebellum by arachidonic acid and its lipoxygenase metabolites. FEBS Lett 243:177–182
124. Onodera H, Araki T, Kogure K (1989) Protein kinase C activity in the rat hippocampus after forebrain ischemia: Autoradiographic analysis by [^{3}H] phorbol 12,13-dibutyrate. Brain Res 481:1–7
125. Cardell M, Bingren H, Wieloch T, Zivin J, and Saitoh T (1990) Protein kinase C is translocated to cell membranes during cerebral ischemia. Neurosci Lett. 119:228–232
126. Wieloch T, Cardell M, Bingren H, Zivin J, Saitoh T (1991) Changes in the activity of protein kinase C and the subcellular redistribution of its isozymes during and following forebrain ischemia. J Neurochem. In press
127. Kaczmarek LK (1987) The role of protein kinase C in the regulation of ion channels and neurotransmitter release. Trends Neurosci 10:30–34
128. Hsu L (1985) Neurite-promoting effects of 12-O-tetradecanoyl-phorbol-13-acetate on chick embryo neurons. Neurosci Lett 62:382–389
129. Magnusson K, Wieloch T (1989) Impairment of protein ubiquitination may cause delayed neuronal death. Neurosci Lett 96:264–270

3

Are Glutamate/Aspartate Antagonists Protective in Cerebral Ischemia?

John C. Drummond[1]

Introduction

The dicarboxylic amino acids glutamate and aspartate are abundant excitatory neurotransmitters in the mammalian nervous system [1]. Glutamate and aspartate and perhaps other endogenous "excitotoxic" substances, eg., quinolinic acid [2,3] have been implicated as contributory or primary causative agents of neuronal injury in several neuropathologic processes including epilepsy [4], Huntington's disease [5,6], Alzheimer's disease [7] olivo-pontocerebellar atrophy [8], the degenerative changes associated with hypoglycemia [9], aging, and hypoxic-ischemic injury [10]. The relevance of excitatory amino acids (EAA's) to hypoxic-ischemic neural injury will be the focus of this presentation. However, an overview of EAA physiology is a necessary back-ground for the discussion.

The Physiology of Excitatory Amino Acids

Of all the EAA's, glutamate is the most widespread and has been the most extensively investigated. There are at least three glutamate receptor sub-types all of which probably act on the same ion channel complex to produce different conductance states [11]. These sub-types have been classified according to the exogenous agonists which preferentially activate them: N-methyl-D-aspartate (NMDA), quisqualate, and kainate. Glutamate is an effective agonist for all three receptor subtypes. Aspartate, the second widely occurring EAA neurotransmitter appears to act preferentially at NMDA receptors [12]. In the rat, NMDA-preferring receptors (Table 1) are numerous in the cerebral cortex with particularly heavy concentration in its superficial layers, the thalamus, the

[1] Department of Anesthesiology, University of California, San Diego, CA 92093-0629, USA

Table 1. The location of glutamate receptors in the brain

NMDA	Quisqualate	Kainate
CORTEX (superficial)	CORTEX (deep)	—
Striatum	Striatum	—
Hippocampus (throughout)	Hippocampus (stratum moleculare-dentate)	Hippocampus (stratum lucidum-CA3)
Thalamus	Thalamus	—
Cerebellum (granule cell)	Cerebellum (molecular layer)	—

striatum, the granule cell layer of the cerebellum, and the hippocampus. There are glutamate receptors throughout the hippocampus. The majority are of the NMDA-preferring type. The exceptions are populations of quisqualate-preferring receptors in the pyramidal cell layer of the hippocampus [13] and in the inner third of the stratum moleculare (dentate gyrus) [12] and kainate-preferring receptors in the stratum lucidum (CA-3) [12]. Quisqualate-preferring receptors have also been identified by autoradiography in the deep layers of the cerebral cortex, the striatum, the thalamus, and the molecular layer of the cerebellum, [12]. The highest concentrations of quisqualate receptors are in the molecular layer of the cerebellum [14]. Kainate receptors are apparently located almost exclusively in the stratum lucidum of the hippocampus [12]. The normal functions of the individual receptor types are not known with certainty. However, it appears that quisqualate and kainate receptors are involved principally in synaptic transmission [15] while NMDA receptors are important in long term potentiation [15], a process thought to be central to learning and memory [16]. Both the NMDA and kainate receptors are implicated in pathologic processes [17,18]. The accumulated evidence principally concerns the former, and there has been relatively little interest in investigations of kainate because of its very limited distribution in the CNS (see above) and because there are no known kainate antagonists.

EAAs are presumed to cause injury as a result of persistent stimulation causing the normal cationic channels to be held in the "open" position with a resultant intracellular accumulation of cations (initially sodium and subsequently calcium). The neuronal insult is thought to be biphasic. The initial process may be osmotic in nature [4] (1986). This theory is consistent with the observation that the initial histopathologic manifestation of the toxicity of a locally applied EAA is swelling in the vicinity of the areas of receptor concentration, i.e., the dendrites [18] with sparing of axons traversing the vicinity [2]. In vitro studies suggest a second EAA-triggered insult process. This process has a longer time course [19,20] and is proposed to be mediated by intracellular accumulation of calcium resulting in activation of lipases and proteases, and the impairment of the function of both mitochondria and membranes. The elimination of calcium from a cultured cortical neuron preparation following glutamate exposure greatly attenuated this delayed neuronal injury [19,20]. The relative contributions to the ultimate outcome of these two processes (early swelling and delayed calcium accumulation) have not been established. However, the early and apparently osmotic component of the injury seen after local administration does not have an obvious counterpart in the histologic picture seen in vivo following hypoxic or ischemic injuries. Furthermore, in a study in which an NMDA antagonist was administered both before and after forebrain ischemia in the rat, Swan et al. [21] observed no difference in the status of neurons in the CA1 area of the hippocampus 24 h post ischemia but noted better neuronal preservation in the treated animals 7 days after the insult. Their observations are consistent with an important contribution by delayed effects.

Glutamate and Hypoxic-Ischemic Injury

Additional evidence supports the notion that glutamate contributes to post-ischemic neuronal injury. This includes the observations that:

1. *The distribution of glutamate receptors (see above) corresponds to a large extent with the recognized areas of "selective neuronal vulnerability"*. According to Greenfield [22], these areas in adults include, the cerebral cortex (layers 3, 5 and 6), hippocampus (CA1 area is more vulnerable than (>) CA3 and CA4 >> CA2 and dentate gyrus), cerebellar cortex (Purkinje and basket cells > granule cells > Golgi cells), amygdaloid nucleus (deep and basolateral portions), outer portions of the caudate and putamen, pallidum, anterior nuclear complex of the thalamus, and, in the brain stem, the substantia nigra, inferior colliculi, and inferior olives.

2. *Ischemia causes enhanced release and/or impaired uptake of glutamate at synapses*. The theory of excitotoxic neuronal injury requires that there be either an increased concentration of glutamate (or other excitant) at the synapse, or that there be an increased sensitivity to normal concentrations. There are data to support the former possibility but not the latter. Benveniste et al. [23] demonstrated an increased concentration of glutamate in the extracellular space in the hippocampus after transient forebrain ischemia in the rat. Diminished uptake has not been demonstrated. However, uptake is an active, energy-consuming process and Rothman and Olney [10] speculate that it might well be impaired during and after ischemia.

3. *In vitro and local microinjection studies confirm EAA toxicity and protection by EAA antagonists*. Glutamate has been shown to cause neuronal injury in cortical cell culture [24], and EAA antagonists have been shown to protect neurons from damage by anoxia in both culture [25,26], and in hippocampal slice preparations [27]. In addition, two groups [21,28] demonstrated that local microinjection of an NMDA antagonist (2-amino-7-phosphonoheptanoic acid) prior to ischemia protected against the neuronal damage caused by temporary carotid artery occlusion in the rat. It should be noted that the results have been inconsistent with respect to the local microinjection studies. Block and Pulsinelli [29]

reported an absence of protection by local microinjection of EAA antagonists into the hippocampus and striatum. However, that investigation employed a 4-vessel occlusion in the rat, and the failure to protect may reflect the greater severity of the ischemic insult yielded by that model. (See below for a discussion of the implications of the severity of ischemia on the protective potential of EAA antagonists.)

4. *The interruption of glutaminergic nerve tracts to areas of glutamate receptor concentration reduces glutamate concentration during ischemia and protects against ischemic neuronal injury.* Diemer et al. [30] produced a reduction in the glutamatergic input to the CA1 area from the CA3 area by intra-cerebroventricular injection of kainic acid in rats. They observed a dramatic reduction in glutamate concentration in the hippocampus during a subsequent ischemic event (the ischemia method was not specified). Wieloch et al. [31] sectioned an afferent glutamatergic pathway (the perforant path) to the hippocampus and observed a reduction in the extent of the histologic injury occurring after ischemia. Johansen et al. [32] performed a comparable study in which the same pathway was interrupted at several levels. They, too, observed a reduction in histologic injury in the CA1 area after ischemia.

This accumulation of evidence provides substantial support for a role for glutamate in hypoxic-ischemic injury. However, there are questions which remain to be answered. For instance, the matching of areas of glutamate receptor concentration with regions of selective vulnerability is not absolutely consistent. There are glutamate receptors in high concentration throughout the hippocampus but one region, the CA1 area, is inexplicably more vulnerable than another, the dentate gyrus. In the cerebral cortex, the greatest density of glutamate receptors is in the superficial layers (1 and 2) [12] but layers 3, 5 and 6 are most vulnerable. Pulsinelli has also criticized the concept of the synaptic accumulation of glutamate as the basis of excitotoxicity [33]. Normal synaptic glutamate levels are on the order of 2 micromolar and have been observed to rise eight-fold during ischemia (neck tourniquet-hypotension) in the rat [23]. The levels necessary to produce injury in mature cortical neuron cultures (which lack a glutamate uptake system) are 50-100 μm [24]. It is unclear whether the concentrations of glutamate that occur at the synapse after ischemia can achieve these levels, although it remains a possibility (unproven) that altered receptor sensitivity renders lesser concentrations toxic after ischemia. This issue requires clarification.

Systemic Administration of EAA Antagonists

The initial investigations of EAA antagonists, including all of those mentioned above, were performed principally with agents that do not cross the blood brain barrier and that were therefore administered locally or investigated in vitro. The promising results and the subsequent availability of agents that cross the blood brain barrier (Table 2) and that can therefore be administered systemically have made EAA antagonists potentially useful clinical therapeutic tools. There are

Table 2. NMDA antagonists that cross the blood brain barrier

Non-competitive	Competitive
MK-801	CGS 19755
PCP	CPP
Ketamine	AP-7 (APH)
SKF 10047	
Dextrorphan	
Dextromethorphan	
MG++	

several specific EAA antagonists that are amenable to systemic administration, including: MK-801 (+-5-methyl-10,11-dihydro-dibenzo-[a,d]cyclohapten-5,10-imine maleate), CGS 19755 (cis-4-phosphonomethyl-2-2-piperidine-carboxylic acid), CPP (4-phosphonopropyl-2-piperazine-carboxylic acid), and AP-7 (2-amino-7-phosphonoheptanoic acid, also abbreviated "APH"). All are highly selective for the NMDA subset of glutamate receptors. MK-801 is a non-competitive antagonist that does not interact with the NMDA recognition site but rather directly blocks the ion channel. CGS 19755, CPP and AP-7 are competitive NMDA antagonists. There are other agents with NMDA antagonist activity including sigma opiates and the opiate derivatives dextrorphan and dextromethorphan [34–37]. There have been numerous investigations of the influence of these agents on the result of cerebral ischemia. These fall into two categories: (1) investigations of incomplete/focal cerebral ischemia, and (2) investigations of severe or complete global cerebral ischemia.

(1) Among the *incomplete/focal ischemia* studies (Table 3) are included investigations in which ischemia was produced by various vessel occlusion techniques (principally carotid or middle cerebral artery occlusion) either with or without reperfusion. In general the insult in this group is characterized by a densely ischemic core with a peripheral penumbral zone wherein cerebral blood flow (CBF) is reduced but not absent because of collateral perfusion. The majority of these studies have demonstrated improved outcome when the antagonist was given either before or shortly after ischemia. Swan et al. [21] administered APH intravenously before and at intervals for 10 h after 10 min of bilateral carotid artery occlusion. They observed significant reduction of neuronal loss in the CA1 region of the hippocampus seven days post insult. Kochhar et al. [38] administered MK-801 to rabbits 5 min after the start of ischemic insults to either the brain (microspheres injected into the carotid artery) or the spinal cord (temporary aortic occlusion). The volume of microspheres necessary to cause a neurologic deficit was larger and the duration of temporary spinal cord ischemia necessary to produce paraplegia was longer in MK-801-treated animals. Ozyurt et al. [39] administered MK-801 5 min prior to permanent middle cerebral artery occlusion (MCAO) in the cat. Six hours post occlusion, the volume of ischemic damage was smaller than that observed in vehicle-treated control animals. Using

Table 3. Investigations of glutamate antagonists in incomplete/focal ischemia

Author	Species	Model	Drug	Timing	Analysis	Protection
Swan et al. [21]	Rat	BCAO	APH	Pre and post	CA1 Hist	+
Kochhar et al. [38]	Rabbit	Mult Emb	MK801	Post	NDS	+
Ozyurt et al. [39]	Cat	MCAO	MK801	Pre	Infarct size	+
Park et al. [40]	Cat	MCAO	MK801	Post	Infarct size	+
Park et al. [41]	Rat	MCAO	MK801	Pre or post	Infarct size	+
Steinberg et al. [37]	Rabbit	UCAO/ ACAO	Dextro- methorphan	Pre and post	Hist/edema/ evoked responses	+

BCAO, Bilateral carotid artery occlusion; APH, +2-amino-7-phosphonoheptanoate; Mult Emb, multiple transcarotid emboli; NDS, neurologic deficit score; MCAO, middle cerebral artery occlusion; Hist, histology; UCAO/ACAO, combined unilateral carotid artery occlusion and anterior cerebral artery occlusion

the same preparation (MCAO in the cat), Park et al. [40] gave MK-801 beginning 2 h after occlusion and also observed a reduction in the volume of ischemic damage. Park et al. [41] observed less histologic evidence of injury in cerebral cortex 3 h after permanent middle cerebral artery occlusion in rats that received MK-801 either 30 min before or after vessel occlusion. Steinberg et al. [37] administered dextromethorphan, an NMDA antagonist, before, during, and after temporary cerebral ischemia (unilateral internal carotid and anterior cerebral artery occlusion) in rabbits. They observed a reduction in the severity of both histologic ischemic changes and edema in the neocortex, and a more rapid recovery of somatosensory evoked potentials in animals treated with dextromethorphan.

(2) Among the *severe or complete global ischemia* studies (Table 4) are included those that entailed complete cessation of CBF by circulatory arrest or neck tourniquet-hypotension techniques, and those that produced severe CBF reduction without the presence of a substantial penumbral flow zone, e.g., 4-vessel occlusion in the rat and bilateral carotid occlusion in the gerbil.

a) Cerebral circulatory arrest. There have been four such studies. All involved the administration of MK-801 following a period of ischemia. The models employed were: aortic arch/vena caval occlusion in the dog [42], ventricular fibrillation in the cat [43], neck tourniquet/hypotension in the monkey [44] and interruption of cardiopulmonary bypass in the dog [45]. In none of these investigations was there a significant improvement in neurologic or histologic outcome.

b) Four-vessel occlusion in the rat. Block and Pulsinelli [29] observed no histologic protection by either competitive or non-competitive EAA antagonists.

c) Carotid artery occlusion in the gerbil. Gill et al. [46] administered MK-801 to gerbils 1 h prior to ischemia. MK-801 resulted in a dose-dependent reduction in the histologic injury seen in the CA1 and CA2 areas 4 days after ischemia produced by 5 min of bilateral carotid artery occlusion, and an increase in survival in animals that underwent 30 min of unilateral carotid artery occlusion. Lawrence et al. [47] performed a similar investigation in which cerebral ischemia in

Table 4. Investigations of glutamate antagonists in severe or complete cerebral ischemia

Author	Species	Model	Drug	Timing	Analysis	Protection
Michenfelder et al. [42]	Dog	Ao X-clamp	MK801	Post	Hist/NDS	0/0
Fleischer et al. [43]	Cat	Vent fib	MK801	Post	Hist/NDS	0/0
Lanier et al. [44]	Monkey	Neck tourn	MK801	Post	Hist/NDS	0/0
Sterz et al. [45]	Dog	Stop CPB	MK801	Pre or post	Hist/NDS	0/0
Block et al. [29]	Rat	4-vess occ	MK801	?Pre	Hist	0
Gill et al. [46]	Gerbil	5' BCAO	MK801	Pre	CA1 Hist	+
Lawrence et al. [47]	Gerbil	7.5' BCAO	MK801	Pre	CA1 Hist	+
Boast et al. [48]	Gerbil	20' BCAO	AP7	Pre	CA1 Hist/NDS	+/+
				Post	CA1 Hist/NDS	+/+
Boast et al. [49]	Gerbil	20' BCAO	CGS19755	Pre	CA1 Hist/NDS	+/+
				Post	CA1 Hist/NDS	+/+
			CPP	Pre	CA1 Hist/NDS	+/+
				Post	CA1 Hist/NDS	+/+

Ao X-clamp, occlusion of proximal aorta and vena cavae; BCAO, Bilateral carotid artery occlusion; 4 Vess Occ, bilateral carotid and bilateral vertebral artery occlusion; MCAO, middle cerebral artery occlusion; Hist, histology; NDS, neurologic deficit score; Vent Fib, ventricular fibrillation; Neck tour, tourniquet with simultaneous hypotension; CPB, cardiopulmonary bypass; +, protection was observed; 0, protection was not observed

gerbils was produced by 7.5 min of bilateral carotid occlusion. Pre-treatment with MK-801 15 min before ischemia resulted in a dose-related protection of the pyramidal cells of the CA1 area of the hippocampus. Boast et al. [48] observed less histologic damage in the hippocampus and better motor function in gerbils treated with AP-7. The improvement was seen when treatment was initiated either before or after 20 min of bilateral carotid occlusion. Boast et al. [49] gave CGS 19755 or CPP to gerbils after a period of bilateral carotid occlusion and observed both histologic protection and a reduction of ischemia-related motor abnormalities. CGS 19755 was more effective than CPP and the former was protective with a post ischemic delay of up to 4 h.

The pattern of the results of these groups of investigations is superficially striking. The incomplete/focal ischemia studies have indicated that EAA antagonists have protective effects, while the studies of severe or complete global ischemia (except for those in the gerbil) have not. What principally distinguishes the two groups is the severity of the ischemic insult. The suggestion has been put forward that EAA receptor antagonists, which can block only one of several mechanisms leading to increased cytosolic calcium, will be ineffective in preventing the calcium-initiated cascade of secondary processes unless there is a sufficient residual energy supply to permit the cell to stave off the intracellular cytosolic calcium accumulation that occurs by routes other than the glutamate-gated channel [50,51]. These conditions may prevail in the event of a focal ischemic insult when there is a peripheral penumbral zone with some collateral CBF. They can not occur with the complete ischemia that accompanies cardiac arrest, and would be unlikely to appear in the severely ischemic circumstances of 4-vessel occlusion in the rat or bilateral carotid artery occlusion in the gerbil. In the gerbil, with its incomplete circle of Willis, bilateral carotid occlusion produces dense forebrain ischemia with minimal collateral flow. The thesis mentioned at the beginning of this paragraph leads to the prediction of an absence of protection in the gerbil, but the data cited above indicate the contrary. However, there may be two confounding variables in the gerbil investigations. Post-ischemic hypothermia may have been a factor in the improved results of gerbils that received EAA antagonists. For instance, in the study of Gill et al. [46], post ischemic temperature was not controlled. Buchan and Pulsinelli performed a similar study, and observed that the protective effect of MK-801 was not apparent if post-ischemic hypothermia was prevented [52]. The reports of Lawrence et al. and Boast et al. [47–49] make no mention of temperature monitoring or control, and hypothermia may therefore have intruded in these investigations as well. The anti-epileptic properties of the EAA antagonists also may have been contributory. Gerbils are particularly seizure-prone in the post-ischemic period, and NMDA antagonists are effective anticonvulsants [53]. The "protection" observed may have been related to seizure prevention rather than other processes. These uncertainties concerning the variables that contributed to the results obtained in gerbils are such that these data are not sufficient to invalidate the hypothesis that EAA antagonists may be protective in circumstances of incomplete ischemia but not with severe complete ischemia.

There is an additional possible explanation for the apparent failure of EAA antagonists to protect in the circumstances of complete cerebral ischemia. In this group of studies (Table 4) the endpoint was often the gross neurologic examination. Protection of selected neuronal populations might well have been masked by the effects of injury to non-protected populations on neurologic deficit scores. In the study of Boast et al. of temporary bilateral carotid occlusion in the gerbil [49], when the competitive NMDA antagonist CGS 19755 was administered beginning one hour after ischemia, there was a reduction in both neuronal loss in the hippocampus and in the hypermotility that is caused by ischemia in gerbils. However, when administration was delayed until 2 or 4 h after the ischemic episode, histologic "protection" was observed but there was **no** reduction in hypermotility. This indicates that there can be a dissociation between the efficacy of an EAA antagonist as simultaneously measured by neurologic and histologic endpoints. Furthermore, it seems reasonable to expect that glutamate antagonists may confer protection on some neuron populations but not on others. Models which involve a greater degree of ischemia in the thalamus and brainstem, where glutamate receptors are much less numerous, might not be expected to show the same protection. Accordingly, the gross neurologic consequences of injury to many types of neurons in multiple locations might mask the evidence of the preservation of selectively protected areas. Nevertheless, the histologic evaluations of at least three of the complete cerebral circulatory arrest studies [42,43,45] have not revealed evidence of selective histologic protection.

The possibility that the glutamate antagonists confer protection by some other and non-specific mechanism should also be considered. In models of incomplete or focal ischemia there is the opportunity for CBF augmentation, and it is possible that some of the benefits of NMDA antagonists might have been the result of favorable redistribution of CBF. There are few relevant data. MK-801 produced a substantial increase in global CBF in normal dogs during total spinal anesthesia [54]. During halothane anesthesia in the rat, MK-801 caused a decrease in CBF in non-ischemic brains and no alteration of CBF around an area of ischemia caused by middle cerebral artery occlusion [55]. However, the effects of glutamate antagonists on CBF and on shunt/steal phenomena have not been extensively investigated and should, pending further investigation, be considered a possible explanation for the beneficial effects of these agents.

The distinction between the results of EAA antagonist administration in focal and global ischemia is reminiscent of the pattern of efficacy of barbiturates in cerebral protection. The parallel is likely to be entirely coincidental. The efficacy of barbiturates is probably the result of metabolic suppression with a possible contribution from favorable redistribution of CBF as a result of vasoconstriction in the normal brain. Neither of these mechanisms is likely to be relevant to EAA antagonists. MK-801 is ketamine/PCP-like in its clinical effect [56,57] and it is probable that all of these agents act at the NMDA receptor [58–60]. NMDA antagonists might therefore be anticipated to influence cerebral metabolic rate (CMR) in a manner similar to ketamine. Ketamine produces either no change or an increase in CMR [61]. In the hippocampus specifically, CMR in the rat was either unchanged or increased depending upon the dose of ketamine administered. In addition, Swan et al. [21] examined the local CBF effects of the NMDA antagonist AP-7 and saw no change, which argues against the occurrence of metabolic suppression.

Summary. The multiplicity of species, investigators, and techniques involved in the various investigations make it difficult to draw definitive conclusions. Nevertheless, the available information suggests that antagonists of the NMDA subset of glutamate receptors are effective in protecting against ischemic neuronal injury. It should be noted that even in the event of proven efficacy as protective agents there may be difficulties in the clinical application of these agents. For the time being, the only available agents require parenteral administration. However, more significant than this minor limitation is the fact that the available agents have a substantial ketamine/PCP-like "anesthetic" effect [59,62,63] which may complicate considerably the clinical management of patients who receive them.

It should also be noted that these agents may prove to be "selectively protective", and that the apparent efficacy of glutamate antagonists in reducing neurologic injury may vary with the identity of the neuron population at risk and with the method used (i.e., histology vs neurologic function) to identify a "protective" effect. It is also possible that the protective potential of these agents will be most apparent clinically in the event of threshold hypoxic/ischemic injuries as opposed to those in which there is severe ischemia with total failure of energy supply. In addition, the benefits may be more apparent when the injury principally involves those areas of greatest selective vulnerability (CA1 area of the hippocampus, neocortex, and striatum) in which there are also conspicuous concentrations of NMDA receptors. However, even in these areas there are glutamate receptors of other subtypes (quisqualate receptors in the striatum and cerebellar cortex, and kainate receptors in the hippocampus) and the available antagonists may not protect all of the neuronal sub-populations at risk for glutamate mediated injury.

It is also probable that excitotoxic injury is only one of several mechanisms of ischemic neuronal damage. While glutamate antagonists will probably prove to be important neural protectants they can ultimately be only one part of any pharmacologic brain protection armamentarium.

References

1. Fagg GE, Foster AC (1983) Amino acid neurotransmitters and their pathways in the mammalian central nervous system Neuroscience 9:701–719
2. Schwarcz R, Whetsell WO, Mangano RM (1983) Quinolinic acid: An endogenous metabolite that produces axon-sparing lesions in rat brain. Science 219:316–318
3. Schwarcz R, Meldrum B (1985) Excitatory amino acid antagonists provide a therapeutic approach to neurological disorders. Lancet II: 140–143
4. Greenamyre JT (1986) The role of glutamate in neurotransmission and in neurologic disease. Arch Neurol 43:1058–1063
5. Beal MF, Kowall NW, Ellison DW, Mazurek MF, Swartz KJ, Martin JB (1986) Replication of the neurochemical characteristics of Huntington's disease by quinolinic acid. Nature 321:168–171
6. Koh J-Y, Choi DW (1988) Cultured striatal neurons containing NADPH-diaphorase or acetylcholinesterase are selectively resistant to injury by NMDA receptor agonists. Brain Res 446:374–378
7. Greenamyre JT, Penney JB, Young AB, D'Amato CJ, Hicks SP, Shoulson I (1985) Alterations in L-glutamate binding in Alzheimer's and Huntington's diseases. Science 227:1496–1499
8. Plaitakis A, Berl S, Yahr M (1982) Abnormal glutamate metabolism in an adult-onset degenerative neurological disorder. Science 216:193–196
9. Wieloch T (1985) Hypoglycemia-induced neuronal damage prevented by an N-methyl-D-aspartate antagonist. Science 230:681–683
10. Rothman SM, Olney JW (1986) Glutamate and the pathophysiology of hypoxic-ischemic brain damage. Ann Neurol 19:105–111
11. Jahr CE, Stevens CF (1987) Glutamate activates multiple single channel conductances in hippocampal neurons. Nature 325:522–525
12. Greenamyre JT, Olson JMM, Penney JB, Young AB (1985) Autoradiographic characterization of N-methyl-D-aspartate, quisqualate and kainate sensitive glutamate binding sites. J Pharmacol Exp Ther 233:254–263
13. Monaghan DT, Holets VR, Toy DW, Cotman CW (1983) Anatomical distributions of four pharmacological distinct H-L-glutamate binding sites. Nature 306:175–179
14. Cha J-HJ, Greenamyre T, Nielsen EO, Penney JB, Young AB (1988) Properties of quisqualate-sensitive L-[^{3}H] glutamate binding sites in rat brain as determined by quantitative autoradiography. J Neurochem 51(2):469–478
15. Robinson MB, Coyle JT (1987) Glutamate and related acidic excitatory neurotransmitters: From basic science to clinical application. FASEB J 1:446–455
16. Collingridge GL, Bliss TVP (1987) NMDA receptors—their role in long-term potentiation. Trends Neurosci 10:288–292

17. Foster AC, Gill R, Kemp JA, Woodruff GN (1987) Systemic administration of MK-801 prevents N-methyl-D-aspartate-induced neuronal degeneration in rat brain. Neurosci Lett 76:307–311
18. Olney JW, Fuller T, deGubareff T (1979) Acute dendrotoxic changes in the hippocampus of kainate-treated rats. Brain Res 176:91–100
19. Choi DW (1985) Glutamate neurotoxicity in cortical cell culture is calcium-dependent. Neurosci Lett 58:293–297
20. Choi DW (1987) Ionic dependence of glutamate neurotoxicity. J Neurosci 7:369–379
21. Swan JH, Evans MC, Meldrum BS (1988) Long-term development of selective neuronal loss and the mechanism of protection by 2-amino-7-phosphonoheptanoate in a rat model of incomplete forebrain ischaemia. J Cereb Blood Flow Metab 8:64–78
22. Greenfield's Neuropathology, 4th edn (1984) Adams JH, Corsellis JAN, Duchen LW (eds) Wiley, New York, pp 131–132
23. Benveniste H, Dreyer J, Schousboe A, Diemer NH (1984) Elevation of the extracellular concentrations of glutamate and aspartate in rat hippocampus during transient cerebral ischemia monitored by intracerebral microdialysis. J Neurochem 43:1369–1374
24. Choi DW, Maulucci-Gedde M, Kriegstein AR (1987) Glutamate neurotoxicity in cortical cell culture. J Neurosci 7:357–368
25. Rothman SM (1984) Synaptic release of excitatory amino acid neurotransmitter mediates anoxic neuronal death. J Neurosci 4:1884–1891
26. Choi DW, Koh J, Peters S (1988) Pharmnacology of glutamate neurotoxicity in cortical cell culture: Attenuation by NMDA antagonists. J Neurosci 8:185–196
27. Robinson MB, Anderson KD, Koerner JF (1984) Kynurenic acid as an antagonist of hippocampal excitatory transmission. Brain Res 309:119–126
28. Simon RP, Swan JH, Griffiths T, Meldrum BS (1984) Blockade of N-Methyl-D-Aspartate receptors may protect against ischemic damage in the brain. Science 226:850–852
29. Block GA, Pulsinelli WA (1987) Excitatory amino acids receptor antagonists: Failure to prevent ischemic neuronal damage. J Cereb Blood Flow Metab 7 (Suppl 1):S149
30. Diemer NH, Sandberg M, Jorgensen MB, Benveniste H (1989) Ischemia-induced release of glutamate in the hippocampal CA1 region is decreased after removal of the excitatory input from the CA3. J Cereb Blood Flow Metab 9 (Suppl 1):S747
31. Wieloch T, Lindvall O, Blomqvist P, Gage FH (1985) Evidence for amelioration of ischaemic neuronal damage in the hippocampal formation by lesions of the perforant path. Neurosci Res 7:24–26
32. Johansen FF, Jorgensen MB, Diemer NH (1987) Ischemia-induced delayed neuronal death in the CA-1 hippocampus is dependent on intact glutamatergic innervation. In: Hicks M, Lodge D, McLennan H (eds) Excitatory amino acid transmission. Liss, New York, pp 245–248
33. Pulsinelli WA (1988) CNS excitation-inhibition and ischemic injury to neurons. Anesthesiology Review 15:50–51
34. Choi DW, Peters S, Viseskul V (1987) Dextrorphan and levorphanol selectively block N-methyl-D-aspartate receptor-mediated neurotoxicity on cortical neurons. J Pharmacol Exp Ther 242:713–720
35. Goldberg MP, Pham P-C, Choi DW (1987) Dextrorphan and dextromethorphan attenuate hypoxic injury in neuronal culture. Neurosci Lett 80:11–15
36. Kemp JA, Foster AC, Wong HF (1987) Non-competitive antagonists of excitatory amino acid receptors. Trend Neurosci 10 :295–298
37. Steinberg GK, George CP, DeLaPaz R, Shibata DK, Gross T (1988) Dextromethor-

phan protects against cerebral injury following transient focal ischemia in rabbits. Stroke 19(9):1112–1118

38. Kochhar A, Zivin JA, Lyden PD, Mazzarella V (1988) Glutamate antagonist therapy reduces neurologic deficits produced by focal central nervous system ischemia. Arch Neurol 45:148–153

39. Ozyurt E, Graham DI, Woodruff GN, McCulloch J (1988) Protective effect of the glutamate antagonist, MK-801 in focal cerebral ischemia in the cat. J Cereb Blood Flow Metab 8:138–143

40. Park CK, Nehls DG, Graham DI, Teasdale GM, McCulloch J (1988) Focal cerebral ischemia in the cat: Treatment with the glutamate antagonist MK-801 in the cat after induction of ischemia. J Cereb Blood Flow Metab 8:757–762

41. Park CK, Nehls DG, Graham DI, Teasdale GM, McCulloch J (1988) The glutamate antagonist MK-801 reduces focal ischemic brain damage in the rat. Ann Neurol 24:543–551

42. Michenfelder JD, Lanier WL, Scheithauer BW, Perkins WJ, Shearman GT, Milde JH (1989) Evaluation of the glutamate antagonist dizocilipine maleate (MK-801) on neurologic outcome in a canine model of complete cerebral ischemia: Correlation with hippocampal histopathology. Brain Res 481:229–234

43. Fleischer JE, Tateishi A, Drummond JC, Scheller MS, Zornow MH, Grafe MR, Zornow MH, Shearman GT, Shapiro HM (1989) MK-801, an excitatory amino acid antagonist, does not improve neurologic outcome following cardiac arrest in cats. J Cereb Blood Flow Metab 9:795–804

44. Lanier WL, Perkins WJ, Ruud B, Milde JH, Michenfelder JD (1988) The effect of the excitatory amino acid antagonist MK-801 on neurologic function following complete cerebral ischemia in primates. Anesthesiology 69:A846

45. Sterz F, Leonov Y, Safar P, Radovsky, Stezoski SW, Reich H, Shearman GT, Greber TF (1989) Effect of excitatory amino acid receptor blocker MK-801 on overall, neurologic, and morphologic outcome after prolonged cardiac arrest in dogs. Anesthesiology 71:907–918

46. Gill R, Foster AC, Woodruff GN (1987) Systemic administration of MK-801 protects against ischemia-induced hippocampal neuro-degeneration in the gerbil. J Neurosci 7:3343–3349

47. Lawrence JJ, Fuller TA, Olney JW (1987) MK-801 and PCP protect against ischemic neuronal degeneration in the gerbil hippocampus. Soc Neurosci Abstr 13:1079

48. Boast CA, Gerhardt SC, Janak P (1987) Systemic AP7 reduces ischemia brain damage in gerbils. In: Hicks TP, Lodge D, McLennan-H (eds) Excitatory amino acid transmission. Liss, New York, pp 249–252

49. Boast CA, Gerhardt SC, Pastor G (1988) The N-methyl-D-aspartate antagonists CGS 19755 and CPP reduce ischemic brain damage in gerbils. Brain Research 442:345–348

50. Siesjö BK, Bengtsson F (1989) Calcium fluxes, calcium antagonists, and calcium-related pathology in brain ischemia, hypoglycemia, and spreading depression: A unifying hypothesis. J Cereb Blood Flow Metab 9:127–140

51. Wieloch T, Gustafson I, Westerberg E (1989) The NMDA antagonist, MK-801, is cerebro-protective in situations where some energy production prevails but not under conditions of complete energy deprivation. J Cereb Blood Flow Metab 9 (Suppl 1):S6

52. Buchan AM, Pulsinelli WA (1990) Hypothermia but not N-Methyl-D-Aspartate antagonist MK-801 attenuates neuronal damage in gerbils subjected to transient global ischemia. J Neurosci 10:311–316.

53. Wong EHF, Kemp JA, Priestley T, Knight AR, Woodruff GN, Iversen LL (1986) The anticonvulsant MK-801 is a potent N-methyl-D-aspartate antagonist. Proc Natl Acad Sci USA 83:7104–7108
54. Perkins WJ, Lanier WL, Ruud B, Milde JH, Michenfelder JD (1988) The cerebral and systemic effects of the excitatory amino acid receptor antagonist MK-801 in dogs: Modification by prior complete cerebral ischemia. Anesthesiology 69:A589
55. Park CK, Nehls DG, Teasdale GM, McCulloch J (1989) Effect of the NMDA antagonist MK-801 on local cerebral blood flow in focal cerebral ischaemia in the rat. J Cereb Blood Flow Metab 9:617–622
56. Boast CA, Pastor G (1987) Characterization of motor activity patterns induced by N-methyl-D-aspartate antagonists in gerbils. Pharmacol Biochem Behav 27:553–557
57. Koek W, Woods JH, Winger GD (1988) MK-801, a proposed noncompetitive antagonist of excitatory amino acid neurotransmission, produces phencyclidine-like behavioral effects in pigeons, rats and Rhesus monkeys. Pharmacol Exp Ther 245:969–974
58. Fagg GE (1987) Phencyclidine and related drugs bind to the activated N-methyl-D-aspartate receptor-channel complex in rat brain membranes. Neurosci Lett 76:221–227
59. Koek W, Woods JH, Mattson MV, Jacobson AE, Mudar PJ (1987) Excitatory amino acid antagonists induce a phencyclidine-like catalepsy in pigeons: Structure-activity studies. Neuropharmacology 26:1261–1265
60. Bowery NG, Wong EHF, Hudson AL (1988) Quantitative autoradiography of [^{3}H]-MK-801 binding sites in mammalian brain. Br J Pharmacol 93:944–954
61. Davis DW, Mans AM, Biebuyck JF, Hawkins RA (1988) The influence of ketamine on regional brain glucose use. Anesthesiology 69:199–205
62. Bennett DA, Bernard BS, Amrick CL, Wilson DE, Hutchinson AJ (1987) The pharmacological profile of an N-methyl-D-aspartate (NMDA) antagonist, CGS 19755. Soc Neurosci Abstr 13:1561
63. Scheller MS, Zornow MH, Shearman GT, Greber TF (1989) Non-competitive N-methyl-D-Aspartate receptor blockade profoundly reduces volatile anesthetic requirement in rabbits. Neuropharmacology: 28:677–681

4

N-Methyl-D-Aspartate Excitotoxicity: Is It Important in Ischemic Neuronal Injury?

Alastair M. Buchan[1] and William A. Pulsinelli[2]

Introduction

The decline in the incidence of stroke is attributable to the detection and modification of risk factors, the treatment of heart disease, and prophylaxis with
aspirin. New technology, particularly through brain and blood vessel imaging,
has allowed prompt assessment of stroke patients but, inspite of this, we are still
without an effective treatment for acute cerebral ischemia. While thrombolytic
therapy has become an established treatment for patients with acute myocardial
ischemia, in order to use it effectively in cerebral ischemia there is a need for
"neuronal cyto-protective agents". Successful brain resuscitation must stabilize
neurons during ischemia, arrest the evolution of delayed ischemic neuronal
death, and prevent reperfusion injury.

Neurons and glia in the core of an infarct remain viable for at most 3–4 h.
Cells in the surrounding area of a cerebral infarct, the so-called "penumbra",
may undergo metabolic changes but maintain the potential for recovery. Following successful cardiac resuscitation certain neurons, particularly those in the hippocampus, are extremely vulnerable and die. Other neurons are characteristically resistant [1]. This selective death of hippocampal neurons occurs after a delay
of up to 72 h [2,3], suggesting that a therapeutic window exists. Ischemic injury,
to both selectively vulnerable neurons following global ischemia and penumbral
neurons surrounding focal ischemia, is postulated to result from an imbalance
between neuronal excitation and inhibition. The most vulnerable cells, the CA1
pyramidal neurons, have a rich concentration of excitatory synapses.

Interest has therefore centered on the role of the excitatory amino-acids, L-
glutamate and L-aspartate. These neurotransmitters are well defined and are

[1]The Laboratory of Cerebral Ischemia, Robarts Research Institute, London, Ontario,
N6A 5A5, Canada
[2]Cerebrovascular Disease Research Center Cornell Medical College, New York, NY
10021, USA

widely distributed in the mammalian central nervous system [4]. At least three sub-types of membrane receptors that bind these neurotransmitters have been identified; N-methyl-D-aspartate (NMDA), Quisqualate, and Kainate [4]. The NMDA receptor is coupled to an ion channel which, when opened, allows an influx of sodium and calcium [5]. Stimulation of this receptor-ion channel complex may pose a lethal threat to neurons in the aftermath or on the edge of ischemia, while affording long-term potentiation and temporal integration in normal circumstance [6] by promoting an excessive intracellular calcium influx [7]. By blocking excitatory neurotransmission or, alternatively, by enhancing inhibitory neurotransmission, selective neuronal damage following transient global ischemia might be prevented, focal infarcts might be shrunken, and time might be gained to allow surgical or medical therapy to irrigate critically under-perfused areas of the brain before damage is inevitable.

In this paper we present our data. The non-competitive NMDA antagonist, MK-801, has been tested in two models of global ischemia: the rat 4-vessel occlusion (4-VO) model, and a gerbil model. Glutamatergic deafferentation lesions have been made in order to see if prior removal of excitatory inputs would ameliorate selective neuronal death in the 4-VO model. We have also tested the effects of MK-801 in focal ischemia models. Lastly, we have explored the possibility that neurotrophic factors may be more important than the inhibiting of glutamate neurotransmission in bringing about cytoprotection for the resuscitated brain.

Scientific Background on Global Ischemia

Animal models of cerebral ischemia that use the rat and the gerbil accurately mimic the pathology seen in humans [8]. Pyramidal neurons in the CA1 zone of the hippocampus are very sensitive to transient forebrain ischemia [1] and periods as brief as 5 min of bilateral carotid artery ligation in the gerbil or 5–15 min of 4-vessel occlusion in the rat are capable of injuring 95% of CA1 pyramidal neurons. This "selective neuronal vulnerability" is characteristically delayed [2,3]. Following the initial reperfusion there is neuronal repolarization, a return of normal electrical activity, and restoration of high energy metabolites and regional cerebral blood flow [9]. However, despite this apparent recovery, the neurons inextricably go on to burst fire and subsequently develop the histological criteria for irreversible neuronal injury including eosinophilia and nuclear pyknosis [2]. The animal models accurately reproduce the topographical and temporal pathological characteristics seen in the human [10] but the molecular mechanism which accounts for this delayed and selective death is unknown.

The excitotoxic hypothesis proposes that endogenous excitatory amino-acids, by stimulating membrane bound receptors which are coupled to ion channels, promote a massive influx of calcium which kills sensitive neurons [5]. In vitro experiments have shown that while immature cultured neurons are insensitive to hypoxia, once they have established synaptic connections they become relatively vulnerable [11]. Even then, the effects of hypoxia can be attenuated if a glutamate or an NMDA antagonist is added to the culture media before the hypoxic exposure [5]. Ionic substitution experiments have demonstrated that post-hypoxic neuronal death can be prevented by exchanging chloride ions [12], while slow death can be attenuated by the replacement of calcium ions in the media [13]. The suggestion has been made that a sodium and chloride influx through the NMDA-ion channel results in osmotic swelling and early neuronal death. If chloride is removed, then the surviving cultured neurons succumb to an excito-toxic mechanism dependent on calcium entry [5]. Calcium influx stimulates pro-teolytic and lipolytic enzymes as well as incapacitating mitochondria [7]. In ma-ture cultures competitive (acting on the receptor) [14] and non-competitive (acting on the ion channel) [15] NMDA antagonists prevent hypoxic neuronal injury.

In vivo transient forebrain ischemia results in an excessive release of gluta-mate [16]. In addition, there is a failure of re-uptake mechanisms [17] so the extracellular concentrations of glutamate remain elevated. The post-synaptic NMDA receptors, however, remain functional in the wake of ischemia [18]. The observation that the highest concentrations of NMDA receptors are in the CA1 pyramidal cell layer [19] links the explanation for selective CA1 necrosis to the excitotoxic hypothesis.

Focal Ischemia

In focal ischemia the core of the infarction is necrotic within 3–4 h (? acute osmotic death), but in the surrounding penumbra cell death may occur selective-ly for periods lasting perhaps days (Ca^{2+}-mediated death). Waves of spreading depression are seen in the penumbra which stimulate the release of glutamate [20], suggesting that glutamate antagonists might play a role in ameliorating the effects of focal ischemia.

Experimental Evidence with NMDA Antagonists

Global Ischemia

Competitive NMDA receptor antagonists such as APH have to be infused in-tracerebrally since they will not cross the blood brain barrier. Simon et al. de-monstrated that intra-hippocampal infusions of d-APH reduced CA1 neuronal injury in an anaesthetized rat model of global ischemia [21]. Following 30 min of ischemia, but only 2 h of reperfusion, rats were perfused and sections of the dorsal hippocampus showed that on the side of the infusion of APH there was a reduction of cell damage from around 50%–12% of CA1 cells [21]. Criticisms of this study include not only the fact that very few animals were used but also the premature termination of the experiments at 2 h of reperfusion. The same laboratory was unable to replicate these findings using a 10 min model of fore-brain ischemia, but they were able to show partial cyto-protection in a group of animals whose ischemic lesions were allowed to mature for a week [22]. Block

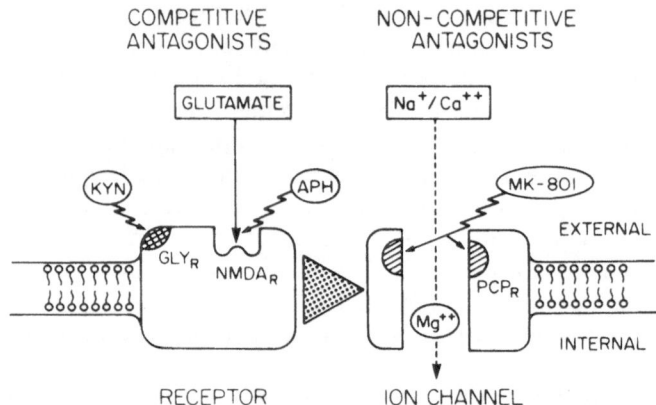

COMPETITIVE
ANTAGONISTS

NON-COMPETITIVE
ANTAGONISTS

Fig. 1. Cartoon of the NMDA receptor/ion channel complex and the sites of action for both competitive and non-competitive antagonists. Glutamate activates the receptor complex allowing the ion channel to open, causing the conductance of sodium (Na^+) and calcium (Ca^{2+}). Glutamate opens the channel from a closed to an open (activated) state. Competitive antagonists such as APH compete with L-glutamate for the NMDA receptor site ($NMDA_R$) preventing the ion channel from opening. Once opened, the non-competitive antagonists, e.g, MK-801 combines to sites within the channel (PCP_R), and block the channel. Non-specific antagonists like Kynurenate (KYN) might antagonize the glycine receptor (GLY_R) or have non-specific antagonist properties. (Modified from [49] with permission)

and Pulsinelli, who used the 4-VO model, were unable to replicate these results despite using large numbers of animals and a variety of dose paradigms of APH (including single doses pre-ischemically, multiple doses, and continuous drug infusions). They concluded that not only was APH ineffective but that its infusion was glial-toxic and worsened the damage [23]. Criticism was made that the ischemic insult was too severe [22].

MK-801 (± 5methyl-10-11-dyhdro-5H-dibenzo (A,D) cyclohepton-5-10 imminmaleate) is the most potent **non-competitive NMDA antagonist** known [24]. It is capable of crossing the blood brain barrier and its site of action is indicated in Fig. 1. It blocks the calcium channel linked to the NMDA receptor and, like PCP, is use-dependent. Claims have been made that the drug prevents CA1 necrosis in rats [25] and gerbils [26,27] following transient forebrain ischemia, and that it reduces infarct volumes in cat [28] and rat models [29] of focal ischemia.

Our first set of experiments was to test the effects of parenterally administered MK-801 in rats subjected to transient forebrain ischemia.

MK801 in the Rat

We have completed a number of studies that test MK-801 in the rat 4-vessel occlusion (4-VO) model [30]. The rats in these experiments were pretreated with

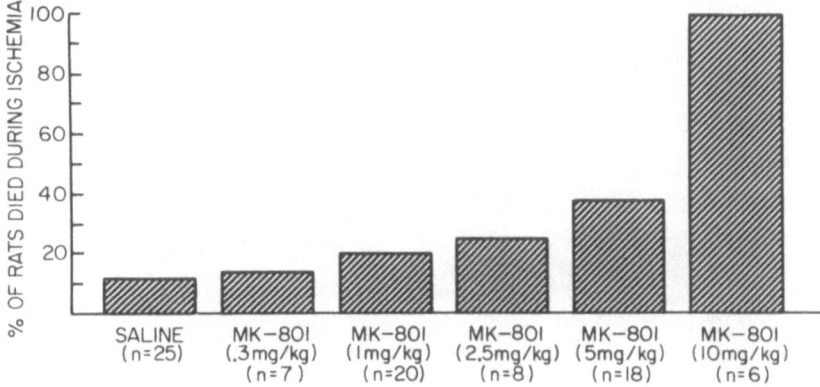

Fig. 2. Histogram showing a dose-dependent increase in death attributed to neurogenic pulmonary edema during 4-vessel occlusion in the rat, pretreated with MK-801

either physiological saline or MK-801 (0.3–10mg/kg), administered intraperitoneally 1 h prior to transient forebrain ischemia. In the first set of experiments, animals sustained 15 min of ischemia and then blood flow was restored for those animals that met full ischemic criteria. Subsequent experiments employed a 5 min model and confirmatory experiments included both 15 min and 5 min with MK-801 administered in multiple doses both before and after the ischemic insult. Survivors were allowed to reperfuse for 72 h prior to fixation. Animals sustaining seizures were excluded. Histological analysis of the hippocampal CA1 zone was performed and a grade determined: zero = normal, $1 = <10\%$, $2 = 10 - 50\%$, and $3 = >50\%$ of CA1 neurons irreversibly damaged. Comparison was made with the non-parametric Mann Whitney U test and, where indicated, a Bonferoni correction was made.

The drug produced behaviourial effects: an initial hyperactivity gave way to sluggish unresponsive behaviour, with pin-point pupils, and excessive salivation. In drug-treated animals, ischemia resulted in dose-dependant mortality (Fig. 2). None of the animals given 10mg/kg of MK-801 were able to maintain respiration during 4-VO ischemia. Figure 3 shows the results for those animals exposed to 15 min of ischemia and pretreated with either saline or increasing doses of MK-801. CA1 injury grades are given only for those animals which met the criteria and subsequently survived 15 min of forebrain ischemia and 72 h of reperfusion. A mean grade of damage to the CA1 is presented for each group. No differences in the amount of damage were detected for any of the doses administered.

Because 15 min of ischemia produced consistent and maximal CA1 damage that was equivalent for both MK-801 and saline treated animals, we reduced the ischemic period to 5 min. The results of the 5 min experiments are presented in Fig. 4. Five min of ischemia resulted in significantly less damage ($P<0.01$, 5 min compared to 15 min) with a mean grade of 1.9. The MK-801–treated groups were not significantly different from the saline-treated group. Despite the fact that 5 min of ischemia resulted in less damage and an increase in variability,

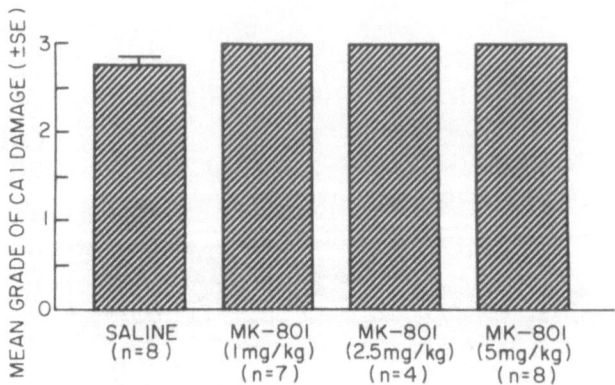

Fig. 3. Mean grades of damage (± SE) for animals treated with saline or increasing doses of MK-801 exposed to 15 min of 4-VO ischemia and 72 h of reperfusion.

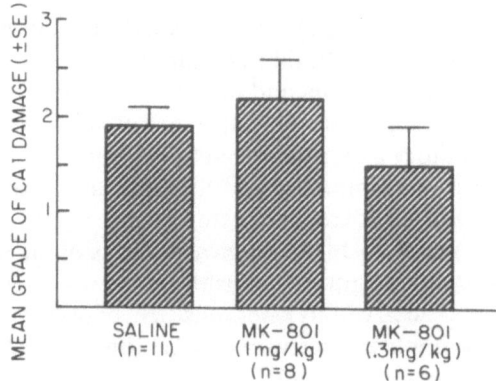

Fig. 4. Mean grades of damage (± SE) for animals pretreated with either saline or MK-801 and exposed to 5 min of 4-VO ischemia with 72 h of reperfusion

MK-801 was unable to favorably influence the outcome for the hippocampal cells.

Subsequent experiments using both 15 min and 5 min of ischemia, with multiple dosages of MK-801 are presented in Figs. 5 and 6. These confirmed our earlier findings, and failed to support the contention that MK-801 acts cytoprotectively, at least for CA1 cells in the rat 4-VO model [30].

MK-801 in the Gerbil

The experiments published by Gill et al. show a dramatic reduction in damage to the CA1 hippocampus following either pretreatment [26] or post-ischemic treatment [27] with MK-801. A dose-dependent effect was shown and we sought to replicate their experiments [31]. Gerbils were given MK-801 1mg/kg (middle of

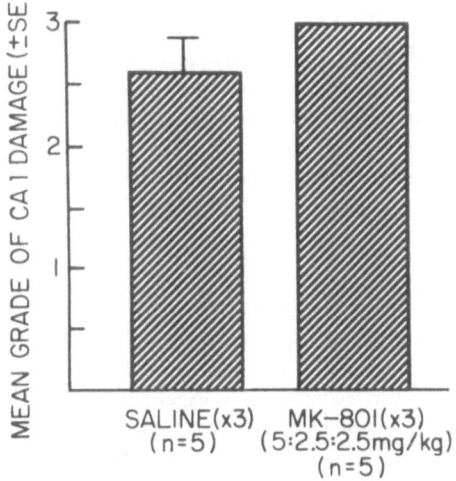

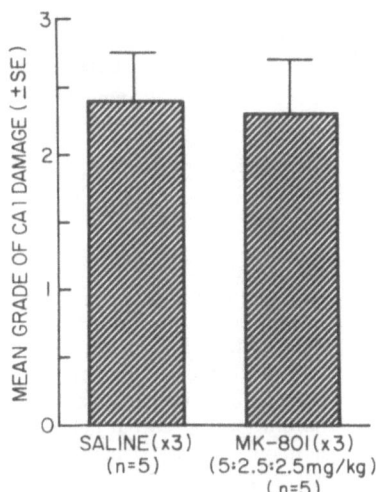

Fig. 5. Mean grades of damage for animals exposed to 15 min of forebrain ischemia and 72 h of reperfusion for animals given multiple doses of either saline or MK-801, at the time of reperfusion (5mg/kg), and at 8 and 20 h of reperfusion (2.5mg/kg)

Fig. 6. Mean grade of CA1 damage for rats exposed to 5 min of forebrain ischemia, followed by 72 h of reperfusion and given multiple doses of MK-801, 1 h prior (5mg/kg) and at 8 and 20 h (2.5mg/kg) following ischemia

the effective dose range [26]) 1 h prior to 5 min of bilateral carotid occlusion. Following 5 days of reperfusion, these drug-treated animals had significantly less damage than those given saline (Table 1). Of eight animals given saline, the mean grade of damage following 5 min of ischemia and 5 days of reperfusion was 2.4 ($\pm$ .2). Of the fourteen animals given MK-801 the mean grade of damage fell to 0.8 ($\pm$ .3). This was significantly less ($p < 0.01$, Mann Whitney U) and confirmed MK-801's capability of protecting neurons from transient forebrain ischemia.

However, while animals given saline were responsive and righting within 10–15 min of reperfusion, the animals which were given MK-801 (and were initially hyperactive and slightly hyperthermic prior to ischemia) remained comatose and hypothermic—a state which lasted for many hours following reperfusion (Fig. 7). The temperature of gerbils treated with MK-801 dropped to a mean of 35.5°C for 3 h following ischemia and remained depressed for at least 8 h. When the drug-treated animals were maintained at normothermic levels during reperfusion, the protective effect was lost: ten animals given saline and kept normothermic had a mean grade of damage of 2.7 ($\pm$ .2) while eleven MK-801 treated animals who were kept normothermic survived 5 min of ischemia and 5 days of reperfusion and ended up with a mean grade of damage of 2.6 ($\pm$ .2). Deliberate post-ischemic hypothermia of 34.5°C in untreated gerbils also resulted in a significant protection of the same order as that produced by animals treated with MK-801 who did not have their temperature maintained (Table 1).

Table 1. Five-minute gerbil model

Treatment	Experiment 1 Temp. monitored		Experiment 2 Temp. maintained		Experiment 3 Temp. controlled	
	Saline	MK-801	Saline	MK-801	Normo	Hypo
Mean temp°C (post isch.)	38.5°C	35.5+	38.5	38.5	38.5	34.5+
Number	8	14	10	11	9	8
Mean grade damage (± SE)	2.4(.2)	0.8(.3)*	2.7(.2)	2.6(.2)	2.2(.3)	0.75(.1)*

Temperatures and mean grades for gerbils treated with either saline or MK-801 and exposed to 5 min of bilateral carotid occlusion. In experiment 1, the drug produced hypothermia; in experiment 2, normothermia was maintained; in experment 3, no drug was given but deliberate hypothermia was induced in one group, while normothermia was maintained in the other. *$P < 0.01$ (MWU); +$P < 0.01$ (Student's t)

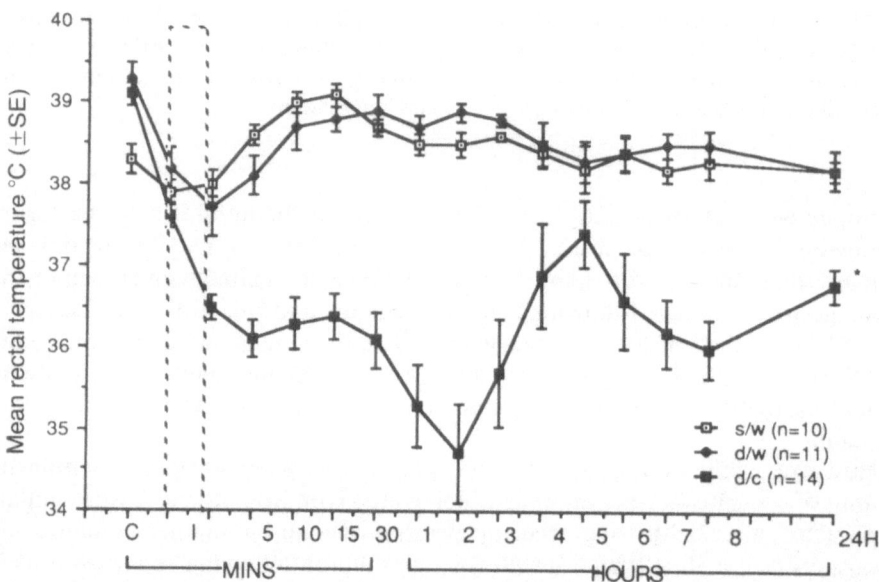

Fig. 7. Temperature curves for gerbils exposed to 5 min of bilateral carotid occlusion: c, control temperatures; I, ischemic period. Each curve represents the mean rectal temperature with standard error bars. S/W, ten animals given saline with temperature maintained; D/C, fourteen animals given 1mg/kg of MK-801 with no attempt made to correct temperatures (temperature-monitored); D/W, eleven gerbils given 1mg/kg of MK-801 1 h prior to ischemia with the temperature maintained at normothermic levels. * = $p < 0.01$

In our 4-VO rat experiments, all of the animals were maintained at nor-mothermic levels, not only during ischemia and the initial post-ischemic reperfusion period, but also when a multiple drug dose paradigm was used and the rats remained sluggish with drug-induced motor retardation. We concluded that the neuroprotective activity of MK-801, seen in the gerbil model of transient global ischemia, appeared to be a consequence of post-ischemic hypothermia rather than a direct action on NMDA receptor channels [31].

Focal Cerebral Ischemia

A wide variety of non-competitive antagonists including kynurenate [32], MK-801 [28,29,33], dextromethorphan [34,35] and dextrorphan [34] have shown pro-tective effects in focal middle cerebral artery occlusion models. These drugs have been given both pre- and post-occlusion and have consistently shown reductions in the volume of neocortical infarction (although not changing sub-cortical in-farction). Models with a large variation in infarct size have shown quite marked reductions in infarct volume, but those models—particularly the SHR rat model which has little variability (perhaps because there is less potential for collateral supply)—have shown much smaller changes in infarct volume [36,37]. One re-port has suggested that the reason for reduction in infarct size might not be related to antagonism of the NMDA receptors, but rather to an increase in re-gional CBF to the ischemic penumbra, and the authors propose a vaso-active explanation for the "cyto-protection" [37].

The results in focal ischemia have been quite consistent when compared to those in global ischemia, hence the "modified excitotoxic hypothesis" put for-ward by Wieloch [38] and Seisjö [7] (see their chapters, this volume).

Deafferentation Experiments

Another way to test the hypothesis that the imbalance of excitation inhibition causes ischemic injury to neurons is to interrupt the afferent input to sensitive neurons, allow time for degeneration of nerve terminals, and then examine the effect of ischemia on the post-synaptic neurons. Several groups using this strategy have reported less ischemic injury to CA1 neurons in rats with lesioned hippocampal afferent pathways. These reports have suggested that gluta-matergic inputs, such as the perforant path [39] and lesions of the entorhinal cortex [40], protect CA1 hippocampal neurons from ischemia. Interruption of the trisynaptic input, the mossy fibre/Schaffer collateral pathway caused by le-sions of the dentate granule cells [41] or lesions of the CA3 neurons [42], have also been reported to protect the rat CA1 zone against ischemic injury. These data all imply that the lesions of importance are glutamatergic.

The purpose of our studies was to cause a unilateral lesion in the afferent fibres travelling in the fimbria/fornix, the perforant path, or in the Schaffer col-lateral either individually or in combination [43] (Fig. 8). Seven to 13 days later, the rats were subjected to 30 min of reversible forebrain ischemia and, following 72 h of cerebral reperfusion, irreversible damage to the CA1 neurons was as-sessed and graded with a light microscope. The left unlesioned hippocampus

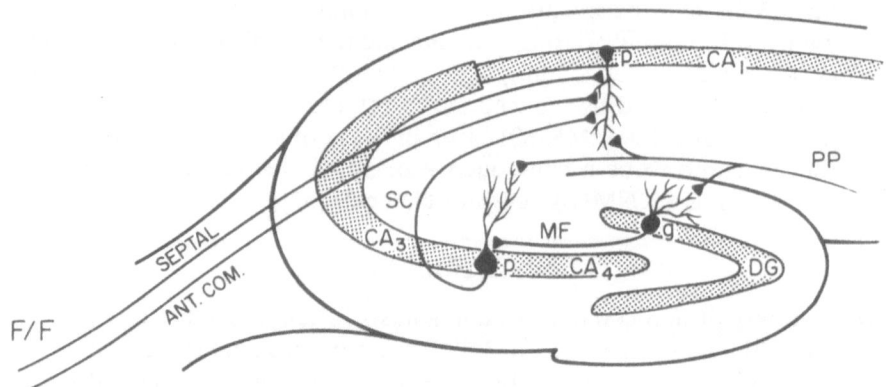

Fig. 8. The afferents to the hippocampal CA1 pyramidal cells. Each input has been the subject either collectively or individually of a deafferentation experiment. PP, perforant path with perforant path fibres rising from the entorhinal cortex; SC, Schaffer collateral with Schaffer collateral fibres rising from CA3 pyramidal cells (p); F/F, fimbria/fornix containing septal and anterior commissural (ant.com.) inputs; Pyramidal cells (p), CA_1, CA_3, and CA_4; DG, dentate gyrus; g, granule cells; MF, mossy fibres (Modified from [50] with permission)

served as a control. Simultaneous lesions of the three major afferents afforded protection against CA1 neuronal damage when compared to the unlesioned left hippocampus or to sham lesioned control animals. Selective lesions of the fimbria/fornix, but not the perforant or Schaffer collateral pathway, protected against ischemic damage to CA1 neurons [43]. These experiments were repeated and the results confirmed (Figs. 9 and 10).

Both sets of experiments showed that the triple lesion and the individual lesions of the fimbria/fornix bestowed protection, whereas lesions involving only the perforant and Schaffer collateral pathway did not. These data are again inconsistent with the hypothesis that glutamatergic inputs have a singular role in the selective injury to CA1 neurons but they do implicate neurotransmitters carried with the fimbria/fornix [43].

Time Course Experiments

We have also reported that there is a time course for this protective effect. When separate groups of fimbria/fornix-lesioned animals were exposed to ischemia at different points following the deafferentation, the cyto-protective effect peaked 9–13 days following the lesion. The first set of animals which were exposed to 30 min of forebrain ischemia at 2 days, 6 days, 9 days, 13 days, and 28–35 days from the time of the fimbria/fornix lesion; protection was significant at the 9–13 day period [44] (Table 2). No time course was discerned for the perforant path lesion, with the animals sustaining ischemia at 4, 9, and 13 days following the perforant path lesion. The existence of this time course was examined in a subse-

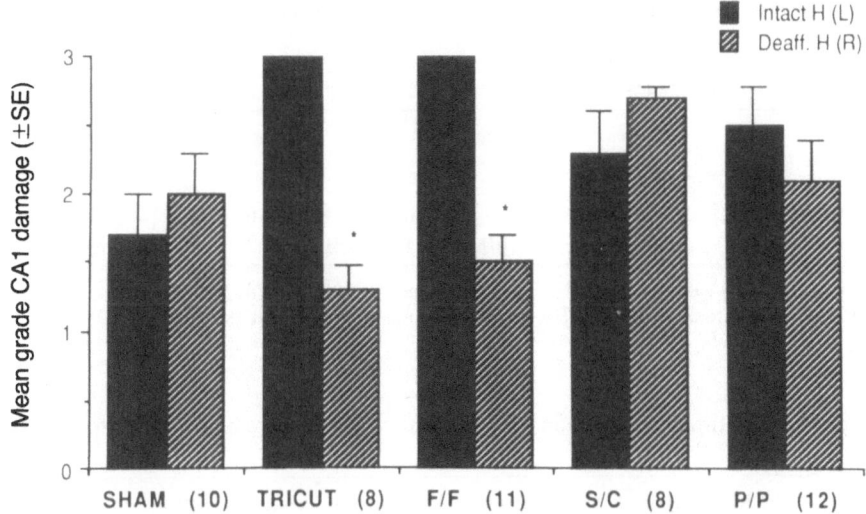

Fig. 9. The effects of deafferentation lesions to the right hippocampus, Deaff, H(R), as compared to the control or intact left hippocampus, intact, H(L). Mean grades of animals (number in brackets) are given for sham lesions, triple lesion, and the individual fimbria/fornix (F/F), Schaffer collateral (SC), and perforant path (P/P) lesions. There is significant protection(*) for the triple lesion, and the fimbria/fornix lesion. *$P<.001$ (MWU)

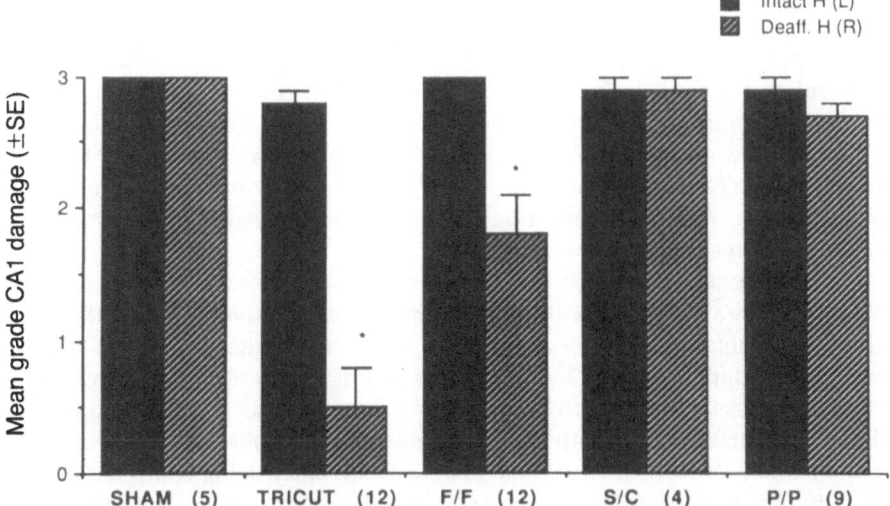

Fig. 10. The confirmatory experiment showing that the triple lesion and the fimbria/fornix lesion are protective while perforant path and Schaffer collateral lesions were not. Deafferentation lesions to the right hippocampus, Deaff, H(R) as compared to the control or intact left hippocampus, intact, H(L). Mean grades of animals (number in brackets) are given for sham lesions, triple lesion, and the individual fimbria/fornix (F/F), Schaffer collateral (SC), and perforant path (P/P) lesions. (*) = $P < .001$ (MWU)

Table 2. The time course for fimbria/fornix lesions in the 30 min 4-vessel occlusion model

	Mean grades of CA1 damage ($\pm$SE) post F/F lesion survival time				
	2 days	6 days	9 days	13 days	28–35 days
	$n = 11$	$n = 10$	$n = 21$	$n = 18$	$n = 9$
Left (intact)	3.0 (0)	2.8 (.1)	3.0 (0)	3.0 (0)	3.0 (0)
Right (F/F lesion)	3.0 (0)	3.0 (0)	2.16 (.19)*	1.75 (.17)**	2.83 (.11)

* $P<0.001$ vs left side (MWU), $P<0.01$ vs right side at 2, 6, and 28–35 d (KW)
** $P<0.001$ vs left side (MWU), $P<0.001$ vs right side at 2, 6, and 28–35 d (KW)

Table 3. The time course when fimbria/fornix animals were exposed to 15 min of ischemia

	Mean grades of CA1 damage ($\pm$SE) post F/F lesion survival time				
	2 days	6 days	9 days	13 days	28–35 days
	$n = 6$	$n = 7$	—	$n = 19$	$n = 8$
Left (intact)	2.92 (.08)	2.79 (.15)	—	2.97 (.03)	2.88 (.13)
Right (F/F Lesion)	2.83 (.17)	2.35 (.14)	—	1.34 (.14)**	2.13 (.34)

** $P<0.001$ vs left side (MWU), $P<0.001$ vs right side at 2, 6, and 28–35 days (KW)

quent experiment with 15 min of 4-VO; significant protection was confirmed ipsilateral to the fimbria/fornix lesion at 13 days (Table 3). While the fimbria/fornix lesion might suggest neurotransmitters other than those activating excitatory amino-acid receptors to be important in determining the sensitivity of specific neurons to ischemia, the temporal profile for this protection suggests that it is a response to this deafferentation rather than the lesion per se that is important in the protective effect [44].

Blood-flow studies have shown that the 13 day fimbria/fornix-lesioned rats have normal blood flow to both hippocampi which is reduced to less than 10%, with no left-right differences during the forebrain ischemia (Table 4). Furthermore, the hippocampal DC potential was lost bilaterally immediately during ischemia in those animals with the fimbria/fornix lesion at 13 days, and returned following reperfusion to both the control and the fimbria/fornix lesion sides simultaneously (Table 5). Biochemical data is currently being compiled to determine whether high energy metabolites are reduced equally in both the control and the lesioned side during ischemia. Further studies are underway to ensure that the lesions successfully remove the neurotransmitter input that are purported to be carried by the fibre pathways that have been transected.

Because the temporal profile for protection suggests that it is a response to the deafferentation, factors other than neurotransmitter depletion may be involved. Such factors might include the transient increase in nerve growth factor,

Table 4. Cerebral blood flows recorded in the 15 min model of 4-vessel occlusion

	Control $n = 6$	15 min ischemia $n = 6$
Unlesioned (L)	95.6 (±10)	9.86 (±2.3)
Unlesioned (R)	97.0 (±11)	9.0 (±2.0)
	$n = 6$	$n = 6$
Unlesioned (L)	103.5 (±10)	7.4 (±1.8)
F/F (13d) (R)	101.0 (±8.4)	9.95 (±2.2)

CA1 blood flow ml $100g^{-1}$ min^{-1} (Mean ± SE)

Table 5. Fimbria/fornix lesions–electrophysiology

	DC potential lost
Sham lesioned ($n = 6$)	5
F/F ($n = 6$)	5

The DC potential is lost in five out of six animals on both the sham and the fornix lesioned sides (day 13). This returned within minutes following reperfusion in both the sham and lesioned hippocampus

known to occur at 7–14 days after fimbria/fornix transection [45]. Ordinarily, hippocampal-manufactured nerve growth factor (NGF) is retrogradely transported to the septum. If the fimbria/fornix is lesioned, NGF accumulates in the hippocampus (maximum levels at 7–14 days) and septal acetylcholine cells die. NGF in high concentration might be sustaining, keeping ischemically-injured cells alive [46]. Another possible explanation is that following septohippocampal deafferentation, vasogenic sympathetic fibres sprout into the hippocampus [47]. Noradrenergic neurotransmission might favorably influence delayed CA1 necrosis [48].

Conclusions

1. In vitro and early in vivo reports suggest an important role for L-glutamate as an excitotoxic cause of cell death following cerebral ischemia [5].
2. No protective effects were found for MK-801, a non-competitive NMDA antagonist, in the rat 4-VO model [30].
3. The protective effect for MK-801 in a gerbil forebrain ischemic model appears to be due to post-ischemic hypothermia [31].
4. Glutamatergic lesions of the perforant path and Schaffer collateral input to CA1 did not prevent selective neuronal damage [43].

5. The fimbria/fornix lesion did protect CA1 neurons from ischemia but only at 9–13 days after the lesion [44], suggesting that it is the response to this deafferentation rather than the deafferentation per se which leads to the protection. Responses include a transient increase in NGF [45] and an ingrowth of noradrenergic fibres [47].
6. Non-competitive NMDA antagonists appears to bring about a reduction of infarct volume in focal models of ischemia [28]. The effect is attributed to NMDA receptor antagonism but it is possible that there is a vasoactive component to this cytoprotection [37].

Summary. The excitotoxic hypothesis implies that the post-ischemic release of the excitatory amino-acid neurotransmitter, L-Glutamate, promotes a lethal neuronal influx of calcium through NMDA receptor-ion channels. In vivo experiments have claimed that prior removal of glutamate afferents or the blockade of the NMDA receptor/ion channel complex can prevent ischemic neuronal injury following global ischemia and that NMDA antagonists can reduce the volume of neocortical infarction in focal ischemia.

Our experiments with the rat 4-vessel occlusion model have not demonstrated a protective effect for the most potent of the NMDA receptor-ion channel antagonists, MK-801. MK-801 did, however, have a protective effect in a gerbil model, but this was associated with post-ischemic hypothermia. When normothermic conditions were maintained the cyto-protective effect of MK-801 was lost. Prior deafferentation of glutamate afferents (perforant path, Schaffer collateral) also failed to protect hippocampal CA1 neurons following transient 4-vessel occlusion in the rat.

In contrast, we have shown that septo-hippocampal deafferentation has a cyto-protective effect, but only if an interval of 7–14 days elapses between the time of the lesion and the exposure to ischemia. This time course suggests new hypotheses to account for the cyto-protection, including a lesion-induced hippocampal accumulation of trophic factors, such as Nerve Growth Factor and vasogenic sympathetic sprouting of noradrenergic fibres into the hippocampus which either singularly or in combination may ameliorate the effects of transient cerebral ischemia.

Acknowledgements. This work was supported by the Mihara Memorial Prize to William Pulsinelli and by NIH grant NS-03346. Alastair Buchan was supported by a Canadian Medical Research Council Centennial Fellowship and grants from the Heart and Stroke Foundation of Ontario.

The authors would like to thank Mrs. Pam Gardner for her expert preparation of the manuscript.

References

1. Spielmeyer W (1925) Zur Pathogenes der Ortlich Elecktiven Gehirnveränderungen. Z Ges Neurol Psychiatr 99:756–777

2. Pulsinelli WA, Brierley JB, Plum F (1982) Temporal profile of neuronal damage in a model of transient forebrain ischemia. Ann Neurol 11:491–498
3. Kirino T (1982) Delayed neuronal death in the gerbil hippocampus following ischemia. Brain Res 237:57–69
4. Cotman CW, Iversen LL (1987) Excitatory amino acids in the brain–focus on NMDA receptors. TINS 10:263–265
5. Rothman SM, Olney JW (1987) Excitotoxicity and the NMDA receptor. TINS 10:299–302
6. Ascher P, Nowak L (1987) Electrophysiological studies of NMDA receptors. TINS 10:284–287
7. Siesjö B, Bengtsson F (1989) Calcium fluxes, calcium antagonists, and calcium-related pathology in brain ischemia, hypoglycemia, and spreading depression: A unifying hypothesis. J Cereb Blood Flow Metab 9:127–140
8. Pulsinelli WA, Buchan A (1989) The utility of animal ischemia models in predicting pharmacotherapeutic response in the clinical setting. In: Ginsberg MD, Dietrich WD (eds) Cerebrovascular diseases. Raven, New York, pp 87–91
9. Pulsinelli WA, Duffy TE (1983) Regional energy balance in rat brain after transient forebrain ischemia. J Neurochem 40:1500–1503
10. Petito CK, Feldmann E, Pulsinelli WA, Plum F (1987) Delayed hippocampal damage in humans following cardiopulmonary arrest. Neurology 37:1281–1286
11. Rothman SM (1983) Synaptic activity mediates death of hypoxic neurons. Science 220:536–537.
12. Rothman SM (1985) The neurotoxicity of excitatory amino acids is produced by passive chloride influx. J Neurosci 5(6):1483–1489
13. Choi DW (1985) Glutamate neurotoxicity in cortical cell culture is calcium dependent. Neurosci Lett 58:293–297
14. Choi DW, Koh J-Y, Peters S (1988) Pharmacology of glutamate neurotoxicity in cortical cell culture: Attenuation by NMDA antagonists. J Neurosci 8:185–196
15. Choi DW, Peters S, Viseskul V (1987) Dextrorphan and Levorphanol selectively block N-methyl-D-aspartate receptor-mediated neurotoxicity on cortical neurons. J Pharmacol Exp Ther 242:713–720
16. Benveniste H, Drejer J, Schousboe A, Diemer NH (1984) Elevation of the extracellular concentrations of glutamate and aspartate in rat hippocampus during transient cerebral ischemia monitored by intracerebral microdialysis. J Neurochem 43:1369–1374
17. Silverstein FS, Buchanan K, Johnston MV (1985) Hypoxia-ischemia causes severe but reversible depression of striatal synaptosomal ^{3}H-glutamate uptake. Ann Neurol 18:122
18. Westerberg E, Monaghan DT, Cotman CW, Wieloch T (1987) Excitatory amino acid receptors and ischemic brain damage in the rat. Neurosci Lett 73:119–124
19. Monaghan DT, Cotman CW (1985) Distribution of N-methyl-D-aspartate-sensitive L-[^{3}H]glutamate-binding sites in rat brain. J Neurosci 5:2909–2919
20. Van Harreveld A, Fifkova E. (1970): Glutamate release from the retina during spreading depression. J Neurobiol 2:13–29
21. Simon RP, Swan JH, Griffiths T, Meldrum BS (1984) Blockade of N-methyl-D-aspartate receptors may protect against ischemic damage in the brain. Science 226:850-852
22. Swan JH, Evans MC, Meldrum BS (1988) Long-term development of selective neuronal loss and the mechanism of protection by 2-amino-7-phosphonoheptanoate in a rat model of incomplete forebrain ischemia. J Cereb Blood Flow Metab 8:64–78

23. Block GA, Pulsinelli WA (1987) N-methyl-D-aspartate receptor antagonists: Failure to prevent ischemia-induced selective neuronal damage. In: Raichle ME, Powers WJ, (eds) Cerebrovascular diseases. Raven, New York, 37–42

24. Wong EHF, Kemp JA, Priestley T, Knight AR, Woodruff GN, Iversen LL (1986) The anticonvulsant MK-801 is a potent N-methyl-D-aspartate antagonist. Proc Natl Acad Sci USA 83:7104–7108

25. Rod MR, Auer RN (1989) Pre- and post-ischemic administration of dizocilpine (MK-801) reduces cerebral necrosis in the rat. Can J Neurol Sci 16:340–344

26. Gill R, Foster AC, Woodruff GN (1987) Systemic administration of MK-801 protects against ischemia-induced hippocampal neurogeneration in the gerbil. J Neurosci 7:3343–3349

27. Gill R, Foster AC, Woodruff GN (1988) MK-801 is neuroprotective in gerbils when administered during the post-ischemic period. Neuroscience 25:847–855

28. Ozyurt E, Graham DI, Woodruff GN, McCulloch J (1988): Protective effect of the glutamate antagonist, MK-801 in focal cerebral ischemia in the cat. J Cereb Blood Flow Metab 8:138–143

29. Park CK, Nehls DG, Graham DI, Teasdale GM, McCulloch J (1988) The glutamate antagonist MK-801 reduces focal ischemic brain damage in the rat. Ann Neurol 24:543–551

30. Buchan AM, Li H, Pulsinelli WA (1991) The N-methyl-D-aspartate antagonist, MK-801, fails to protect against neuronal damage caused by transient severe forebrain ischemia in adult rats. J Neurosci in press

31. Buchan AM, Pulsinelli WA (1990) Hypothermia but not the N-methyl-D-aspartate antagonist, MK-801, attenuates neuronal damage in gerbils subjected to transient global ischemia. J Neurosci 10:311–316

32. Germano IM, Pitts LH, Meldrum BS, Bartkowski HM, Simon RP (1987) Kynurenate inhibition of cell excitation decreases stroke size and deficits. Ann Neurol 22:730–734

33. Park CK, Nehls DG, Graham DI, Teasdale GM, McCulloch J (1988) Focal cerebral ischemia in the cat: Treatment with the glutamate antagonist MK-801 after induction of ischemia. J Cereb Blood Flow Metab 8:757–762

34. Steinberg GK, Saleh J, Kunis D (1988) Delayed treatment with dextromethorphan and dextrorphan reduces cerebral damage after transient focal ischemia. Neurosci Lett 89:193–197

35. George CP, Goldberg MP, Choi DW, Steinberg GK (1988) Dextromethorphan reduces neocortical ischemia neuronal damage in vivo. Brain Res 440:375–379

36. Dirnagl U, Tanabe J, Pulsinelli WA (1990) MK-801, an NMDA receptor antagonist protects against focal cerebral infarction. Brain Res

37. Buchan AM, Xue D, Slivka A, Zhang C, Hamilton J, Gelb A (1989) MK-801 increases cerebral blood flow in a rat model of temporary focal cortical ischemia (abstract). Soc Neurosci Abstract 15:804.

38. Wieloch T, Gustafson I, Westerberg E, (1989) The NMDA antagonist, MK-801, is cerebro-protective in situations where some energy production prevails but not under conditions of complete energy deprivation. J Cereb Blood Flow Metab 9 (Supp 1):S6

39. Wieloch T, Lindvall O, Blomquist P, Gage FH (1985) Evidence for amelioration of ischemic neuronal damage in the hippocampal formation by lesions of the perforant path. Neurol Res 7:24–26

40. Jorgensen MB, Johansen FF, Diemer NH (1987) Removal of the entorhinal cortex protects hippocampal CA-1 neurons from ischemic damage. Acta Neuropathol (Berl) 73:189–194

41. Johansen FF, Jorgensen MB, Diemer NH (1986) Ischemic CA-1 pyramidal cell loss is prevented by preischemic colchicine destruction of dentate gyrus granule cells. Brain Res 377:344–347
42. Onodera H, Sato G, Kogure K (1986) Lesions to Schaffer collaterals prevent ischemic death of CA1 pyramidal cells. Neurosci Lett 68:169–174
43. Buchan AM, Pulsinelli WA (1990) Septo-hippocampal deafferentation protects CA1 neurons against ischemic injury. Brain Res 512:7–14
44. Buchan AM, Pulsinelli WA (1989) Fimbria/fornix lesions: The temporal profile for protection of CA1 hippocampus against ischemic injury. J Cereb Blood Flow Metab 9(Supp 1):S749
45. Korsching S, Heumann R, Thoenen H, Hefti F (1986) Cholinergic denervation of the rat hippocampus by fimbrial transection leads to a transient accumulation of nerve growth factor (NGF) without change in mRNA NGF content. Neurosci Lett 66: 175–180
46. Buchan AM, Williams L, Bruederlin B (1990) Nerve growth factor: Pretreatment ameliorates ischemic hippocampal neuronal injury. Stroke 21:177
47. Stenevi U, Bjorklund A (1978) Growth of vascular sympathetic axons into the hippocampus after lesions of the septo-hippocampal pathway: A pitfall in brain lesions studies. Neurosci Lett 7:219–224
48. Gustafson I, Miyauchi Y, Wieloch TW (1989) Postischemic administration of idazoxan: An α-2 adrenergic receptor antagonist, decreases neuronal damage in the rat brain. J Cereb Blood Flow Metab 9:171–174
49. Kemp JA, Foster AC, Wong HF (1987) Non-competitive antagonists of excitatory amino acid receptors. Trends Neurosci 10:294–298
50. Wieloch T (1985) Neurochemical correlates to selective neuronal vulnerability. In: Kogure K, Hossman KA, Siesjö BK, Welsh FA (eds) Prog Brain Res 63:69–85

5

Metabolic Derangement and Cell Damage in Cerebral Ischemia with Emphasis on Protein and Nucleic Acid Metabolism

Takehiko Yanagihara[1]

Introduction

Brain resuscitation is an important issue in a clinical practice dealing with patients with cardiac arrest, asphyxia and stroke. Since an increasing number of patients with cardiac arrest have been resuscitated successfully, prevention and treatment of postischemic damage associated with cardiac arrest should be addressed critically. If the thrombolytic agents such as a tissue-type plasminogen activator and streptokinase prove to be safe and effective to recanalize the thrombosed cerebral arteries, prevention of post-reperfusion damage will become an important issue. While the advances in modern technology, such as positron emission tomography and nuclear magnetic resonance spectroscopy, have provided us opportunities to study metabolic derangements associated with cerebral ischemia and hypoxia, in some instances involving human patients, such investigations immediately after cardiac arrest or stroke are often technically difficult and may be unethical. Therefore, we have to rely on animal models to clarify the molecular mechanism for selective cerebral tissue vulnerability and to find rational ways to resuscitate damaged brains. In this presentation, I will first review animal experimental models and metabolic derangements associated with cerebral ischemia and hypoxia, and then review protein and RNA metabolism during and following cerebral ischemia and hypoxia, as well as their relationship to various metabolic derangements. Finally, I will discuss some aspects of structural damage which may be related to suppression of protein synthesis and intracellular transport.

[1] Department of Neurology, Mayo Clinic and Mayo Medical School, Rochester, MN 55905, USA

Experimental Models for Cerebral Ischemia and Hypoxia

Global cerebral ischemia has been produced experimentally in large animals such as cats [1], dogs [2] and monkeys [3] by occlusion of major blood vessels in the thoracic cavity, in small animals such as rats by induction of intracranial hypertension [4], and by occlusion of major cervical arteries [5]. Focal cerebral ischemia has been produced by occlusion of intracranial arteries, usually a middle cerebral artery, with monkeys [6], squirrel monkeys [7], baboons [8], dogs [9], cats [7], rabbits [10] and rats [11]. Since the first report by Levine and Payan in 1966, mongolian gerbils (*Meriones unguiculatus*) have been widely used for production of experimental cerebral ischemia. The advantage of this species is that we are able to produce not only unilateral and bilateral global ischemia by occlusion of one or both common carotid arteries [12], but also regional cerebral ischemia in the hippocampus or cerebral cortex by occlusion of a posterior communicating artery or a middle cerebral artery [13]. It is also possible to produce selective ischemia in the vertebrobasilar circulation with this species [14]. While the heterogeneity in the severity of ischemia after unilateral carotid occlusion may be a drawback of this experimental model [15], the severity of ischemia can be predicted by the appearance of the common carotid artery distal to the site of brief occlusion; gerbils with severe, moderate and mild cerebral ischemia can be selected [16].

Cerebral hypoxia has been produced both in large [17] and small animals [18] by alteration of an inhaled gas mixture. Cerebral anoxia and hypoxia can be produced in vitro with brain tissue slices [19,20]. Nerve and glia cells isolated by Ficoll density gradient ultracentrifugation have been used for investigation of cerebral hypoxia [21] and, more recently, nerve cells in tissue culture have been used for this purpose as well [22,23]. In this presentation, I will primarily review metabolic derangements, protein and RNA synthesis, and structural damage seen in gerbils after unilateral or bilateral carotid occlusion, and protein and RNA synthesis seen with the in vitro model of cerebral anoxia/hypoxia.

Metabolic Derangements in Cerebral Ischemia and Hypoxia

In cerebral ischemia, the blood supply is reduced or eliminated, and deprivation of oxygen occurs in cerebral hypoxia. Metabolic derangement can be quite different in each case. Since Lowry et al. [24] reported a rapid decline of high energy phosphates and a rise of lactic acid after decapitation in mice, many investigators examined these compounds, as well as other nucleotides and nucleosides, and intermediates of carbohydrate metabolism. After complete compression ischemia caused by intracranial hypertension in rats, the ATP level was about one-half of the control level and lactic acid rose to 10 times the control level in 1 min [4]. On the other hand, the ATP and phosphocreatine levels were well retained during cerebral hypoxia for 30 min when mice were kept at 5% oxygen atmosphere, although lactic acid increased to 3 times the control level [25]. A more recent study with mice at 4.4% oxygen atmosphere, however, revealed a

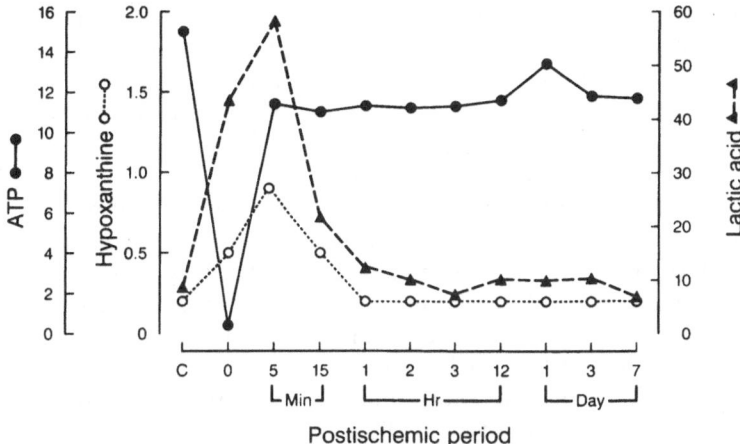

Fig. 1. The tissue ATP, lactate and hypoxanthine levels in the cerebral cortex after bilateral carotid occlusion for 10 min and subsequent reperfusion for 7 days in gerbils. The results are expressed as nmoles/mg dry tissue weight for ATP (•—•), lactate (▲--▲) and hypoxanthine (○--○) based on 4 experiments for each ischemic and postischemic period, respectively. ATP and hypoxanthine were measured by an isocratic HPLC method simultaneously, while lactic acid was measured by a conventional enzymatic method

failure of oxidative phosphorylation and a gradual decline of aerobic ATP synthesis which led to animal death in a few minutes [26].

The energy state has been investigated extensively in cerebral ischemia after unilateral or bilateral carotid occlusion in gerbils. After unilateral occlusion for 1 h, the ATP level declined to 10% of the control while the lactate level was over 8 times the control value. After reperfusion, the tissue ATP level rose to normal or near-normal and the lactate level returned to normal in 4–24 h [27–29]. After bilateral carotid occlusion, high energy phosphates and lactic acid returned to normal within 30 min when the ischemic period was 5 min [30]. There is general agreement that high energy phosphates decline sharply and lactic acid increases during ischemia, but uncertainties remain regarding the recovery of ATP and the decline of lactic acid following reperfusion. Therefore, we re-evaluated these unsettled questions by using an isocratic HPLC method [31]. An advantage of this technique is that we are able to measure not only phosphocreatine and adenine nucleotides (ATP, ADP and AMP) but also guanine nucleotides (GTP, GDP and GMP) and subsequent metabolites (adenosine, inosine and hypoxanthine) down to uric acid. After bilateral carotid occlusion for 10 min, there was a sharp decline of high energy phosphates (Fig. 1). Following reperfusion, the tissue ATP level recovered in 5 min but did not return to normal during reperfusion for 7 days (Fig. 1). The same was true for GTP. Phosphocreatine returned to normal but declined again when the morphologic evidence of postischemic damage was expected to become widespread [32]. While a transient rise in adenosine, inosine, and hypoxanthine occurred after reperfusion, they returned to normal within 1 h (Fig. 1). While the persistently subnormal ATP level in various areas of the brain may reflect the extent of tissue damage incurred during ischemia, this may induce various metabolic derangements and postischemic damage during reperfusion.

Lactic acid increases several-fold during and after cerebral ischemia and hypoxia, and we observed transient rise for 3 h in our investigation (Fig. 1). This is believed to be the cause of tissue acidosis with low pH in acute cerebral ischemia. After occlusion of the middle cerebral artery in rabbits, brain pH dropped from 7.01 to 6.64 in 10 min, reflecting the increase of lactic acid [33]. After a prolonged ischemia, brain pH went down further to 6.0 [33], which may be a combined effect of lactic acid and free fatty acids [34]. Tissue acidosis can affect a variety of cellular functions and can cause tissue damage as observed after intracerebral injection of an acidic solution [35]. While excessive acidosis may be the cause of extensive cerebral infarcts encountered in cerebral ischemia in the presence of hyperglycemia [36], a beneficial effect of lactic acidosis has been observed with an in vitro model of hypoxia [37]. The discrepancy may be due to the difference in pathophysiologic conditions and residual circulation [38].

Sodium-potassium and sodium-calcium pumps in plasma membranes are ATP-dependent. When ATP is depleted, an influx of sodium and calcium ions and an efflux of potassium ions are expected to occur. After occlusion of the middle cerebral artery in baboons, there was a dramatic increase of potassium ions and a decrease of calcium ions in the extracellular space, implying influx of calcium ions into cells and efflux of potassium ions out of cells [39]. The work from our laboratory by atomic absorption spectrometry revealed a decrease of potassium and an increase of calcium in gerbil brains after unilateral carotid occlusion and reperfusion. These shifts were transient in clinically reversible ischemia but were progressive after a prolonged ischemia (Fig. 2) [40]. This study suggested correlation between progressive accumulation of calcium and irreversible damage of the brain tissue. Subsequently, accumulation of calcium in the areas without biochemical recovery was observed in cat brains [41], and another autoradiographic investigation following transient ischemia and reperfusion in the rat brain also showed accumulation of radioactive calcium in the areas with postischemic damage [42]. A more recent investigation using proton microprobe also revealed accumulation of calcium in the areas with neuronal necrosis [43]. It is still difficult to determine whether calcium accumulated because of progressive cellular damage or accumulation of calcium caused cellular damage [44].

Another mechanism for the accumulation of calcium in the neuronal structure is membrane depolarization by excitatory neurotransmitters. An influx of calcium ions into dendrites can be induced by glutamate or N-methyl-D-aspartate [45], and delayed neuronal damage has been observed in nerve cells in tissue cultures after exposure to excitatory amino acids in the presence of calcium ions [46,47].

An influx of calcium ions may activate phospholipase A_2 and C resulting in release of free fatty acids—particularly arachidonic acid—from phosphatidyl

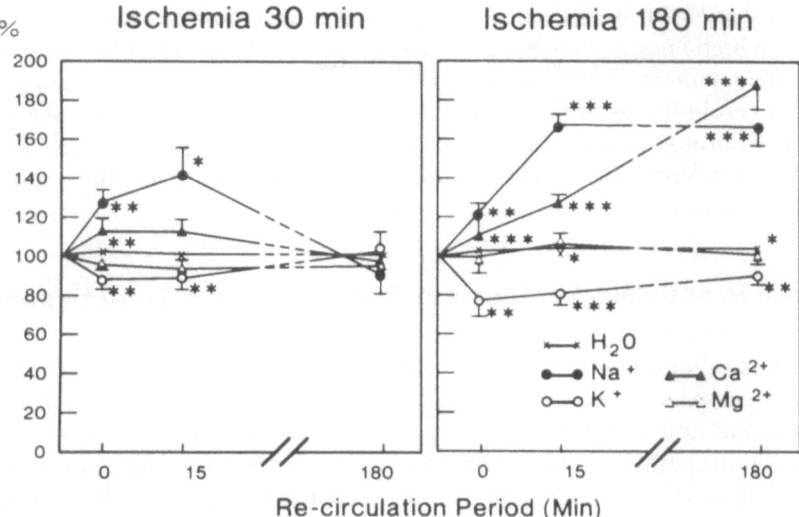

Fig. 2. The tissue water and electrolyte levels in the cerebral cortex after unilateral carotid occlusion in gerbils for 30 min or 3 h and subsequent reperfusion. The results are expressed as percent of the control value based on 4 experiments at each ischemic and postischemic period. *** $P < 0.01$; ** $0.01 < P < 0.05$; * $0.05 < P < 0.1$. (Reproduced by permission from Pergamon Press plc. [40])

ethanolamine and phosphoinositides [48–50]. Loss of these phospholipids may result in damage in plasma membranes and intracytoplasmic subcellular organelles. Once oxygen is resupplied after reperfusion, arachidonic acid is converted to prostaglandins [51] and leukotrienes [52] by cyclooxygenase and lipoxygenase, respectively. Since many of them, particularly prostaglandin $F_{2\alpha}$ and leukotriene C_4, have vasoconstrictive effects, cerebral edema and hypoperfusion may be induced by these compounds.

When oxygen is resupplied to previously ischemic tissues, oxygen radicals such as superoxides and hydroxyl radicals may be formed. Specifically, superoxides may be formed from hypoxanthine or xanthine by xanthine oxidase [53]. They are also formed during conversion of arachidonic acid to vasoactive prostanoids [54]. Hydroxyl radicals may be formed from a superoxide reaction in the presence of Fe^{3+}. The conversion of xanthine dehydrogenase to xanthine oxidase may be mediated by Ca^{2+}-activated protease, where influx of calcium ions and oxygen may trigger the process [53]. Oxygen and hydroxyl radicals cause alterations of proteins and lipids which may cause cell damage and death [55–57]. The presence of xanthine oxidase action has been observed after unilateral common carotid occlusion for 3 h in gerbils [58]. However, our investigation revealed only transient increase of hypoxanthine after bilateral ischemia for 10 min and reperfusion, which returned to normal within 1 h (Fig. 1) [31] while no notable rise in xanthine or uric acid was observed during that time.

Thus, major biochemical derangements during severe ischemia and hypoxia are associated with inefficient oxidative phosphorylation resulting in a marked decline of high energy phosphates and a rise of lactic acid. The brain may sustain irreversible damage if the ischemic or hypoxic condition is not reversed promptly. After cerebral reperfusion, high energy phosphates may or may not return to normal. Neuronal damage may occur because of a worsening of tissue acidosis, an influx of calcium ions, and/or tissue peroxidation. However, those processes appear to be transient if the ischemic period is short.

Protein and RNA Synthesis in Cerebral Ischemia and Hypoxia

Regardless of whether it is decline of the energy state, tissue acidosis, calcium influx, or peroxidation, each of them individually or combined can exert damage to functional and structural proteins as well as to membrane lipids, and may affect protein synthesis in the brain. Protein synthesis is essential for the survival of cells and microorganisms but is a very complex process [59]. It requires two different ribosomal subunits, messenger RNA, transfer RNA, and about 10 initiation factors (eIF). Initially, eIF-2 and GTP form a binary complex and then form a ternary complex with methionine tRNA, an aminoacyl tRNA essential for initiation of polypeptide synthesis. This complex, in turn, forms a 43S preinitiation complex with a 40S ribosomal subunit before the initiation messenger RNA joins to form a 48S preinitiation complex. This process is ATP-dependent. Finally, a 60S ribosomal subunit joins to form a 80S initiation complex. The whole process clearly requires ATP and GTP.

Protein synthesis can be measured by several different methods. Although each component for formation of the 80S initiation complex can be isolated and reassembled, it is very difficult to do such an experiment in pathophysiologic conditions, thus we are obliged to rely on more crude methods. A commonly used method is to quantitatively measure incorporation of radioactive amino acids into proteins after systemic or intraventricular administration. A second method is to measure incorporation of radioactive amino acids into proteins by quantitative autoradiography after intravenous administration. The third is to measure amino acid incorporation during incubation of tissue slices or cells, and the fourth is to measure radioactive amino acid incorporation into polypeptides by purified microsomes or polyribosomes. Each method has its advantages and disadvantages.

As early as 1965, suppression of protein synthesis was observed during hypoxia and anoxia-ischemia using biochemical and autoradiographic methods [60,61]. While reduction of protein synthesis occurred with each method, precise measurements and interpretations are difficult in the absence of the information regarding the precursor availability or when the precursor availability was altered because of the pathophysiologic condition. This problem may have existed in the study of protein synthesis in vivo during hypoxia in rabbits [62], or after prolonged global ischemia in cats [63,64]. An alternative approach (which we used) is an in vitro model of cerebral anoxia [65], and another is the measurement of polypeptide synthesis by isolated microsomes or polyribosomes [66]. RNA synthesis can be measured in vitro by measurement of nucleoside incorporation into RNA [67] or the assay of DNA-dependent RNA polymerase activity [66].

It is difficult to measure protein synthesis during complete cerebral ischemia because the radioactive precursor does not adequately reach the ischemic area. In any event, it is expected to be very low. When protein synthesis and the acid-soluble fraction (precursor availability) were actually measured after occlusion of the posterior communicating artery in gerbils using quantitative autoradiography, they were extremely low [68]. We also evaluated protein synthesis in vivo during hypoxia, by exposure of rats to 5% oxygen atmosphere [69]. While protein synthesis was reduced by 25% during hypoxia, there was an increase in the acid-soluble fraction and an accurate measurement for protein synthesis was difficult. With the same experimental model, polypeptide synthesis with microsomes was reduced by 30%–40% after hypoxia for 20 to 30 min. The polyribosomal size distribution, an index for intactness of polysomes, revealed an increase of the monomer and dimer peaks, indicating disintegration of polysomes.

We have also measured protein synthesis during hypoxia in vitro with brain slices [20]. In one series of experiments, rabbit brain slices were incubated under constant flow of 10%–80% oxygen in the presence of radioactive leucine. In another series of experiments, incubation was carried out without radioactive leucine first, and polypeptide synthesis was then carried out with polyribosomes. As shown in Fig. 3, the acid-soluble fraction did not change until the oxygen concentration was reduced to 10%. Protein synthesis in the tissue homogenate started to decline when the oxygen concentration was reduced to 50% after which it declined linearly. Polypeptide synthesis with purified polyribosomes also showed steady decline after reduction of the oxygen concentration to 50%. It should be noted that there was a close agreement between protein synthesis with tissue slices and polypeptide synthesis with polyribosomes when there was no change in the acid-soluble fraction. While polyribosomal size distribution profiles have not been studied with this in vitro model of cerebral hypoxia, we expect progressive increase of the monomer and dimer peaks and decline of polysomes. Disaggregation of polyribosomes has been observed with the in vivo model of cerebral hypoxia [69] and with the in vitro model of cerebral anoxia [70].

The brain is composed of heterogenous cellular components. Therefore, it is important to study protein synthesis under pathophysiologic conditions in nerve and glia cells separately. For this purpose, we have utilized the Ficoll density gradient ultracentrifugation technique [70] to separate the neuron- and neuroglia-enriched fractions after cerebral hypoxia in vitro [20]. As the oxygen concentration was lowered in vitro, protein synthesis steadily decreased in the neuronal and neuroglial fractions to a similar extent (Fig. 4). The effect of hypoxia was also evaluated with isolated nerve and glia cells during incubation under low oxygen atmosphere in the presence of radioactive leucine or uridine (Table 1) [21]. Again, hypoxia affected protein synthesis in isolated nerve and glia cells to a similar extent. This was also true for RNA synthesis. On the other hand, hypoglycemia, simulated by removal of glucose from the incubation medium, affected nerve cells more than glia cells.

Protein synthesis is rigidly controlled by messenger RNA and it is highly con-

Takehiko Yanagihara

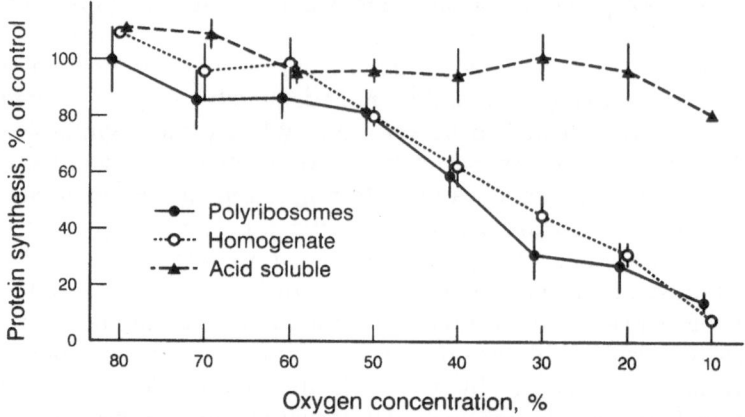

Fig. 3. Protein synthesis with brain slices and polypeptide synthesis with polyribosomes during hypoxia in vitro. Brain slices were incubated under constant flow of 10–80% oxygen for 30 min. For protein synthesis with brain slices, [³H]leucine was added during incubation and the specific radioactivity (DPM/mg protein) was measured after washing with trichloroacetic acid. The acid-soluble radioactivity (DPM/mg wet tissue weight) was measured from the supernatant after precipitation of a brain slice with trichloroacetic acid. Polypeptide synthesis with polyribosomes was measured by incubation of purified polyribosomes in the presence of [³H]leucine. The results for polypeptide synthesis was expressed as DPM/µg protein. The result from the experimental group was further expressed as percent of the corresponding control group and presented as mean ± SEM based on 4–6 experiments for each oxygen concentration

Table 1. Protein and RNA synthesis with isolated cell fractions

	Neuron	Glia
Protein synthesis		
Control	100	100
Hypoxia	68.7 ± 9.3	68.5 ± 14.6
Hypoglycemia	61.9 ± 17.4	79.6 ± 13.2
RNA synthesis		
Control	100	100
Hypoxia	56.4 ± 9.1	50.0 ± 6.6
Hypoglycemia	61.7 ± 5.5	85.1 ± 8.2

Results are expressed as percent ± S.D. of the specific radioactivity (DPM/µg protein) of controls based on 4 experiments for each cell fraction and pathophysiologic condition, respectively. (Modified from [21])

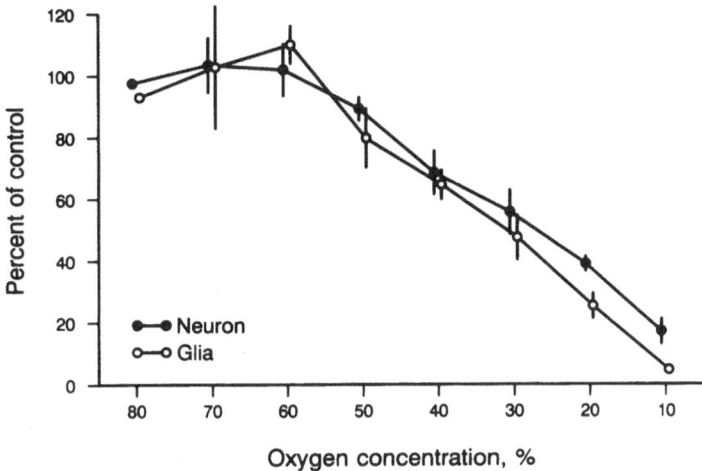

Fig. 4. Protein synthesis in the neuron- and neuroglia-enriched fractions during cerebral hypoxia in vitro. Protein synthesis was carried out as described in Fig. 3 with brain slices and then the neuron- and neuroglia-enriched fractions were separated using Ficoll density gradient ultracentrifugation. Each cell fraction was precipitated by trichloroacetic acid and the specific radioactivity (DPM/μg protein) was measured. The results from the experimental group was expressed as percent of the corresponding control group and presented as mean ± SEM based on 4–6 experiments for each oxygen concentration except at the oxygen concentration of 80%, where only 2 experiments were carried out

ceivable that protein synthesis is affected by alteration of RNA synthesis. We therefore investigated the relationship between protein and RNA synthesis by measuring polypeptide synthesis with isolated microsomes and DNA-dependent RNA polymerase activity (oligonucleotide synthesis) in cell nuclei (Fig. 5). With the in vitro model of cerebral anoxia, protein synthesis promptly declined to 60% of control in 10 min but no change was noted in RNA polymerase activity in isolated cell nuclei [67]. The discrepancy between protein synthesis and RNA synthesis was also present in gerbil brains during the first 2 h after unilateral carotid occlusion [71]. However, a direct comparison between cerebral anoxia and cerebral ischemia is difficult since the experimental models are quite different. Together with suppression of protein synthesis, disaggregation of polyribosomes occurred promptly after cerebral anoxia in vitro [70] but more slowly in cerebral hypoxia in vivo [69] and cerebral ischemia (Fig. 6) [72].

As reviewed earlier, biochemical derangements after reperfusion or reoxygenation are quite different from those during complete cerebral ischemia. Therefore, the effect on protein and RNA synthesis may be quite different between these two distinct conditions. Due to the nature of the experimental model, we have not carried out an investigation to measure the effect of prolonged re-oxygenation after cerebral anoxia. However, the in vitro model of cerebral anoxia [19,64] is actually equivalent to brief re-oxygenation after anoxia in vivo if subsequent protein or RNA synthesis is carried out under a well-oxygenated condition. In such a condition, protein synthesis was affected in the neuron-

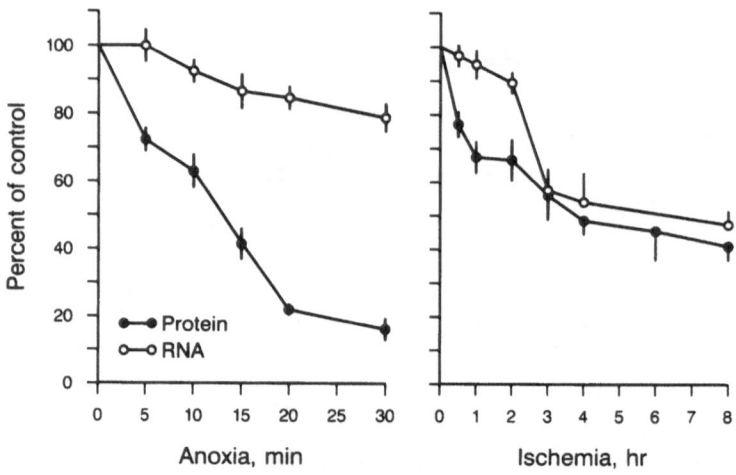

Fig. 5. Comparison of polypeptide synthesis by microsomes and oligonucleotide synthesis by cell nuclei in cerebral anoxia in vitro and cerebral ischemia in vivo. For production of cerebral anoxia, rabbit brain slices were incubated under constant flow of nitrogen for 5–30 min. For production of cerebral ischemia, the right common carotid artery was occluded for 30 min to 8 h in gerbils. Polypeptide synthesis was carried out by incubation of microsomes in the presence of [³H]leucine. The specific radioactivity was measured as described in Fig. 3. Oligonucleotide synthesis was carried out by incubation of purified nuclei in the presence of [³H]GTP. The assay methods for Mn-dependent RNA polymerase activity (for cerebral anoxia) and RNA polymerase II activity (for cerebral ischemia) were used as representing messenger RNA synthesis. The radioactivity and the DNA content were measured and the specific radioactivity was expressed as DPM/μg DNA. The results from the experimental group was expressed as percent of the corresponding control group and presented as mean ± SEM based on 6–8 experiments for cerebral anoxia and 4–6 experiments for cerebral ischemia. (Data from [66,71])

enriched and neuroglia-enriched fractions to a similar degree [70] and the extent of suppression was similar to that observed during exposure to 10% oxygen [20]. This indicated that protein synthesis did not show any tendency for recovery during re-oxygenation following anoxia for 30 min. Although we have not examined protein synthesis following exposure of rats to 5% oxygen atmosphere, judging from their behavior, we expect it to recover promptly. The situation was also similar in cerebral ischemia in gerbils if protein or RNA synthesis was carried out in vitro with brain slices after cerebral ischemia in vivo [15,66]. In this case, protein synthesis did not recover promptly and was affected in the neuron- and neuroglia-enriched fractions to a similar extent [15].

The integrity of polyribosomes during reperfusion has been investigated by others. Disaggregation of polyribosomes was demonstrated with the polyribosomal size-distribution profile and electron microscopy in cat brains during reperfusion following prolonged complete ischemia [63]. Disaggregation of polyribosomes was also observed during reperfusion after complete ischemia caused by intracranial hypertension in rat brains [73]. We studied polypeptide synthesis

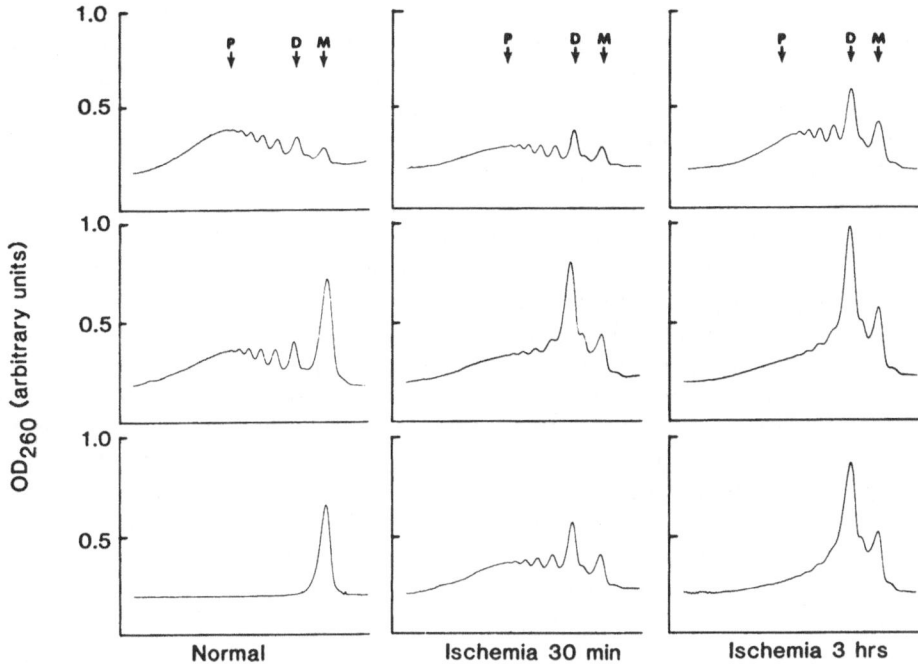

Fig. 6. The polyribosomal size distribution profiles from normal brain (left column), after ischemia for 30 min (center column) and after ischemia for 3 h (right column) following unilateral carotid occlusion in gerbils. For normal brain (left column), the middle profile was prepared after addition of the 80S marker and the bottom profile was prepared with the 80S marker alone. For ischemic brains (center and right column), the profiles at the top were prepared from the brains without reperfusion, while the profiles in the middle and the bottom were prepared from the brains with reperfusion for 15 min and 3 h, respectively. M, D, and P designate the location of the monomer, dimer and polysomes, respectively. (By permission of the American Heart Association Inc. [72])

and the polyribosomal size-distribution profile in gerbil brains during reper-fusion following unilateral carotid occlusion [72]. Marked disintegration of polyribosomes occurred after reperfusion for 15 min, regardless of whether the ischemic period was for 30 min or 3 h (Fig. 6). While the polyribosomal size-distribution profile returned to near normal after reperfusion for 3 h if the ische-mic period was 30 min, no trend for recovery was observed when the ischemic period was extended to 3 h (Fig. 6). A drastic suppression of polypeptide synthe-sis also occurred after reperfusion for 15 min. Electron microscopic examina-tion revealed disaggregation of polyribosomes.

Since messenger RNA is required for integrity of polysomes, we compared the effect of reperfusion on polypeptide synthesis by polyribosomes and DNA-dependent RNA polymerase II activity in cell nuclei [74]. Suppression of oligo-nucleotide synthesis was only 40% compared to 80% for polypeptide synthesis, and recovered promptly if the ischemic period was 30 min.

It is interesting to note that synthesis of a stress protein occurred in gerbil brains 2 h after reperfusion following bilateral carotid occlusion for 5 min [75]. Further immunohistochemical investigation localized synthesis of this protein extensively in the brain and in the nerve cell bodies [76]. However, the CA1 region of the hippocampus—the site of delayed neuronal death—showed only weak or no synthesis of this protein [76].

Quantitative autoradiography has been used for measurement of protein synthesis. The obvious advantage of this method is the regional presentation of protein synthesis. While this method may provide an accurate measurement of the rate of protein synthesis under certain conditions [77], a precise quantitation may not be achieved in various pathophysiologic conditions where the precursor pool, precursor incorporation into proteins, and degradation of these proteins may not be affected uniformly. The measurement may be more complicated if multiple radiotracers are utilized [78]. Protein synthesis has been measured during occlusion of the posterior communicating artery in gerbils [68]. There was virtually no protein synthesis, and radioactive precursors were also markedly reduced. Protein synthesis also has been measured autoradiographically after embolic infarction in monkey brains [79]. Markedly reduced protein synthesis was observed 2 h after embolization, but did not correlate well with the morphologic abnormalities in those areas.

Quantitative autoradiography may be better suited for measurement of protein synthesis after re-establishment of cerebral circulation because of easier access of radiotracers to the ischemic areas. After complete ischemia in the rat brain for 30 min, protein synthesis was reduced in many areas even after reperfusion for 48 h [80]. On the other hand, recovery of protein synthesis occurred in most areas except for the pyramidal neurons in the paramedian and CA1 region of the hippocampus if the ischemic period was only for 10 min [80]. Protein synthesis did not recover in the CA1 region of the hippocampus of gerbil brains after bilateral carotid occlusion for 5 min and subsequent reperfusion up to 72 h [81]. We have also measured protein synthesis autoradiographically after unilateral carotid occlusion for 10 min with subsequent reperfusion in gerbils and observed similar results.

The molecular mechanism for suppression of protein synthesis is uncertain. We have evaluated the possibility of degradation of polyribosomes by proteases or ribonucleases, since these degradative enzymes may be activated by acidic pH and released during anoxia or ischemia. The experiments with the soluble fraction from anoxic or ischemic brain tissues and those with inhibitors for these enzymes indicated that release or activation of the enzymes was unlikely the cause of the disintegration of polyribosomes [82]. A more direct evidence against ribonuclease activation was subsequently provided, when the acidic ribonuclease activity was examined using radioactive cerebral RNA during ischemia for 3 h in gerbils [83]. The effect of various biochemical derangements on protein synthesis was also evaluated with an in vitro system [84]. Low pH and low sodium ions did not affect protein synthesis, but high extracellular calcium ions induced disaggregation of polyribosomes. When the mitochondrial function was inhibited by 2,4-dinitrophenol, marked suppression of protein synthesis and of radioactive precursor transport occurred, but no effect was noted on the integrity of polyribosomes. Therefore, biochemical derangements which occur during ischemia and reperfusion, as reviewed earlier, may contribute significantly to the suppression of protein synthesis and disintegration of polyribosomes. Although we have not evaluated the effect of peroxidation, the deleterious effect of oxygen radicals on proteins [55] suggests that free radicals may also affect protein synthesis.

Accumulation of the monomers and dimers in the polyribosomal size-distribution profile soon after reperfusion suggested a block at the level of the re-initiation step [63,72,73]. As mentioned earlier, initiation of protein synthesis in eukaryotic cells is a very complex molecular event [59]. ATP is required for formation of the 48S preinitiation complex, formation of aminoacyl transfer RNA, and initiation of translation. GTP is required for formation of the binary complex. Both messenger and transfer RNAs are necessary to synthesize polypeptides. The subnormal tissue ATP and GTP levels during reperfusion [31] may thus affect protein synthesis. Structural alterations of various initiation factors or ribosomal subunits caused by tissue acidosis or free radicals may halt the initiation step for a sustained period. If messenger and transfer RNAs are depleted by accelerated degradation or suppressed synthesis, or if their structures are altered by miscoding, the initiation step of protein synthesis will not proceed during reperfusion. At the present time, it is not possible to precisely localize the site of the blockade in the initiation step in cerebral ischemia. If the mechanism is similar to inhibition of protein synthesis in heat shock, the blockade could be at 43S initiation complex formation [85].

The molecular mechanism for suppression of the RNA polymerase activity is not clear either. In our study with gerbils, suppression immediately after reperfusion was transient if the ischemic period was short but became persistent if the ischemic period was prolonged, coinciding well with transient and persistent suppression of polypeptide synthesis [74]. Since both ATP and GTP serve as precursors for oligonucleotide synthesis, the subnormal tissue levels during reperfusion [31] may affect RNA synthesis. Anoxia and ischemia may also alter DNA and protein structures within nuclei, which may result in suppression of RNA synthesis or alteration of RNA structures. Phosphorylation of chromatin proteins, tightly bound to DNA, was suppressed in cerebral anoxia and ischemia, and was qualitatively altered [86]. Since those qualitative and quantitative changes could not be seen until after anoxia for 15 min or ischemia for 3 h, it is uncertain whether or not these changes were directly responsible for suppression of RNA synthesis. We have also observed qualitative alterations of newly synthesized messenger RNA after cerebral ischemia [87]. However, this also occurred after prolonged ischemia and the significance is uncertain at the present time. In order to examine other parts of the nuclear structure, we evaluated the nuclear binding sites for triiodothyronine after unilateral carotid occlusion in gerbils [88]. This hormone receptor is known to be present in the neuronal nuclei. During cerebral ischemia, there was a steady increase in the binding sites, which was reversed abruptly upon reperfusion. If the ischemic period was 30 minutes, the binding sites and the affinity normalized in 24 h but no recovery occurred when the ischemic period was extended to 3 h. Thus, the functional elements in neuronal nuclei also appear to sustain "jolts" or "shocks" when cerebral perfusion is re-established. This may be directly or indirectly caused by tissue acidosis, influx of calcium, peroxidation, or other biochemical derangements but the exact mechanism is uncertain.

Thus a series of investigations by us and others thus indicated that the reversibility of protein synthesis may depend upon the duration and severity of ischemia or hypoxia as well as upon selective tissue vulnerability. In complete ischemia or severe hypoxia, cell death may ensue because of the failure of oxidative phosphorylation and depletion of high energy phosphates where protein or RNA synthesis has no role for cell survival. If cerebral perfusion is re-established early enough, protein and RNA synthesis may recover although further deterioration may occur temporarily. If the ischemic period is too long, neither protein nor RNA synthesis recovers even after reperfusion. In incomplete cerebral ischemia or in the areas surrounding the infarcted core of focal cerebral ischemia, the recovery of protein and RNA synthesis may play a significant role in the recovery of partially damaged nerve cells or in the prevention of delayed neuronal death.

Structural Damage in Cerebral Ischemia

In the past, most biochemical analyses for cerebral ischemia and reperfusion have been carried out with relatively large amounts of brain tissues. This is also true for most investigations to elucidate the molecular mechanism of ischemic and postischemic damage. However, it has been shown by various morphologic investigations that selective tissue vulnerability exists not only in gross anatomical regions but also at the microscopic level. This requires us to study the molecular mechanism of ischemic and postischemic damage at the microscopic level.

In the past few years, we have been engaged in immunohistochemical investigations of cerebral ischemia. We have found that early ischemic and postischemic damages occur in certain anatomic locations in the hippocampus, cerebral cortex, caudoputamen, and thalamus, and that early immunohistochemical lesions occur as early as 3 min after unilateral or bilateral carotid occlusion in gerbils [32,89–91]. Since only a handful of brain-enriched and brain-specific proteins have been evaluated immunohistochemically in cerebral ischemia, it is premature to assess the contribution of this technique for investigation of ischemic and postischemic brains. However, a series of immunohistochemical investigations clearly pointed out the presence of selective tissue vulnerability and the location of the vulnerable sites at the microscopic level. They also clearly demonstrated that dendrites are as vulnerable as neuronal cell bodies (Fig. 7). These findings have led us to investigate by the electron microscopic technique one of the most vulnerable sites in the hippocampus of gerbils, from the pyramidal cell bodies to the periphery of the apical and basal dendrites [92]. The most striking finding was a prompt swelling of the distal part of the dendrites as early as 5 min after unilateral carotid occlusion (Fig. 8). Inside the swollen

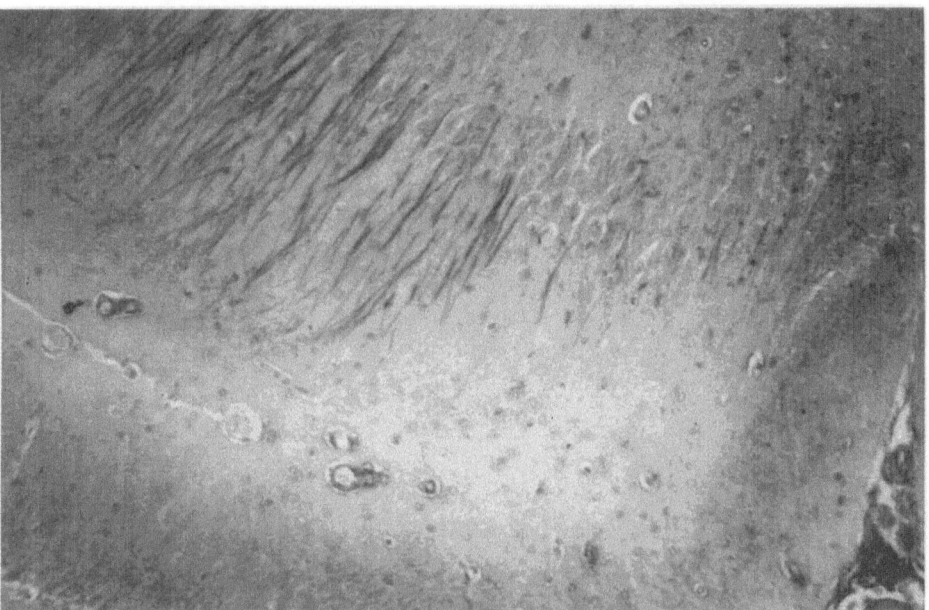

Fig. 7. Loss of the immunohistochemical reaction for creatine kinase BB-isoenzyme in the subiculum-CA1 region of the hippocampus 10 min after unilateral carotid occlusion in a gerbil. Cell nuclei were counter-stained with hematoxylin (X 100). Note the disappearance of the reaction in the apical and basal dendrites

dendrites, swelling of mitochondria and disintegration of microtubules were observed. At longer ischemic periods, further disintegration of mitochondria and disruption of plasma membranes occurred (Fig. 8), and the morphologic evidence of ischemic damage spread proximally to the pyramidal cell bodies. We have also observed central propagation of postischemic damage during reperfusion [93].

Although an earlier report described a breakdown of ribosomal units and swelling of rough endoplasmic reticulum as the first neuronal abnormalities 15 min after unilateral carotid occlusion in gerbils [94], we observed ultrastructural damage occurring earlier in the periphery of the dendrites both during progressive cerebral ischemia and following reperfusion. Regardless of whether or not massive membrane depolarization by excitatory neurotransmitters and/or influx of calcium ions are responsible for these ultrastructural changes, disintegration of mitochondria would result in tissue death in that particular area. The role of microtubules in the dendrites is not well defined but they may be important in dendritic transport.

In contrast to the well-established concept of axonal transport which was proposed over 40 years ago [95], the idea of dendritic transport has not been well recognized. However, the spreading of newly synthesized proteins from neuronal perikarya to dendrites [96] and the disruption of dendritic spreading by colchi-

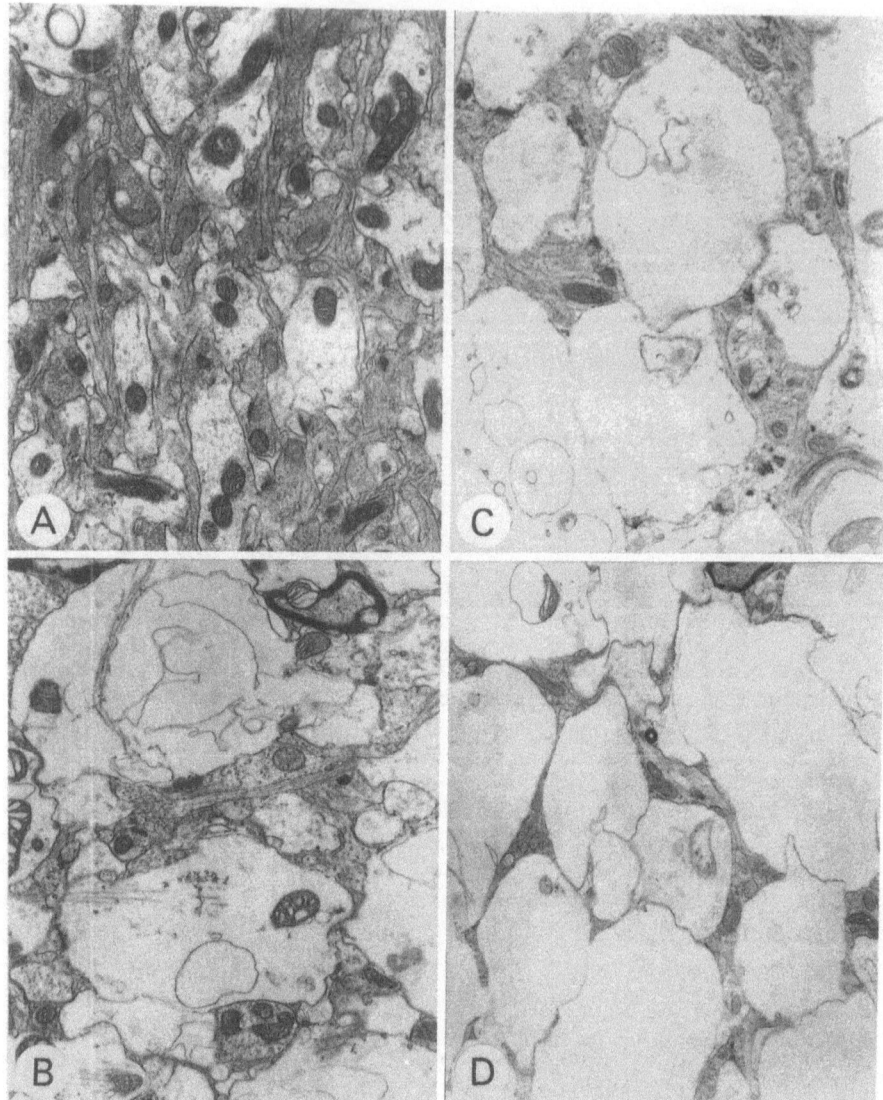

Fig. 8. The stratum moleculare of the subiculum-CA1 region of the hippocampus under electron microscopic examination after unilateral carotid occlusion for 5 (B), 10 (C) and 30 min (D) in gerbils. Note marked swelling of the distal parts of the apical dendrites with progressive loss of the cytoplasmic content and disruption of plasma membranes as compared to the control section (A). (X 17,500) (By permission from Elsevier Science Publishers BV [92])

cine [97] have hinted at the presence of dendritic transport and the role of microtubules. Therefore, it is conceivable that disintegration of microtubules may result in disruption of dendritic transport of soluble proteins and possibly other molecules to the peripheral part of the dendrites. Suppression of protein synthesis may also disrupt dendritic transport of proteins. Therefore, sustained suppression of protein synthesis and disruption of dendritic transport may delay repair of ischemic damage and may facilitate progressive postischemic damage, thus leading to eventual neuronal death.

Summary. Biochemical derangement occurs in the brain during and after cerebral ischemia/hypoxia where conditions such as loss of high energy phosphates, tissue acidosis, influx of calcium ions, free radical formation, and release of excitatory neurotransmitters, among others, have been observed. Biochemical derangements may be different during ischemia/hypoxia and after reperfusion/re-oxygenation. They can affect vital intracellular molecular metabolism as shown in protein and RNA synthesis, where clear differences have been observed during progressive ischemia and after reperfusion. Morphologically, selective vulnerability exists in cerebral ischemia both during progression and after reperfusion. In both conditions, structural damage occurs first in the periphery of dendrites and propagate centrally. Sustained suppression of RNA and protein synthesis and disruption of dendritic transport may result in delay of the repair process in partially damaged neurons and may facilitate postischemic damage. Resuscitation of the brain following global ischemia, rescue of partially damaged tissue in focal cerebral ischemia, and prevention of postischemic damage may be possible by preventing or minimizing the observed biochemical and molecular derangements which eventually cause structural damage.

Acknowledgment. This series of investigations have been supported by grant NS-06663 from the National Institutes of Health, the U.S. Public Health Services. The author thanks Mrs. Joan M. Brengman and all research associates involved in the present series of investigations for their excellent works and thanks Mrs. Gail Sim for the preparation of this manuscript.

References

1. Hossmann K-A, Sato K (1970) Recovery of neuronal function after prolonged cerebral ischemia. Science 168:375–376
2. Kramer RS, Sanders AP, Lasage AM, Woodhall B, Sealy WC (1968) The effect of profound hypothermia on preservation of cerebral ATP content during circulatory arrest. J Thorac Cardiovasc Surg 56:699–709
3. Hossmann K-A, Kleihues P (1973) Reversibility of ischemic brain damage. Arch Neurol 29:375–384
4. Ljunggren B, Schutz H, Siesjö BK (1974) Changes in energy state and acid-base parameters of the rat brain during complete compression ischemia. Brain Res 73:277–289
5. Pulsinelli WA, Brierley JB (1979) A new model of bilateral hemispheric ischemia in the unanesthetized rat. Stroke 10:267–272

6. Harvey J, Rasmussen T (1951) Occlusion of the middle cerebral artery: An experimental study. Arch Neurol Psychiat 66:20–29
7. Sundt TM Jr, Waltz AG (1966) Experimental cerebral infarction: Retro-orbital, extradural approach for occluding the middle cerebral artery. Mayo Clin Proc 41:159–168
8. Symon L, Dorsch NWC, Crockard HA (1975) The production and clinical features of a chronic stroke model in experimental primates. Stroke 6:476–481
9. Crowell RM, Olsson Y (1973) Effect of extracranial intracranial vascular bypass graft on experimental acute stroke in dogs. J Neurosurg 38:26–31
10. Yamamoto K, Yoshimine T, Yanagihara T (1985) Cerebral ischemia in rabbit: A new experimental model with immunohistochemical investigation. J Cereb Blood Flow Metab 5:529–536
11. Tamura A, Graham DI, McCulloch J, Teasdale GM (1981) Focal cerebral ischemia in the rat: 1. Description of technique and early neuropathological consequences following middle cerebral artery occlusion. J Cereb Blood Flow Metab 1:53–60
12. Levine S, Payan H (1966) Effects of ischemia and other procedures on the brain and retina of the gerbil (Meriones unguiculatus). Exp Neurol 16:255–262
13. Yoshimine T, Yanagihara T (1983) Regional cerebral ischemia by occlusion of the posterior communicating artery and the middle cerebral artery in gerbils. J Neurosurg 58:362–367
14. Yamada K, Hayakawa T, Yoshimine T, Ushio Y (1984) A new model of transient hindbrain ischemia in gerbils. J Neurosurg 60:1054–1058
15. Yanagihara T (1978) Experimental stroke in gerbils: Correlation of clinical, pathological and electroencephalographic findings and protein synthesis. Stroke 9:155–159
16. Matsumoto M, Hatakeyama T, Akai F, Brengman JM, Yanagihara T (1988) Prediction of stroke before and after unilateral occlusion of the common carotid artery in gerbils. Stroke 19:490–497
17. Gurdjian ES, Stone WE, Webster JE (1944) Cerebral metabolism in hypoxia. Arch Neurol Psychiat 51:472–477
18. Siesjö BK, Nilsson L (1971) The influence of arterial hypoxemia upon labile phosphates and upon extracellular and intracellular lactate and pyruvate concentrations in the rat brain. Scand J Clin Lab Invest 27:83–96
19. Yanagihara T (1973) Cerebral anoxia: An improved in vitro model for biochemical study. Stroke 4:409–411
20. Yanagihara T (1974) Protein metabolism in the neuronal and neuroglial fractions of rabbit brain during hypoxia. Trans Am Soc Neurochem 5:108
21. Yanagihara T (1979) Protein and RNA synthesis and precursor uptake with isolated nerve and glia cells. J Neurochem 32:169–177
22. Rothman S (1983) Synaptic activity mediates death of hypoxia neurons. Science 220:536–537
23. Goldberg MP, Weiss JH, Phuong-Chi P, Choi DW (1987) N-methyl-D-aspartate receptors mediate hypoxic neuronal injury in cortical culture. J Pharmacol Exp Ther 243:784–791
24. Lowry OH, Passonneau JV, Hasselberger FX, Schultz DW (1964) Effect of ischemia on known substrates and cofactors of the glycolytic pathway in brain. J Biol Chem 239:18–30
25. Duffy TE, Nelson SR, Lowry OH (1972) Cerebral carbohydrate metabolism during acute hypoxia and recovery. J Neurochem 19:959–977
26. Ueda H, Hashimoto T, Furuya E, Tagawa K, Kitagawa K, Matsumoto M, Yoneda S, Kimura K, Kamada T (1988) Changes in aerobic and anaerobic ATP-synthesizing

activities in hypoxic mouse brain. J Biochem 104:81–86

27. Mršulja BB, Mršulja BJ, Ito U, Walker JTJr, Spatz M, Klatzo I (1975) Experimental cerebral ischemia in Mongolian gerbils II: Changes in carbohydrates. Acta Neuropathol (Berl) 33:91–103

28. Mršulja BB, Lust WD, Mršulja BJ, Passonneau JV, Klatzo I (1976) Post-ischemic changes in certain metabolites following prolonged ischemia in the gerbil cerebral cortex. J Neurochem 26:1099–1103

29. Levy DE, Duffy TE (1977) Cerebral energy metabolism during transient ischemia and recovery in the gerbil. J Neurochem 28:63–70

30. Nowak TS Jr, Fried RL, Lust WD, Passonneau JV (1985) Changes in brain energy metabolism and protein synthesis following transient bilateral ischemia in the gerbil. J Neurochem 44:487–494

31. Brengman JM, Morimoto K, Yanagihara T (1988) Purine nucleotides and phosphocreatine in postischemic gerbil brain. Trans Am Soc Neurochem 19:147

32. Hatakeyama T, Matsumoto M, Brengman JM, Yanagihara T (1988) Immunohistochemical investigation of ischemic and postischemic damage after bilateral carotid occlusion in gerbils. Stroke 19:1526–1534

33. Meyer FB, Anderson RE, Sundt TM Jr, Yaksh TL (1986) Intracellular brain pH, indicator tissue perfusion, electroencephalography and histology in severe and moderate focal cortical ischemia in the rabbit. J Cereb Blood Flow Metab 6:71–78

34. Kawashima J, Nakamura K, Fujitani B, Kadokawa T, Yoshida K, Shimizu M (1978) Relationship between cerebral energy failure and free fatty acid accumulation following prolonged brain ischemia. Jpn J Pharmacol 28:277–287

35. Petito CK, Kraig RP, Pulsinelli WA (1987) Light and electron microscopic evaluation of hydrogen ion-induced brain necrosis. J Cereb Blood Flow Metab 7:625–632

36. Siemkowicz E, Hansen AJ (1978) Clinical restitution following cerebral ischemia in hypo-, normo- and hyperglycemic rats. Acta Neurol Scand 58:1–8

37. Schurr A, Dong W-Q, Reid KH, West CA, Rigor BM (1988) Lactic acidosis and recovery of neuronal function following cerebral hypoxia in vitro. Brain Res 438:311–314

38. Prado R, Ginsberg MD, Dietrich WD, Watson BD, Busto R (1988) Hyperglycemia increases infarct size in collaterally perfused but not end-arterial vascular territories. J Cereb Blood Flow Metab 8:186–192

39. Harris RJ, Symon L, Branston NM, Bayhan M (1981) Changes in extracellular activity in cerebral ischemia. J Cereb Blood Flow Metab 1:203–209

40. Yanagihara T, McCall JT (1982) Ionic shift in cerebral ischemia. Life Sci 30:1921–1925

41. Hossmann K-A, Paschen W, Csiba L (1983) Relationship between calcium accumulation and recovery of cat brain after prolonged cerebral ischemia. J Cereb Blood Flow Metab 3:346–353

42. Dienel GA (1984) Regional accumulation of calcium in postischemic rat brain. J Neurochem 43:913–925

43. Martins E, Inamura K, Themner K, Malmqvist KG, Siesjö BK (1988) Accumulation of calcium and loss of potassium in the hippocampus following transient cerebral ischemia: A proton microprobe study. J Cereb Blood Flow Metab 8:531–538

44. Farber JL (1981) The role of calcium in cell death. Life Sci 29:1289–-1295

45. Connor, JA, Wadman WJ, Hockberger PE, Wong RKS (1988) Sustained dendritic gradients of Ca^{2+} induced by excitatory amino acids in CA1 hippocampal neurons. Science 240:649–653

46. Choi DW, Maulucci-Gedde MA, Kriegstein AR (1987) Glutamate neurotoxicity in

cortical cell culture. J Neurosci 7:369–379
47. Choi DW (1987) Ionic dependence of glutamate neurotoxicity in cortical cell culture. J Neurosci 7:380–390
48. Bázan NG Jr (1970) Effects of ischemia and electroconvulsive shock on free fatty acid pool in the brain. Biochim Biophys Acta 218:1–10
49. Yoshida S, Inoh S, Asano T, Sano K, Kubota M, Shimazaki H, Ueta N (1980) Effect of transient ischemia on free fatty acids and phospholipids in the gerbil brain: Lipid peroxidation as a possible cause of postischemic injury. J Neurosurg 53:323–331
50. Abe K, Kogure K, Yamamoto H, Imazawa M, Miyamoto K (1987) Mechanism of arachidonic acid liberation during ischemia in gerbil cerebral cortex. J Neurochem 48:503–509
51. Gaudet RJ, Levine L (1980) Effect of unilateral common carotid artery occlusion on levels of prostaglandins D_2 $F_{2\alpha}$ and 6-keto-prostaglandin $F_{1\alpha}$ in gerbil brain. Stroke 11:648–652
52. Moskowitz MA, Kiwak KJ, Hekimian K, Levine L (1984) Synthesis of compounds with properties of leukotrienes C_4 and D_4 in gerbil brains after ischemia and reperfusion. Science 224:886–889
53. McCord JM (1985) Oxygen-derived free radicals in postischemic tissue injury. N Engl J Med 312:159–163
54. Kontos HA (1985) Oxygen radicals in cerebral vascular injury. Circ Res 57:508–516
55. Davies KJA (1987) Protein damage and degradation by oxygen radicals. 1. General aspects. J Biol Chem 262:9895–9901
56. Burton KP (1988) Evidence of direct toxic effects of free radicals on the myocardium. Free Radical Biol Med 4:15–24
57. Zweier JL, Kuppusamy P, Lutty GA (1988) Measurement of endothelial cell free radical generation: Evidence for a central mechanism of free radical injury in post-ischemic tissues. Proc Natl Acad Sci USA 85:4046–4050
58. Patt A, Harken AH, Burton LK, Rodell TC, Piermattel D, Schorr WJ, Parker NB, Berger EM, Horesh IR, Terada LS, Linas SL, Cheronis JC, Repine JE (1988) Xanthine oxidase-derived hydrogen peroxide contributes to ischemia reperfusion-induced edema in gerbil brains. J Clin Invest 81:1556–1562
59. Pain VM (1986) Initiation of protein synthesis in mammalian cells. Biochem J 235:625–637
60. Sanders AP, Hale DM, Miller AT Jr (1965) Some effects of hypoxia on respiratory metabolism and protein synthesis in rat tissues. Am J Physiol 209:443–446
61. Yap S-L, Spector RG (1965) Cerebral protein synthesis in anoxic-ischaemic brain injury in the rat. J Path Bact 90:543–549
62. Blomstrand C (1970) Effect of hypoxia on protein metabolism in neuron and neuroglia cell-enriched fractions from rabbit brain. Exp Neurol 29:175–188
63. Kleihues P, Hossmann K-A (1971) Protein synthesis in the cat brain after prolonged cerebral ischemia. Brain Res 35:409–418
64. Kleihues P, Hossmann K-A (1973) Regional incorporation of $L-[3-{}^3H]$ tyrosine into cat brain proteins after 1 hour of complete ischemia. Acta Neuropathol (Berl) 25:313–324
65. Yanagihara T (1972) Cerebral anoxia: Protein metabolism during recovery in in vitro model. Stroke 3:733–738
66. Yanagihara T (1974) Cerebral anoxia: Effect on transcription and translation. J Neurochem 22:113–117
67. Yanagihara T (1976) Cerebral ischemia in gerbils: differential vulnerability of protein, RNA, and lipid synthesis. Stroke 7:260–263

68. Yoshimine T, Hayakawa T, Kato A, Yamada K, Matsumoto K, Ushio Y, Mogami H (1987) Autoradiographic study of regional protein synthesis in focal cerebral ischemia with TCA wash and image subtraction techniques. J Cereb Blood Flow Metab 7:387–393

69. Metter E J, Yanagihara T (1979) Protein synthesis in rat brain in hypoxia, anoxia and hypoglycemia. Brain Res 161:481–492

70. Yanagihara T (1976) Cerebral anoxia: effect on neuron-glia fractions and polysomal protein synthesis. J Neurochem 27:539–543

71. Yanagihara T (1978) Experimental stroke in gerbils: effect on translation and transcription. Brain Res 158:435–444

72. Morimoto K, Yanagihara T (1981) Cerebral ischemia in gerbils: polyribosomal function during progression and recovery. Stroke 12:105–110

73. Cooper HK, Zalewska T, Kawakami S, Hossmann K-A, Kleihues P (1977) The effect of ischemia and recirculation on protein synthesis in the rat brain. J Neurochem 28:929–934

74. Morimoto K, Mogami H, Brengman JM, Yanagihara T (1981) Molecular dysfunction following re-circulation in cerebral ischemia. Excerpta Medica International Congress Series No.548, p 73

75. Nowak TS Jr (1985) Synthesis of a stress protein following transient ischemia in the gerbil. J Neurochem 45:1635–1641

76. Nowak TS Jr, Vass K, Welch WJ (1987) Localization of 70 KDa heat shock protein induction in gerbil brain after transient ischemia. Soc Neurosci Abstracts 13:1688

77. Smith CB, Crane AM, Kadekaro M, Agranoff BW, Sokoloff L (1984) Stimulation of protein synthesis and glucose utilization in the hypoglossal nucleus induced by axotomy. J Neurosci 4:2489–2496

78. Mies G, Bodsch W, Paschen W, Hossmann K-A (1986) Triple-tracer autoradiography of cerebral blood flow, glucose utilization, and protein synthesis in rat brain. J Cereb Blood Flow Metab 6:59–70

79. Xie Y, Munekata K, Seo K, Hossmann K-A (1988) Effect of autologous clot embolism on regional protein biosynthesis of monkey brain. Stroke 19:750–757

80. Dienel GA, Pulsinelli WA, Duffy TE (1980) Regional protein synthesis in rat brain following acute hemispheric ischemia. J Neurochem 35:1216–1226

81. Thilmann R, Xie Y, Kleihues P, Kiessling M (1986) Persistent inhibition of protein synthesis precedes delayed neuronal death in postischemic gerbil hippocampus. Acta Neuropathol (Berl) 71:88–93

82. Morimoto K, Brengman J, Yanagihara T (1978) Further evaluation of polypeptide synthesis in cerebral anoxia, hypoxia and ischemia. J Neurochem 31:1277–1282

83. Albrecht J, Yanagihara T (1979) Effect of anoxia and ischemia on ribonuclease activity in brain. J Neurochem 32:1131–1133

84. Yoshimine T, Brengman J, Yanagihara T (1981) Correlation of the integrity of polyribosomes and protein synthesis. Trans Am Soc Neurochem 12:206

85. Austin SA, Pollard JW, Jagus R, Clemens MJ (1986) Regulation of polypeptide chain initiation and activity of initiation factor eIF-2 in Chinese-hamster-ovary cell mutants containing temperature-sensitive aminoacyl-tRNA synthetases. Eur J Biochem 157:39–47

86. Yanagihara T (1980) Phosphorylation of chromatin proteins in cerebral anoxia and ischemia. J Neurochem 35:1209–1215

87. Albrecht J, Yanagihara T (1978) Effect of cerebral anoxia and ischemia on messenger RNA metabolism. Trans Am Soc Neurochem 9:147

88. Thrall CL, Yanagihara T (1984) Nuclear binding sites for triiodothyronine in the

gerbil brain following ischemia and recirculation. Brain Res 301:179–183

89. Yanagihara T, Yoshimine T, Morimoto K, Yamamoto K, Homburger HA (1985) Immunohistochemical investigation of cerebral ischemia in gerbils. J Neuropath Exp Neurol 44:204–215

90. Yoshimine T, Morimoto K, Brengman JM, Homburger HA, Mogami H, Yanagihara T (1985) Immunohistochemical investigation of cerebral ischemia during recirculation. J Neurosurg 63:922–928

91. Matsumoto M, Yamamoto K, Homburger HA, Yanagihara T (1987) Early detection of cerebral ischemic damage and repair process in the gerbil by use of an immunohistochemical technique. Mayo Clin Proc 62:460–472

92. Yamamoto K, Morimoto K, Yanagihara T (1986) Cerebral ischemia in the gerbil: Transmission electron microscopic and immunoelectron microscopic investigation. Brain Res 384:1–10

93. Akai F, Yanagihara T (1986) Electron microscopic study of regional cerebral ischemia in the gerbil after occlusion of a posterior communicating artery (abstract). Xth Internat Congress Neuropath, p 119

94. Dodson RF, Chu LWF, Welch KMA, Achar VS (1977) Acute tissue response to cerebral ischemia in the gerbil: An ultrastructural study. J Neurol Sci 33:161–170

95. Weiss P (1944) Damming of axoplasm in constricted nerve: a sign of perpetual growth in nerve fibers. Anat Rec 88:464

96. Globus A, Lux HD, Schubert P (1968) Somadendritic spread of intracellularly injected tritiated glycine in cat spinal motoneurons. Brain Res 11:440–445

97. Schubert P, Kreutzberg GW, Lux HD (1971) Neuroplasmic transport in dendrites: Effect of colchicine on morphology and physiology of motoneurones in the cat. Brain Res 47:331–343

6

Protein and Polyamine Metabolism in Reversible Cerebral Ischemia of Gerbils

Wulf Paschen, Yaxia Xie, Gabriele Röhn, Joachim Hallmayer, and
Konstantin-Alexander Hossmann[1]

Introduction

The interruption of blood flow to the brain induces disturbances of brain metabolism [1] which are not immediately reversed after restoration of cerebral blood flow. The timecourses of ischemia-induced disturbances do vary, however, for different biochemical events. The recovery of energy metabolism is a fast process following short-term cerebral ischemia and the content of high energy phosphates is rapidly replenished after the onset of recirculation [2]. Inhibition of protein biosynthesis, in contrast, persisted for hours or even days after ischemia. [3–9]. The consequences of prolonged disturbances in protein biosynthesis are not known; however, it is obvious that persistent inhibition of protein biosynthesis must effect the integrity of the cell as long as the degradation of proteins is not reduced to the same extent as the biosynthesis.

Three different time periods must be distinguished when studying the ischemia-induced inhibition of protein biosynthesis: a) the ischemic period itself; b) the early recirculation (up to 24 h), and c) the late recirculation period (24 h or longer). Protein biosynthesis ceases during ischemia because it is an energy dependent process and energy stores are depleted. However, the protein biosynthesis machinery remains intact during ischemia because incorporation of amino acids into proteins is normal during ischemia when studied in an in vitro system to which ATP and an ATP regenerating system are added [4]. In contrast, protein biosynthesis is severely depressed during the early recirculation period both in vivo and in vitro. The cause of this inhibition is thought to be the disaggregation of polyribosomes which occurs shortly after the onset of recirculation [3,4]. With ongoing recirculation, protein biosynthesis slowly recovers to control values in the resistant brain structures, such as the neocortex, but remains

[1] Max-Planck-Institute for Neurological Research, Department of Experimental Neurology, Gleneler Straße 50, 5000 Cologne 41, Federal Republic of Germany

almost completely suppressed in the vulnerable CAl-subfield of the hippocampus in which neuronal necrosis is manifested after a few days [8,9].

In vivo protein biosynthesis is usually evaluated by autoradiographic measurement of the global incorporation of radioactive amino acids into brain proteins. However, the results, do not provide information about individual proteins unless the reaction products are further characterized (e.g. by two dimensional gel electrophoresis). This information may be critical for the interpretation of the observed disturbances because there are indications that certain proteins are preferentially synthesized at a time when the overall protein biosynthesis is severely depressed. Indeed, it has been shown that the synthesis of the heat-shock proteins [10–12] or the enzyme ornithine decarboxylase (ODC) [3,13,14] is increased after cerebral ischemia at the same time global protein biosynthesis is suppressed. Measurement of the content of these proteins therefore provides an opportunity to determine whether the biosynthesis machinery is broken (global suppression of the synthesis of all proteins) or switched to a different pattern (individual changes of different proteins).

In the present study post-ischemic temporal profiles of protein synthesis were studied in the cortex and the selectively vulnerable CAl-subfield of the hippocampus, and compared with changes in the activity of ODC, the key enzyme in polyamine metabolism which exhibits a half-life of only 10–20 min [15]. In addition, changes in the concentration of putrescine, the product of ODC reaction, were measured during and after ischemia because putrescine exhibits activities which might be of interest for the understanding of the mechanisms of ischemic cell injury (see Discussion). Protein biosynthesis, ODC activity, and putrescine levels were evaluated in untreated animals subjected to 5 min reversible cerebral ischemia and in animals treated post-ischemically with pentobarbital. Barbiturates are known to inhibit ischemic cell damage, [16–19] and we were interested to know if this treatment exerts a similar effect on protein biosynthesis, ODC activity, and putrescine levels.

Material and Methods

Animal Preparation

Experiments were performed in male Mongolian gerbils (*meriones unguiculatus*) weighing 50–75 g. The animals were anesthetized with 1.2% halothane, a neck incision was made, and both common carotid arteries were exposed. Reversible cerebral ischemia was produced by occluding both common carotid arteries with aneurysm clips. The aneurysm clips were removed after 5 min ischemia and the brains spontaneously recirculated. The skin incision was sutured and the animals were put back to their cages. Brains were recirculated for 2 or 24 h (protein biosynthesis study) or for 8, 24 and 96 h (polyamine metabolism study). Animals were re-anesthetized with 1.2% halothane at the end of the experiments and processed for autoradiographic measurement of regional protein biosynthesis or analysis of ODC activity and putrescine levels. Barbiturate-treated animals received a single intraperitoneal injection of pentobarbital (50 mg/kg) immediately after removal of the aneurysm clips.

Evaluation of Protein Biosynthesis

Regional protein biosynthesis was studied autoradiographically by measuring the incorporation of amino acids into proteins. A mixture of tritium-labeled amino acids (phenylalanine [phe], tyrosine [tyr], methionine [met], leucine [leu] and isoleucine [ile], specific activity 77 Ci/mmol, 1 mCi/kg) was injected intraperitoneally 45 min prior to the chosen recirculation time. Arterial blood samples were taken at increasing time intervals during tracer circulation to insure similar blood radioactivity in control and ischemic animals. At the end of the experiments the brains were quickly removed and frozen in 2-methylbutane pre-cooled with liquid nitrogen. The brains were then cut at $-20°C$ in a cryostat, and coronal sections of 20 μm thickness were dried on a warm plate at about 45°C. Every second section was incubated with 10% trichloroacetic acid overnight to remove free amino acids. Incubated and non-incubated sections (representing radioactivity incorporated into proteins and total tissue radioactivity, respectively) were exposed together with standards on X-ray film. The resultant autoradiograms were scanned with a rotating microdensitometer and further analyzed using an image processing system. Protein biosynthesis was expressed as the fraction of total radioactivity incorporated into TCA-insoluble proteins. Values given are means $\pm$ SEM.

Biochemical Analysis

Brains were frozen in situ at the end of the experiments for measurements of ODC activity and putrescine levels. They were cut in a cryostat and 20 μm thick coronal sections were taken for morphological localization and correlation with the density of ischemic cell damage. Small tissue samples of about 2–4 mg were taken from the cortex and the hippocampal CAl-subfield of the right and left hemispheres, and used for measuring ODC-activity and putrescine levels, respectively.

ODC activity was evaluated by measuring the release of $^{14}CO_2$ from L-[1-^{14}C]ornithine (spec. activity 52.6 mCi/mmol) as described by Pegg et al.[15] with modifications [14].The assay was carried out in sealed tubes containing center wells. The tissue was homogenized in Tris-buffer supplemented with 2 mM DTT. The test mix was composed of brain homogenate in Tris/DTT-buffer, supplemented with pyridoxalphosphate (50 μM). The reaction was started by adding [^{14}C]ornithine. After 60 min incubation at 37°C the reaction was terminated by injecting 100 μl perchloric acid (0.6 M) into the test mix. The $^{14}CO_2$ released by ODC was trapped on a Hyamine-impregnated filter paper in the center well. The radioactivity on the filter was measured using a liquid scintillation counter.

Quantification of putrescine levels was performed as recently described [20]. In short, putrescine was extracted from tissue samples with perchloric acid and measured after derivatization with o-phthalaldehyde using HPLC and a fluorescence detector. Quantification of the peak area was carried out with external standard solutions.

The data obtained were analyzed using analyzis of variance (ANOVA). Dif-

ferences between groups were calculated using the Scheffe's F-test. Values given
are means ± SEM.

Results

Protein Biosynthesis

Autoradiograms of the regional incorporation of tritiated amino acids into brain
proteins of control animals and animals subjected to 5 min forebrain ischemia
and 2 or 48 h recirculation are illustrated in Fig. 1. The quantitative results are
given in Table 1. Amino acid incorporation into proteins in control animals
reflected neuronal density; $72.4 \pm 3.9\%$ (range 59–84) of total tissue radioacti-

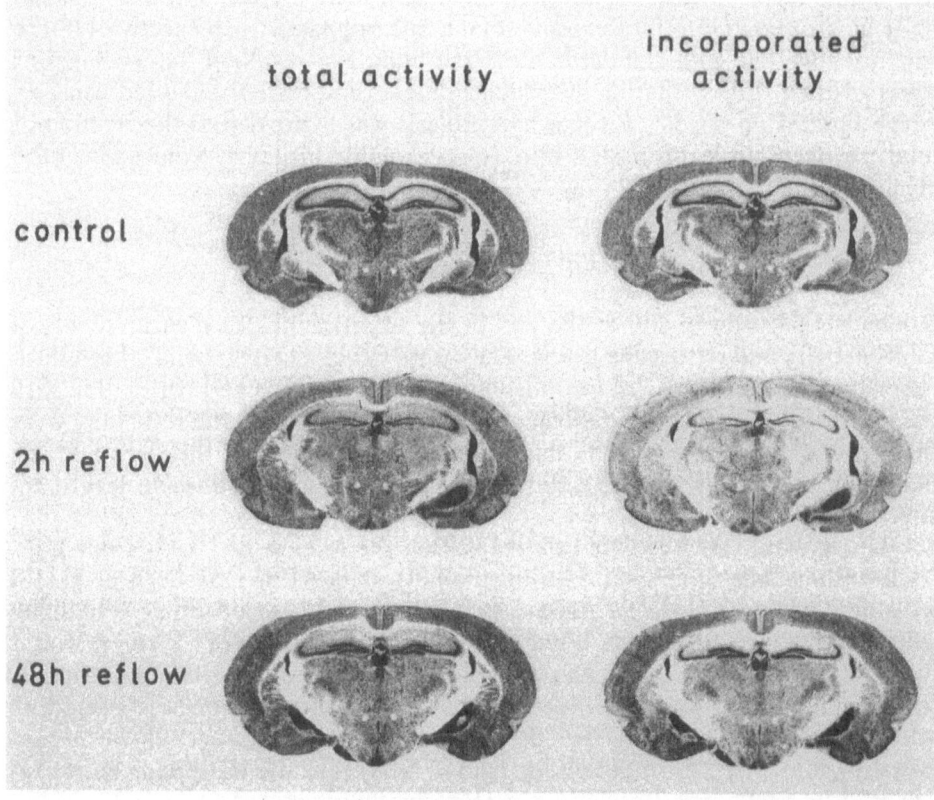

Fig. 1. Regional amino acid incorporation into proteins before and after reversible ischemia of gerbil brain. Images represent total tissue radioactivity (before TCA-wash) and
radioactivity incorporated into proteins (after TCA-wash) in control animals and animals
subjected to 5 min cerebral ischemia and 2 h or 48 h recirculation. Amino acid incorporation is severely suppressed after 2 h recirculation. It recovers to near control values after
48 h recirculation in cerebral cortex and basal ganglia but remains depressed in the CAl-
subfield of the hippocampus. (Modified from [9])

Table 1. Protein biosynthesis (% amino acid incorporation into proteins) after 5 min cerebral ischemia of Mongolian gerbils

| | Control | 2 h recirculation | | 48 h recirculation | |
		Untreated	Treated	Untreated	Treated
Cortex	72.4 ± 3.9	39.5 ± 8.5^a	29.3 ± 7.5^a	63.6 ± 5.5	73.6 ± 4.6
Hippocampus (CAl)	69.0 ± 5.7	13.3 ± 5.6^b	11.6 ± 4.0^b	31.5 ± 12.1	60.1 ± 13.8

Brain sections were exposed to X-ray films before and after TCA-wash for measurement of total radioactivity and the radioactivity incorporated into proteins, respectively (see Figs. 2, 3). From these autoradiograms the percent of amino acid incorporated into proteins was calculated. Treated animals received a single dose of pentobarbital (50 mg/kg) immediately after removal of the aneurysm clips. Values given are means $\pm$ SEM ($n = 6$ in each group). Statistically significant differences between experimental groups and control are indicated by : a, $P \leq 0.05$; b, $P \leq 0.01$

vity were incorporated into proteins of cerebral cortex and $69.0 \pm 5.7\%$ (range 49–82) in the CAl-subfield of hippocampus.

The incorporation of amino acids into proteins was markedly reduced after 5 min cerebral ischemia and 2 h recirculation. The incorporation rate decreased to $39.5 \pm 8.5\%$ (range 7–61) in the cortex and to $13.3 \pm 5.6\%$ (range 0–32) in the CAl-subfield indicating that the vulnerable CAl-sector was more affected than the more resistant cortex. Protein biosynthesis of cerebral cortex recovered to near control values following 48 h recirculation, but in the CAl-subfield incorporation rate varied considerably in different animals: it returned to control values in two out of six animals (66 and 77% incorporation) but it was severely inhibited in the other 4 animals (range 6–19% incorporation). The mean incorporation rate amounted to $31 \pm 12.1\%$ or 50% of control.

To study the relationship between post-ischemic inhibition of protein biosynthesis and the development of ischemic neuronal necrosis, animals were treated after cerebral ischemia with pentobarbital. Post-ischemic barbiturate treatment of gerbils has been previously shown to reduce the density of ischemic cell damage in the CAl-subfield of the hippocapus [16–19]. The results are illustrated in Fig. 2 and summarized in Table 1. Barbiturate treatment of animals did not only not prevent post-ischemic inhibition of protein biosynthesis it further enhanced the disturbance after 2 h recirculation. Following 48 h recirculation, however, the barbiturate resulted in a marked improvement in the vulnerable CAl-subfield: in 4 out of 6 treated animals amino acid incorporation returned to or were above control values (range of incorporation 54–95%) remaining suppressed only in two animals (18 and 23% incorporation).

Polyamine Metabolism

Ornithine decarboxylase. Changes in ODC activity in reversible cerebral ischemia are summarized in Table 2. ODC activity in control animals amounted to 0.30 ± 0.03 and 0.60 ± 0.06 nmol/g/h in the cerebral cortex and CAl-subfield of

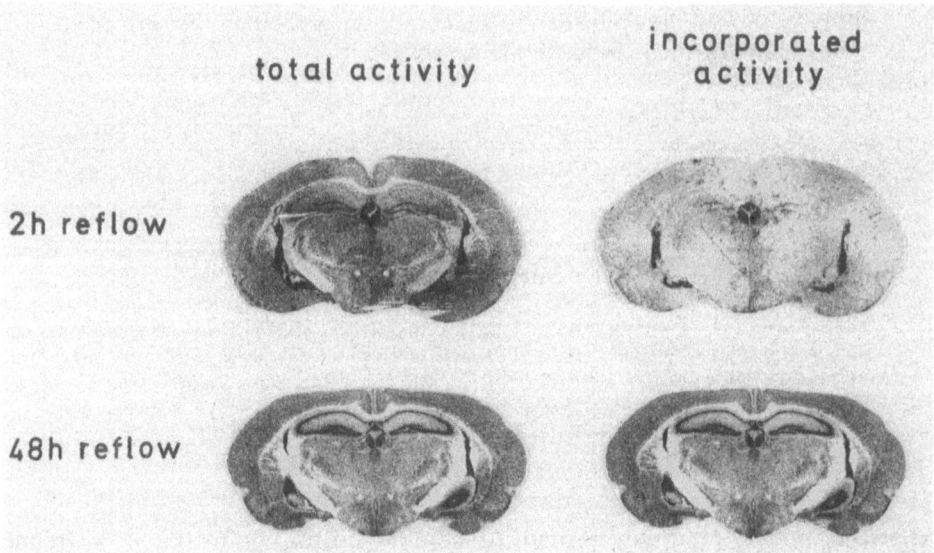

total activity

incorporated activity

2h reflow

48h reflow

Fig. 2. Regional amino acid incorporation into proteins in barbiturate-treated animals subjected to 5 min ischemia and 2 h or 48 h recirculation. Animals received an intraperitoneal injection of pentobarbital (50 mg/kg) immediately after restoration of blood flow. Images represent total tissue radioactivity (unwashed) and the radioactivity incorporated into proteins (washed). Note the severe depression of amino acid incorporation after 2 h followed by complete recovery after 48 h recirculation, even in the vulnerable CA1-subfield of the hippocampus. (Modified from [9])

Table 2. Ornithine decarboxylase activity after 5 min cerebral ischemia of Mongolian gerbils

	Control	8 h recirculation		24 h recirculation	
		Untreated	Treated	Untreated	Treated
Cortex	0.30 ± 0.03	4.15 ± 1.18	4.60 ± 1.61[a]	0.75 ± 0.11	0.56 ± 0.11
Hippocampus (CA1)	0.60 ± 0.06	5.08 ± 0.63[a]	7.66 ± 1.62[c]	1.37 ± 0.25	1.35 ± 0.32

Ornithine decarboxylase activity is given in nmol/g/h. Treated animals received a single dose of pentobarbital (10 mg/kg) injected intraperitioneally immediately after removal of the aneurysm clips. Statistically significant differences between controls and experimental groups are indicated by: a, $P \leq 0.05$; c, $P \leq .001$

the hippocampus, respectively. ODC activity increased considerably after 8 h recirculation (to 4.15 ± 1.18 in the cortex and 5.08 ± 0.63 nmol/g/h in the hippocampus). ODC activity returned to just over twice the control value following 24 h recirculation.

Post-ischemic barbiturate treatment produced only minor effects in the cortex

Table 3. Putrescine levels after 5 min cerebral ischemia of Mongolian gerbils

	Control	8 h recirculation		24 h recirculation	
		Untreated	Treated	Untreated	Treated
Cortex	7.50 ± 0.45	22.26 ± 3.09[a]	19.58 ± 2.57	30.45 ± 5.15[c]	23.05 ± 2.56[a]
Hippocampus (CAl)	10.14 ± 0.61	31.75 ± 1.29[c]	26.23 ± 1.79	40.01 ± 4.98[c]	28.17 ± 2.44[c]

Putrescine is given in nmol/g. Treated animals received a single dose of pentobarbital (10 mg/kg) injected intraperitioneally immediately after removal of the aneurysm clips. Statistically significant differences between controls and experimental groups are indicated by: a, $P \leq 0.05$; b, $P \leq 0.01$; c, $P \leq 0.001$

after 8 h recirculation but it induced a further increase in ODC activity in the CAl-subfield of the hippocampus (Table 2, from 5.08 ± 0.63 to 7.66 ± 1.62 nmol/g/h in untreated and treated animals, respectively). No major effect of barbiturate on ODC activity was observed following 24 h recirculation.

Putrescine. Changes in putrescine levels in reversible cerebral ischemia are summarized in Table 3. In control animals putrescine levels amounted to 7.50 ± 0.45 in the cerebral cortex and 10.14 ± 0.61 nmol/g in the CAl-subfield of the hippocampus. Putrescine increased considerably after 8 h recirculation and even more after 24 h recirculation (to 30.5 ± 5.15 nmol/g in the cortex and 40.01 ± 4.98 nmol/g in the CAl-subfield).

The effects of post-ischemic barbiturate treatment on regional putrescine levels are illustrated in Table 3. The barbiturate produced a decrease rather than an increase in putrescine. The effect of barbiturates was most evident in the CAl-subfield: in 10 out of 12 treated animals putrescine levels after 8 h or 24 h recirculation were found to be lower than those in untreated animals.

Relationship between putrescine levels and the density of ischemic cell damage. In order to study the relationship between putrescine levels and the density of ischemic cell damage in the CAl-subfield of the hippocampus, both parameters were measured within the same animals following 5 min ischemia and 96 h recirculation. The density of ischemic neuronal necrosis was evaluated by counting the number of intact neurons/mm stratum pyramidale. It appeared that the density of neuronal necrosis correlated with the putrescine levels in a threshold relationship (Fig. 3). Less than 10% of neurons were necrotic in animals in which putrescine levels were below 20 nmol/g whereas in animals with putrescine levels above 25-30 nmol/g, more than 90% of the neurons were severely injured.

Post-ischemic barbiturate treatment of gerbils had a significant effect on both the increase in putrescine and the density of ischemic neuronal necrosis (Table 4). Pentobarbital reduced the mean putrescine level measured in the CAl-subfield after 96 h recirculation from 41.8 to 23.2 nmol/g and increased the number of intact neurons from 10.3 ± 2.6 cells/mm to 137 ± 34.6 cells/mm stratum pyramidale.

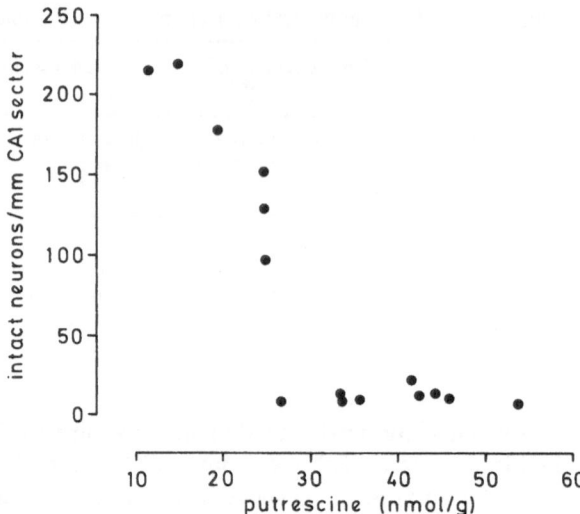

Fig. 3. Relationship between putrescine levels in the CAl-subfield of the hippocampus after 5 min ischemia and 94 h recirculation and the corresponding number of intact cells/ mm stratum pyramidale. The picture illustrates a close threshold-relationship between the two parameters: only minor cell injury was observed in animals with putrescine levels below about 25 nmol/g, whereas nearly all neurons were necrotic in animals exhibiting putrescine levels above 30 nmol/g. (From [19])

Discussion

In the present study, regional changes in global protein biosynthesis were compared with changes in the activity of a protein with a short half-life, ornithine decarboxylase (ODC, half-life about 10–20 min) [15] and the tissue content of putrescine, the product of ODC reaction. In addition, untreated and pentobarbital-treated animals were used to investigate whether the barbiturate, which is known to reduce the ischemic cell injury in hippocampal CAl-subfield [16–19], has any effect on the parameters analyzed. The results illustrate that barbiturate treatment further reduced protein biosynthesis following 2 h recirculation and increased ODC activity after 8 h recirculation in comparison to untreated animals, but considerably ameliorated the postischemic increase in putrescine. In the following discussion, these observations will be dealt with separately.

Protein Biosynthesis

Our findings of a severe reduction in overall protein biosynthesis after cerebral ischemia of Mongolian gerbils corroborates earlier observations from our own laboratory [7] and is in line with results reported by Thilmann et al. [8]. The latter authors observed a marked inhibition of amino acid incorporation

Table 4. Effect of barbiturate on postischemic cell damage and putrescine levels in the CA1-subfield of the hippocampus

	Control	Untreated	Treated
Putrescine (nmol/g)	11.3 ± 5.7	41.8 ± 3.6[a]	23.2 ± 3.1[b]
Intact cells/mm	211.2 ± 5.7	10.3 ± 2.6[a]	137.0 ± 34.6[b]

Ischemic cell damage was quantified in the CA1-subfield of the hippocampus by counting the number of intact neurons per mm stratum pyramidale. In treated animals 50 mg/kg pentobarbital was injected intraperitoneally immediately after removal of the aneurysm clips. Values given are means ± SEM. Statistically significant differences are indicated by: a, $P \leq 0.001$ (cf. control animals); b, $P \leq 0.05$ (cf. untreated animals). (From [18])

throughout the brain after 5 min cerebral ischemia and 2 h recirculation, as indicated by the low radioactivity of sections after the removal of free amino acids by TCA-wash. Amino acid incorporation was even more inhibited after 30 min recirculation and was hardly detectable by autoradiography [8]. This confirms that global cerebral ischemia of short duration produces long-lasting global inhibition of protein biosynthesis. There are, however, marked regional differences in the recovery after ischemia. The amino acid incorporation into proteins recovered within 24 h almost completely in the cerebral cortex but remained severely depressed in the CA1-subfield of the hippocampus.

This finding is in line with previous studies in different animal species as well as after different durations of ischemia, [5,8,21,22]. However, some caveats have to be noted concerning the interpretation of these data in terms of the quantitative protein synthesis rate. A reliable estimate of protein synthesis from tracer incorporation studies requires a precise determination of the tissue-integrated specific activity of the precursor pool, i.e., the amino acyltransfer RNA. In terms of methodology, this is easily done, and the pathological process can be standardized to such a degree that tissue samples from multiple experiments can be combined to establish the time course of the precursor pool. A kinetic approach, by which the time course of the precursor is calculated from blood radioactivity, has to consider up to 10 rate constants which may vary not only in different brain regions but also under different pathological conditions. [23]. Quantification of the post-ischemic protein synthesis rate by in vivo tracer incorporation studies, therefore, is very problematic. It is possible, however, to determine the direction of the error which occurs when conventional measurements of amino acid incorporations are carried out. If at the end of the incorporation period the specific activity of either the amino acyl-tRNA or the free amino acid pool is higher than under control conditions, the changes of amino acid incorporation into proteins overestimate the actual protein synthesis rate. Conversely, a decline of the specific activity of the precursor indicates an underestimation of protein synthesis. Application of this simple approach to the study of post-ischemic protein synthesis revealed a reciprocal relationship, i.e., an increase of the precursor specific activity when fractional incorporation of amino acids into proteins was inhibited, and a decrease of the precursor activity when protein synthesis recovered [22]. The described biochemical or autoradiographic determinations of post-ischemic amino acid incorporation can, therefore, be confidently interpreted as representing similar or even more pronounced alterations in the protein synthesis rate.

Post-ischemic inhibition of protein biosynthesis has been shown to be paralleled by a disaggregation of polyribosomes [3,4], and is supposedly caused by a post-ischemic inhibition of polypeptide chain initiation [4]. Since the protein biosynthesis rate is high in brain tissue, prolonged inhibition will influence the content of cellular proteins if it is not coupled to a similar inhibition of degradation. Consequently, the disturbance of protein biosynthesis is an important target for therapeutical intervention.

The development of a rational therapeutical concept requires knowledge of the mechanisms of this inhibition. A possible approach is the comparison of the temporal profile of the disturbance in vulnerable and non-vulnerable structures. Following 5 min ischemia, protein biosynthesis is completely suppressed throughout the brain during the first 30 min of recirculation [8], i.e., in both vulnerable and non-vulnerable structures. It is, therefore, unlikely that the degree of inhibition during early recirculation determines the extent of ischemic cell damage after longer recirculation times. This interpretation is corroborated by the results of the present study. A barbiturate which is known to reduce ischemic cell damage did not reverse the ischemia-induced disturbances in protein biosynthesis but produced an even further reduction of amino acid incorporation after 2 h recirculation. This finding clearly indicates that it is the capacity for recovery from the post-ischemic inhibition of protein biosynthesis rather than the extent of inhibition during early recirculation which predicts the development of neuronal necrosis. In order to study this process in more detail, autoradiographic measurement of global protein biosynthesis may not be an appropriate technique because it does not provide information about the integrity of the synthesizing machinery. Our findings, therefore, do not reveal whether or not the machinery is injured (which should result in reduction of biosynthesis of all proteins) or is switched to a state in which some proteins are synthesized at the expense of others.

In fact, a change in the pattern of newly synthesized proteins is a common response to various types of stress. This has been studied in detail after heat-shock stress, which induces the synthesis of so-called heat-shock proteins [24–26]. The synthesis of the same class of proteins has been recently shown to be activated after cerebral ischemia [10–12]. It is interesting to note that the temporal profile of heat-shock proteins following cerebral ischemia coincides with that of ODC, both of which peak at about 8 h recirculation [10,13].

The observation that the synthesis of heat-shock proteins [10–12] or ODC [3,13,14] is activated after cerebral ischemia at a time when the overall protein biosynthesis is markedly reduced, clearly illustrates that the synthesizing machinery is not impaired but switched to a post-stress state. Regional changes in the recovery of the process of amino acid incorporation into proteins after cerebral ischemia may therefore depend on the capacity of the cells to switch back the protein biosynthesis process from the post-stress to the physiological state. This reversal is obviously possible in the resistant cortex in which protein biosynthesis recovers to near control values but not in the vulnerable CAl-sector of the hippocampus.

Polyamine Metabolism

A scheme of the biosynthesis and interconversion of the polyamines (putrescine, spermidine, and spermine) is given in Fig. 4. The only metabolic route to putrescine in mammals is by decarboxylation of the amino acid ornithine. This reaction is catalyzed by ODC. Two aminopropyl moieties are then added to form spermidine and spermine, respectively. These aminopropyl groups are taken from decarboxylated S-adenosylmethionine which is produced from S-adenosylmethionine by S-adenosylmethionine decarboxylase (SAMDC). The polyamine metabolism is controlled by the activities of the two key enzymes, ODC and SAMDC [27–30].

Ornithine Decarboxylase

In the present study the activity of ODC was measured in reversible cerebral ischemia because this enzyme has a very short half-life (10–20 min) [15]. An increase in the activity of ODC after cerebral ischemia reflects de novo synthesis of the enzyme because the post-ischemic increase in ODC activity is considerably reduced by conventional blockers of protein biosynthesis [31]. ODC activity can therefore be taken as an indicator of the post-ischemic integrity of the protein biosynthesis machinery.

The transition of the protein biosynthesis machinery from the physiological to a post-stress state influences the two key enzymes in polyamine metabolism in different ways. The activity of ODC is markedly increased parallel to the increase in heat-shock protein biosynthesis [3,13,14], while the activity of SAMDC is severely reduced corresponding to the global reduction in protein biosynthesis [13]. The post-ischemic increase in ODC activity produces an increase in the synthesis of putrescine, the product of ODC reaction. In contrast, conversion of putrescine to spermidine and spermine is considerably inhibited because this reaction is catalyzed by SAMDC, the activity of which is severely depressed following cerebral ischemia. The overshoot in putrescine formation observed following cerebral ischemia may, therefore, be a result of the global reduction of protein biosynthesis and the activation of the biosynthesis of stress proteins.

The specific role of heat-shock proteins and ODC in cellular function following stress is still unknown. It has been suggested that heat-shock proteins are necessary for the preservation of cell structure after stress [32]. It is also possible that induction of ODC activity reflects proliferative changes [13] that are known to occur in glial cells after cerebral ischemia [33]. Preliminary results, however, indicate that during the first hours after ischemia increase in ODC activity is localized in neurons rather than in glial cells [34,35]. It is therefore conceivable

Wulf Paschen et al.

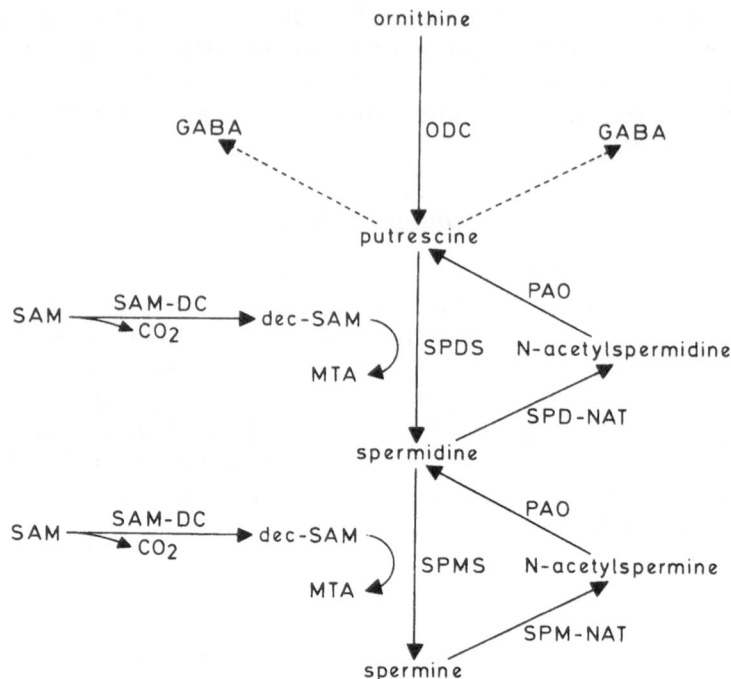

Fig. 4. Schematic representation of polyamine metabolism. In mammals the only route to putrescine is the decarboxylation of ornithine by ornithine decarboxylase (ODC). Two aminopropyl moieties are then added to putrescine to form spermidine and spermine, respectively. These aminopropyl moieties are taken from decarboxylated S-adenosylmethionine (dec-SAM) which is formed from S-adenosylmethionine (SAM) by S-adenosylmethionine decarboxylase (SAMDC). SPDS, Spermidine synthase; SPMS, spermine synthase; MTA, methylthioadenosine; SPM-NAT, spermine N-acetyltransferase; SPD-NAT, spermidine N-acetyltransferase; PAO, polyamine oxidase. (From [18] with permission)

that activation of ODC synthesis after cerebral ischemia is the response of neurons to cellular stress and may play a protective role similar to that suggested for the heat-shock proteins. This assumption is corroborated by the observation that a barbiturate treatment of gerbils, which is known to prevent neuronal necrosis, did not reduce post-ischemic ODC activity but caused an even further increase (see Results).

One very interesting change observed after cerebral ischemia was the marked increase in putrescine. During early recirculation putrescine formation was activated in all brains structures studied. [14,18,19,36,37]. It remained high with longer recirculation times in selectively vulnerable brain structures, such as the striatum and the CA1-subfield of the hippocampus, but returned to near control in the less vulnerable cortex and thalamus [37]. In this respect, changes in putrescine levels resemble the protein biosynthesis rate: a recovery to near control in the resistant structures and persistent depression in the vulnerable ones. It is therefore possible that the persistent inhibition of protein biosynthesis in the vulnerable CAl-subfield of the hippocampus is responsible for the high putrescine levels, as discussed above.

It has been recently shown that post-ischemic putrescine levels correlate closely with the density of ischemic cell damage [18,19]. A threshold-like relationship was observed in the CAl-subfield of the hippocampus between the two parameters after 5 min ischemia and 4 days recirculation [19], the threshold of putrescine being in the range of 25–30 nmol/g. Interestingly, putrescine was already markedly increased during early recirculation at a time when the cells of the CAl-subfield were still intact. A similar relationship could be observed [37] in the vulnerable striatum but the threshold was considerably higher than in the hippocampus and amounted to 50–60 nmol/g (Paschen et al., in preparation). Consequently, different regions of the brain exhibit different cell-inherent thresholds for the induction of cell injury.

The effects of barbiturate-treatment on protein biosynthesis, ODC activity, and putrescine levels differed considerably. The barbiturate markedly reduced the post-ischemic increase in putrescine levels but had either only minor effects or produced an even further inhibition of protein biosynthesis and a further increase in ODC activity. The reduction of putrescine levels by barbiturate treatment correlated closely with the inhibition of cell necrosis, but it remains unclear whether this is the reason or the consequence of reduced injury. However, certain properties of putrescine suggest that it may, in fact, be a mediator of ischemic cell necrosis.

For a long time putrescine was thought to be just a precursor for the polyamines spermidine and spermine. Recently, however, it has been shown to play a significant role in membrane transport processes. These findings are supported by the observation that putrescine is bound to cell membranes [38] and that it increases intracellular calcium by stimulating the influx from the extracellular compartment and the efflux from mitochondria. It also triggers the calcium- or depolarization-dependent release of neurotransmitters from synaptosomes, a reaction that can be inhibited by blocking putrescine formation [39–41]. In addition, putrescine injected into the cerebral ventricle has been shown to reduce GABA levels and to produce electrical hyperactivity in the cortex [42]. All of these processes have been related to the manifestation of ischemic cell injury. It is therefore conceivable that in addition to the prolonged inhibition of protein synthesis, the equally prolonged post-ischemic increase in putrescine is involved in the pathological sequel of events leading to irreversible ischemic cell injury.

Summary. Changes in cerebral protein biosynthesis, ornithine decarboxylase (ODC) activity, and putrescine levels were studied before and after global brain ischemia in untreated and barbiturate-treated Mongolian gerbils. Reversible cerebral ischemia of 5 min duration was produced under halothane anesthesia by occluding both common carotid arteries with aneurysm clips. Brains were recirculated after ischemia for 2 h and 48 h (protein biosynthesis study), or for 8,

24 and 96 h (polyamine metabolism study). Barbiturate-treated animals received a single dose of pentobarbital (50 mg/kg) immediately after removal of the aneurysm clips.

Amino acid incorporation into proteins in untreated animals was markedly depressed after 2 h recirculation (to about 50% and 20% of control values in the cerebral cortex and CAl-subfield, respectively). Protein biosynthesis recovered to near control levels in the cortex following 48 h recirculation but remained severely suppressed in the selectively vulnerable hippocampus. Amino acid incorporation was even further suppressed in barbiturate treated animals after 2 h recirculation but it recovered to near control values after 48 h recirculation in both the cortex and CAl-subfield of the hippocampus.

The ODC activity of untreated animals was markedly increased (to about 10-fold) after 8 h recirculation in the cortex and CAl-subfield of hippocampus. ODC activity declined in both regions after 24 h recirculation but remained about twice as high as in controls. Putrescine increased 3-fold after 8 h and continued to rise after 24 h recirculation in both the cortex and hippocampus. Barbiturate treatment produced an even further increase in ODC activity after 8 h but not after 24 h recirculation. Putrescine levels, in contrast, were markedly reduced by barbiturate after 8 h and 24 h recirculation in the hippocampus and after 24 h recirculation in the cortex.

It is concluded that the capacity of neurous to tolerate the metabolic stress induced by cerebral ischemia correlates with the changes of polyamine metabolism and the duration—but not the extent—of post-ischemic suppression of protein biosynthesis. The observation that synthesis of ODC recovers at a time when the overall protein biosynthesis is still markedly reduced suggests that the protein biosynthesis machinery remains intact after ischemia and is switched to a post-stress pattern.

Acknowledgements. The excellent technical assistance of Christine Magendanz and Anne Pribliczki is gratefully acknowledged. This work was supported by the Deutsche Forschungsgemeinschaft, Grant Pa 266/3-1.

References

1. Lowry OH, Passonneau JV, Hasselberger FY, Schulz DW (1964) Effects of ischemia on known substrates and cofactors of the glycolytic pathway in brain. J Biol Chem 239:18–30
2. Siesjö BK (1978) Brain energy metabolism. Wiley, New York
3. Kleihues P, Hossmann K-A, Pegg AE, Kobayashi K, Zimmermann V (1975) Resuscitation of the monkey brain after one hour complete ischemia: III. Indications of metabolic recovery. Brain Res 95:61–73
4. Cooper HK, Zalewska T, Kawakami S, Hossmann K-A (1977) The effect of ischemia and recirculation on protein synthesis in the brain. J Neurochem 28:929–934
5. Dienel GA, Pulsinelli WA, Duffy TE (1980) Regional protein synthesis in rat brain following acute hemispheric ischemia. J Neurochem 35:1216–1226
6. Nowak TS, Fried RL, Lust WD, Passonneau JV (1985) Changes in brain energy

metabolism and protein synthesis following transient bilateral ischemia in the gerbil. J Neurochem 44:487–494

7. Bodsch W, Takahashi K, Barbier A, Grosse Ophoff B, Hossmann K-A (1985) Cerebral protein synthesis and ischemia. Prog Brain Res 63:197–210

8. Thilmann R, Xie Y, Kleihues P, Kiessling M (1986) Persistent inhibition of protein synthesis precedes delayed neuronal death in postischemic gerbil hippocampus. Acta Neuropathol (Berl) 71:88–93

9. Xie Y, Hossmann KA, Munekata K, Seo K (1987) Prolonged suppression of protein synthesis after brief cerebral ischemia in gerbils. In: Cervos-Navarro J, Ferszt R (eds) Stroke and microcirculation. Raven, New York pp 135–141

10. Nowak TS (1985) Synthesis of a stress protein following transient ischemia in the gerbil. J Neurochem 45:1635–1641

11. Dienel GA, Kiessling M, Jacewicz M, Pulsinelli WA (1986) Synthesis of heat shock proteins in rat brain cortex after transient ischemia. J Cereb Blood Flow Metab 6:505–510

12. Kiessling M, Dienel GA, Jacewicz M, Pulsinelli WA (1986) Protein synthesis in post-ischemic rat brain: A two-dimensional electrophoretic analysis. J Cereb Blood Flow Metab 6:642–649

13. Dienel GA, Cruz NF, Rosenfeld SJ (1985) Temporal profiles of proteins responsive to transient ischemia. J Neurochem 44:600–610

14. Paschen W, Röhn G, Meese CO, Djuricic B, Schmidt-Kastner R (1988) Polyamine metabolism in reversible cerebral ischemia: Effect of α-difluoromethylornithine. Brain Res 453:9–16

15. Pegg AE, Lockwood DH, Williams-Ashman HG (1970) Concentration of polyamines and their enzymic synthesis during androgen-induced prostatic growth. J Biochem 117:17–31

16. Hallmayer J, Hossmann KA, Mies G (1985) Low dose of barbiturates for prevention of hippocampal lesions after brief ischemic episodes. Acta Neuropathol (Berl) 68:27–31

17. Kirino T, Tamura A, Sano K (1986) A reversible type of neuronal injury following ischemia in the gerbil. Stroke 17:455–459

18. Paschen W, Hallmayer J, Röhn G (1988) Relationship between putrescine content and density of ischemic cell damage in the brain of Mongolian gerbils: Effect of nimodipine and barbiturate. Acta Neuropathol (Berl) 76:388–394

19. Paschen W, Hallmayer J, Röhn G (1988) Regional changes of polyamine profiles after reversible cerebral ischemia in Mongolian gerbils: Effects of nimodipine and barbiturate. Neurochem Pathol 8:27–41

20. Djuricic BM, Paschen W, Schmidt-Kastner R (1988) Polyamines in the brain: HPLC analysis and its application in cerebral ischemia. Jugosl Physiol Pharmacol Acta 24:9–17

21. Kleihues P, Hossmann K-A (1973) Regional incorporation of L-(3-^{3}H)tyrosine into cat brain proteins after 1 hour of complete ischemia. Acta Neuropathol (Berl) 25:313–324

22. Bodsch W, Barbier A, Oehmichen M, Grosse Ophoff B, Hossmann K-A (1986) Recovery of monkey brain after prolonged ischemia: II. Protein synthesis and morphological alterations. J Cereb Blood Flow Metab 6:15–21

23. Smith CB, Deibler GE, Eng N, Schmidt K, Sokoloff L (1988) Measurement of local cerebral protein synthesis in vivo: Influence of recycling of amino acids derived from protein degradation. Proc Natl Acad Sci USA 85:9341–9345

24. Hightower LE, White FP (1981) Cellular response to stress: comparison of a family of

71–73 kilodalton proteins rapidly synthesized in rat tissue slices and canavanine treated cells in culture. J Cell Physiol 108:261–275

25. Schlesinger MJ, Ashburner M, Tissueres A (eds) (1982) Heat shock from Bacteria to Man. Cold Spring Harbor Laboratory, New York

26. Cosgrove JW, Brown IR (1983) Heat shock protein in mammalian brain and other organs after physiologically relevant increase in body temperature induced by D-lysergic acid diethylamide. Proc Natl Acad Sci 80:569–573

27. Jänne J, Poso H, Raina A (1978) Polyamines in rapid growth and cancer. Biochim Biophys Acta 473:241–293

28. Canellakis ES, Viceps-Madore D, Kyriakidis DA, Heller JS (1979) The regulation and function of ornithine decarboxylase and of the polyamines. Curr Top Cell Reg 15:155–202

29. Seiler N (1981) Polyamine metabolism and function in the brain. Neurochem Int 3:95–110

30. Pegg AE, McCann PP (1982) Polyamine metabolism and function. Am J Physiol 243: C212–C221

31. Dienel GA, Cruz NF (1984) Induction of brain ornithine decarboxylase during recovery from metabolic, mechanical, thermal or chemical injury. J Neurochem 42:1053–1061

32. Pelham HRB (1984) HSP 70 accelerates the recovery of nuclear morphology after heat shock. EMBO J 3:3095–3100

33. Petito CK, Babiak T (1982) Early proliferative changes in astrocytes in postischemic noninfarcted rat brain. Ann Neurol 11:510–518

34. Dempsey RJ, Maley BE, Cowen DE, Olson JW (1988) Ornithine decarboxylase activity and immunohistochemical location in postischemic brain. J Cereb Blood Flow Metab 8:843–847

35. Nemeth G, Cintra A, Mayer G, Fuxe K, Hoyer S (1988) Characteristics of neurological deficits and behavioral impairments of rats induced by bilateral incomplete cerebral ischemia. In: Meyer JS, Lechner H, Reivich H, Ott EO (eds) Cerebrovascular disease 7. Excerpta Medica, pp 267–270

36. Paschen W, Schmidt-Kastner R, Djuricic B, Meese C, Linn F, Hossmann K-A (1987) Polyamine changes in reversible cerebral ischemia. J Neurochem 49:35–37

37. Paschen W, Hallmayer J, Mies G (1987) Regional profile of polyamines in reversible cerebral ischemia of Mongolian gerbils. Neurochem Pathol 7:143–156

38. Frydman B, Frydman RB, De Los Santos C, Alonso Garrids D, Goldemberg SH, Algranti ID (1984) Putrescine distribution in Escherichia coli studied in vivo by ^{13}C nuclear magnetic resonance. Biochim Biophys Acta 805:337–344

39. Iqbal Z, Koenig H (1985) Polyamines appear to be second messengers in mediating Ca^{2+} fluxes and neurotransmitter release in potassium-depolarized synaptosomes. Biochem Biophys Res Commun 133:563–573

40. Bondy SC, Walker CH (1986) Polyamines contribute to calcium-stimulated release of aspartate from brain particulate fractions. Brain Res 371:96–100

41. Komulainen H, Bondy SC (1987) Transient elevation of intrasynapto-somal free calcium by putrescine. Brain Res 401:50–54

42. Nistico G, Jentile R, Rotiroti D, Di Giorgio RM (1980) GABA depletion and behavioral changes produced by intraventricular putrescine in chicks. Biochem Pharmacol 29:954–957

7

Protective Effect of Microinjury in Brain Ischemic Damage

Koki Shimoji, Yoshio Takahata, Naoshi Fujiwara, Kiichiro Taga, and Satoru Fukuda[1]

Introduction

Ischemic brain damage is one of the most serious concerns associated with post-resuscitation sequelae and other hemodynamic complications seen in clinical anesthesia and critical medicine. Although numerous preventive measures for the treatment of brain ischemia have been reported in recent years, none of these is promising in clinical practice. This is a report on the induction of certain endogenous protective mechanisms in the artificially injured brain to promote survival of mice subjected to incomplete brain ischemia.

Methods

Experiment A

Eighty-six DDY mice weighing 37.8 ± 0.3 g (mean $\pm$ S.E.M.) were randomly divided into three groups: the anesthesia group which received 80 mg/kg pento-barbital with 0.1 mg/kg atropine (i.p. injection) without surgery (group I), the sham-operated group which received surgical incision on the scalp without brain injury under the same anesthesia (group II) and the brain injury group (group III). The brain injury was induced by vertically inserting a 25 gauge hypodermic needle, 5 mm deep from the skull into the brain at 4 sites. All surgical procedures were carried out aseptically. After anesthesia (group I) or surgery (groups II and III), the animals were kept in a plastic box ($20 \times 30 \times 20$ cm) at a constant temperature of 32°C until full recovery from anesthesia (24 h). All the animals of one experiment were then put in the same cage under the ambient temperature of 25°C.

---

[1] Department of Anesthesiology, Niigata University School of Medicine, 1-757 Asahi-machi, Niigata, 951 Japan

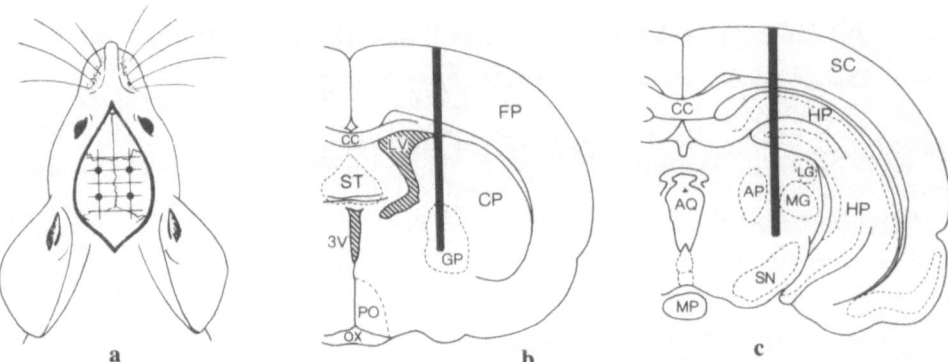

Fig. 1. a Schematic presentation of the sites of the minor brain damage inflicted by a needle (25 gauge) through the skull (1/4 caudal from the bregma and 1/4 rostral from the lambda between the two sutures, 1.7 mm lateral from the midline). **b,c** Camera lucida drawings showing the extent of lesions reconstructed from the Klüver-Barrera-stained sections. Bilateral lesions were extended up to the **b** globus pallidus and **c** thalamus, through the cortex. AP, anterior pretectal area; AQ, cerebral aqueduct; CC, corpus callosum; CP, caudate putamen; FP, frontoparietal cortex; GP, globus pallidus; HP, hippocampus; LV, lateral ventricle, MG, medial geniculate nucleus; MP, mammillary peduncle; OX, optic chiasm; PO, preoptic area; SC, striate cortex; SN, substantia nigra; ST,septal nucleus; 3V, third ventricle. (From [5] with permission)

On the 7th day after these procedures, the bilateral carotid arteries of all animals were exposed and clamped for 1 h with small clips with the aid of a microscope under pentobarbital anesthesia, 80 mg/kg i.p. with 0.1 mg/kg atropine. After the ischemic insult, the wounds were closed and the animals were placed in the original plastic box.

Three to five animals randomly selected in each group were subjected to one experiment (the initial as well as the subsequent surgical interventions). The same experiments were then carried out 7 times. The survival rate was calculated during the postischemic course in each group up to the 7th postischemic day, since no death occurred thereafter in any of the groups. Two-by-three chi-squared tests were used for analysis of distribution of the number of animals that survived or died among the 3 groups. Severity scores were ranked, and non-parametric correlation coefficients were calculated. For all statistical comparisons, P values of less than 0.05 were considered significant. Histological examination was undertaken 2 weeks after the ischemic insult. The artificial lesions extended to the globus pallidus and thalamus through the frontal and striate cortices, respectively (Fig. 1).

Experiment B

The two experiments were carried out in DDY male mice weighing 38.9 ± 0.4 g (mean ± S.E.M.). The initial surgical procedures, brain injury or sham operation, were undertaken under ketamine anesthesia (200 mg/kg, i.p.) with atropine (0.1 mg/kg, i.p.) in both experiments I and II. The second procedure to induce brain ischemia was carried out under pentobarbital anesthesia (80 mg/kg, i.p.) with atropine (0.1 mg/kg, i.p.) in experiment I and under ketamine anesthesia (200 mg/kg, i.p.) with atropine (0.1 mg/kg, i.p.) in experiment II. The animals were randomly divided into two groups: the sham-operated group being given a surgical incision on the scalp without brain injury, and the brain injury group. The brain lesions were induced by vertically inserting a 25 gauge hypodermic needle into the brain to a depth of 5 mm from the skull at 4 sites, as described in experiment A. All surgical procedures were carried out aseptically. After surgery the animals were kept in a plastic box ($20 \times 30 \times 20$ cm) at a constant temperature of 32°C until full recovery from anesthesia and then placed in a cage under an ambient temperature of 25°C. On the 7th day after the initial procedures, the bilateral carotid arteries were exposed and clamped with small clips under microscopical observation for 60 min under pentobarbital anesthesia, 80 mg/kg i.p., with 0.1 mg/kg atropine (experiment I), or for 30 min under ketamine anesthesia, 200 mg/kg i.p., with 0.1 mg/kg atropine (experiment II). In our preliminary experiment, all animals ($\bar{n} = 8$) died following bilateral clamping of the carotid arteries for 60 min under ketamine anesthesia. Therefore, we reduced the clamping time to 30 min with ketamine. After the ischemic insult, the wounds were closed and the animals were placed in the original plastic box.

Three to five animals randomly selected in both groups were subjected to the initial (brain injury or sham operation) as well as subsequent (brain ischemia) surgical interventions. The initial and subsequent surgical procedures were carried out 4 times in experiment I, and 7 times in experiment II. The survival rate was calculated during the postischemic course in both groups up to the 7th postischemic day, since no death occurred thereafter. Two-by-two chi-squared tests were used for analysis of distribution of the number of animals that survived or died between the two groups. Severity scores were ranked, and non-parametric correlation coefficients were calculated. For all statistical comparisons, P values of less than 0.05 were considered significant. Histological examination was undertaken 2 weeks after the ischemic insult. The artificially inflicted lesions extended to the globus pallidus and thalamus through the frontal and striate cortices, respectively. There were no significant differences in extent or degree of lesions between experiments I and II.

Results

Experiment A

No animals died in any group following the initial procedures. Body weight tended to increase in both groups I and II on the 7th day following the initial insult from 37.9 ± 1.06 (mean $\pm$ S.E.M.) to 39.7 ± 0.9 g, and from 38.8 ± 1.5 to 39.4 ± 1.7 g in groups I and II, respectively (not significant), whereas it tended to decrease, from 37.2 ± 0.9 to 36.7 ± 0.8 g in group III (not significant). After the ischemic insult, 8, 9 and 6 animals died within 24 h in groups I, II and III,

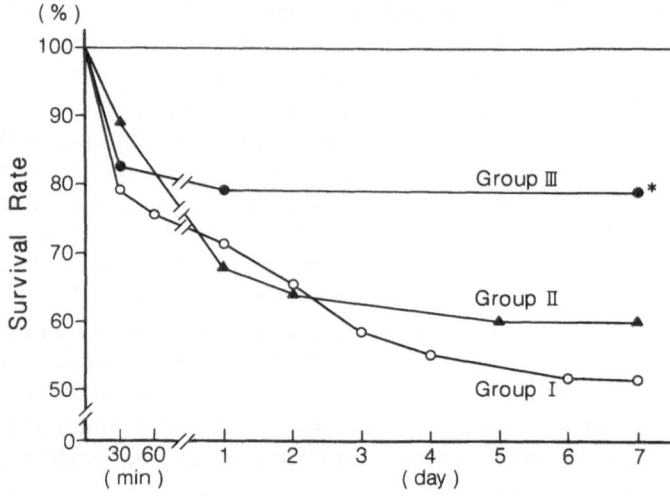

Fig. 2. Changes in the survival rate after brain ischemia (bilateral clamping of the carotid arteries for 1 h) in each group in experiment A. Groups I, II, and III consisted of 29, 28, and 29 mice, respectively. Note that no death occurred in group III after 24 h following the brain ischemia. The asterisk denotes a significant difference from groups I and II. The survival rate in group III became significantly high on the third postischemic day versus that in group I ($P < 0.05$) and on the fifth day versus that in group II ($P < 0.05$). (From [4] with permission)

respectively, and there were no significant differences in the survival or death rates among the groups. An additional 6 and 2 animals died within 7 days after the ischemic episode in groups I and II, respectively. By contrast, however, no death occurred thereafter in group III (Fig. 2). Thus, the survival rate at the 7th post-ischemic day in group III was significantly higher ($P < 0.05$) than that in groups I or II, and there was no significant difference in the survival rates between the latter two groups. No death or behavioral abnormality was observed in the animals which survived more than 7 days following the brain ischemia.

Experiment B

In experiment I (a prior brain injury under ketamine anesthesia and subsequent brain ischemia under pentobarbital anesthesia), 3 animals died immediately after the initial intervention in the brain-injured group ($\bar{n} = 20$), but no animals died in the sham-operated group ($\bar{n} = 21$). In the subsequent ischemic insult, 9 out of 21 animals (43%) survived and 12 died in the sham-operated group. In this group, 9 of the dead animals died within 1 h, and the rest (two animals) died by the 4th post-ischemic day. By contrast, all animals except one (96%) survived the ischemic episode in the brain-injured group (Fig. 3a). Thus, the survival rate of the brain-injured group was significantly high (p < 0.01) when compared to that of the sham-operated group.

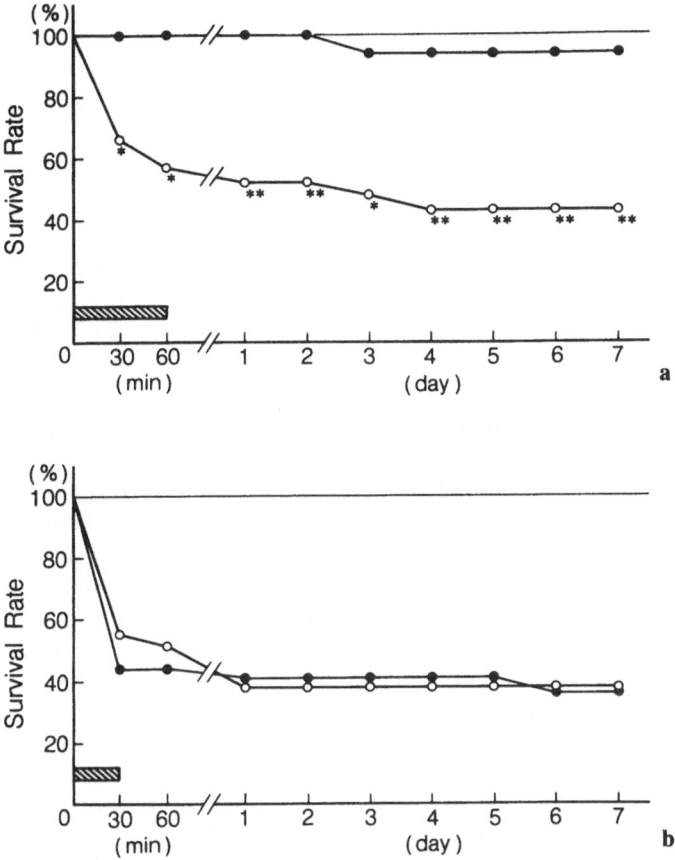

Fig. 3. a Changes in the survival rate after brain ischemia produced by bilateral clamping of the carotid arteries for 1 h (*hatched bar*) in experiment I (brain lesions and ischemia were produced under ketamine and pentobarbital anesthesia, respectively). The survival rate in the brain-injured group (*black circle*, $n = 17$) became significantly higher ($P < 0.01$) within 30 min following brain ischemia than that in the sham-operated group (*white circle*, $n = 21$). The two *asterisks* denote a significant difference between the two groups. **b** Changes in the survival rate after brain ischemia produced by bilateral clamping of the carotid arteries for 30 min (*hatched bar*) in experiment II (both procedures for inducing brain lesions and ischemia were carried out under ketamine anesthesia). There were no significant differences in the survival rates between the two groups during the entire postischemic course. (From [4] with permission)

In experiment II (both surgical interventions, a prior brain injury and subsequent brain ischemia, carried out under ketamine anesthesia), 9 out of 24 (38%) and 9 out of 25 (36%) animals survived the brain ischemic insult in the sham-operated and brain-injured groups, respectively, and there were no significant differences in the survival rates between the two groups (Fig. 3b).

Discussion

The present results have demonstrated that a certain protective mechanism becomes active by 24 h following the ischemic episode, since there was no death in group III after the first postischemic day in experiment A.

The existence of substances with neurotrophic, neurite-promoting, and guiding activities has been known for about 3 decades. One that has been characterized is a nerve growth factor, which exhibits all 3 types of activity [1]. Furthermore, it has been suggested that neurotrophic and neurite-promoting factors become available to facilitate repair following injury to the nervous system [2]. More recently, Nieto-Sampedro et al. [3] have demonstrated that the fluid collected by day 6 after injury in brain wound cavities promotes neuron survival in tissue culture in young adult rats. Such assays have been based on the survival of cultured sensory neurons, but not on the ischemic survival of the mammalian brain in situ.

An anti-ischemic activity such as that demonstrated in the brain injured animals in the present study, could be one of several biological activities afforded by these factors. Another explanation could be that a certain unknown substance with anti-ischemic activity is also released simultaneously with neurotrophic factors from the injured brain site. Alternatively, it could be that other unknown protective mechanisms become active against ischemia following injury within the brain [4].

The protective mechanism developed in the injured brain could not be demonstrated when brain ischemia (the second procedure in experiment B) was induced under ketamine anesthesia [5]. Ketamine has several characteristic pharmacological effects on the central nervous system such as producing increases in cerebral blood flow (CBF) [6] and evoked potentials [7]. These pharmacological features of ketamine may not be advantageous for protecting the brain from ischemic insult. on the other hand, the drug is known as an NMDA receptor antagonist [8] and may influence the sequelae following ischemic brain insult [9]. By contrast, pentobarbital decreases CBF and CMR_{O_2} with a scavenging effect on free radicals [10,11] and an agonistic effect on GABA receptor [12].

The mechanisms of the differential effects of the two anesthetics on the survival rates of mice following brain ischemia remain unknown. The results from the two experiments in experiment B allow us to presume that a masking (or suppressing) effect of ketamine, an activating (or potentiating) effect of pentobarbital, or a combination of these effects on the anti-ischemic activity could be responsible for the differential effects of the two drugs on the survival rates following brain ischemia.

The results also show that a protective mechanism against brain ischemia is active within several hours following the ischemic episode, since a majority of the animals died within 1 h following brain ischemia even though the durations of the two anesthetics differ. However, the differences in duration of action between pentobarbital, which acts for more than several hours, and ketamine, whlch ceases its action within several hours, might be taken into consideration with regard to the present results.

Summary. Minor brain injury was inflicted with a small needle at four sites 1 week before the production of incomplete brain ischemia in the mouse. A bilateral carotid clamp was applied for 60 min under pentobarbital anesthesia, and the number of survivors at 1 week after the ischemic insult was compared with those in animals only anesthetized and those in a sham-operated group. The number of survivors in the brain-injured group was significantly higher than in the other two groups. The results suggest that anti-ischemic factors are released by the injured brain or that certain unknown protective mechanisms against ischemia become active following brain injury. Improvement of the survival rates in mice with brain injury, however, became insignificant when brain ischemia was imposed during ketamine anesthesia, suggesting that the actions of certain factors or protective mechanisms against brain ischemia developed by brain injury are antagonized by ketamine and/or potentiated by barbiturate anesthesia.

References

1. Varon S, Adler R (1981) Trophic and specifying factors directed to neuronal cells. In: Fedoroff S, Hertz L (eds) Advances in Cellular Neurobiology vol 2. Academic, New York, pp 115–163
2. Thoenen H, Edgar D (1985) Neurotrophic factors. Science 229:238–242
3. Nieto-Sampedro M, Needles DL, Cotman CW (1985) A simple, objective method to measure the activity of factors that promote neuronal survival. J Neurosci Methods 15:37–48
4. Takahata Y, Shimoji K (1986) Brain injury improves survival of mice following brain ischemia. Brain Res 381:368–371
5. Shimoji K, Takahata Y, Fujiwara N, Endoh H, Taga K, Ohama E (1987) Effects of pentobarbital and ketamine on brain injury-induced anti-ischemic activity. Brain Res 408:385–388
6. Takeshita H, Okuda Y, Sari A (1972) The effects of ketamine on cerebral circulation and metabolism in man. Anesthesiology 36:69–75
7. Kayama Y, Iwama K (1972) The EEG, evoked potentials, and single unit activity during ketamine anesthesia in cats. Anesthesiology 36:316–328
8. Snell LD, Johnson KM (1985) Antagonism of N-methyl-D-aspartate induced transmitter release in the rat striatum by phencyclidine-like drugs and its relationship to turning behavior. J Pharmacol Exp Ther 235:50–57
9. Olney JW, Price MT, Fuller TA, Labruere J, Samson L, Carpenter M, Mahan K (1986) The anti-excitotoxic effects of certain anesthetics, analgesics and sedative-hypnotics. Neurosci Lett 68:29–34
10. Shapiro H-M (1985) Barbiturates in brain ischemia. Br J Anaesth 57:82–95
11. Messick J-M Jr, Milde LN (1987) Brain protection. In: Stoelting RK, Barash PG, Gallagher TJ (eds) Advances in anesthesia vol 4. Year Book Medical Publishers, Chicago. pp 47–87
12. Higashi H, Nishi S (1982) The effect of barbiturates on the GABA receptor of cat primary afferent neurones. J Physiol 332:299–314

8

Behavioral and Memory Impairment Following Cerebral Ischemia in Rats

Takefumi Sakabe, Yoshitoyo Miyauchi, Hideto Nakayama,
Akio Tateishi, Takanobu Sano, Toshizoh Ishikawa,
and Hiroshi Takeshita[1]

Introduction

Although the mortality and morbidity of victims of cardiac arrest have been improved with recent progress in the techniques of cardiopulmonary resuscitation and intensive care, many of them remain hospitalized in a vegetative state or die of brain death. Even in those who have regained consciousness within a relatively short period of time after an ischemic insult, neuropsychological and memory impairments may persist either transiently or permanently. Studies on the mechanisms responsible for the development of these conditions may provide a therapeutic window. Thus, there are an increasing number of investigations in animal models that examine the behavioral and memory impairment after cerebral ischemia [1–4]. The present study was undertaken to examine the behavioral and memory changes after forebrain ischemia in rats.

Materials and Methods

Forty male Wistar rats were used for the experiment. All rats were housed in standard rat cages and were maintained on a 12-h light-dark cycle. They were fasted overnight preceding the operation but had free access to tap water.

Surgical preparation was done under halothane/nitrous oxide/oxygen anesthesia. The rats were intubated and mechanically ventilated. The surgical preparation consisted of cannulation of the tail artery, tail vein, and external jugular vein, and exposure of both carotid arteries. Halothane was then discontinued and a 30 min period was allowed to elapse in order to obtain a steady state. Assessment of ambulatory behavior was made in 23 rats, 9 in the 10-min ischemia group, 6 in the 15-min ischemia group, and 8 in the sham-operated group.

[1]Department of Anesthesiology-Resuscitology, Yamaguchi University, School of Medicine, 1144 Kogushi, Ube, 755 Japan

Assessment of memory function was made in the remaining 17 rats, 11 in the 10 min-ischemia group, and 6 in the sham-operated group.

Cerebral ischemia was produced by clamping both carotid arteries and inducing hypotension by exsanguination to a mean arterial blood pressure (MAP) of 50 mmHg, preceded by an injection of a small dose of trimethaphan (2.5 mg) [5]. After completion of ischemia for the scheduled period (10 or 15 min in the groups assigned for the assessment of ambulatory behavior and 10 min in the groups assigned for the assessment of memory function), the shed blood was rapidly reinfused and the carotid clamps were released. When the rats regained normal spontaneous respiration, nitrous oxide was discontinued and the endotracheal tubes were removed. The sham-operated rats underwent the same procedures as the rats rendered ischemic except for the carotid occlusion and induced hypotension. Physiological variables including MAP, body temperature (BT), PaO_2, $PaCO_2$, pHa, hemoglobin (Hb), and blood glucose (BG) were determined at appropriate intervals. Pre- and postischemic values shown in the tables (see Results) are those determined at 10 min before clamping and at 10 min after the end of ischemia in the ischemia group, respectively, and at the corresponding time (only one sample) in the sham-operated group.

Ambulatory behavior consisting of locomotion, preening, and grooming was assessed using the open field method [6] for 3 min at each examination. Assessment was done one day before and 2, 3, 5, and 7 days after ischemia of 10 min or 15 min duration, or after the sham operation. For the assessment of memory function the rats were trained on a conditioned avoidance schedule once a day using a two-way shuttle box [7]. This schedule consists of 100 consecutive trials (active avoidance response). Visual and auditory signals lasting 5 sec were used as the conditioning stimuli. When the rats failed avoidance, they were foot-shocked electrically. The rats that retained an avoidance rate of more than 60% for two consecutive days were selected to undergo ischemia or the sham operation followed by repeat memory studies. Assessment of memory function was performed 3, 5, and 7 days after ischemia or the sham operation.

Seven days after the ischemic insult or sham operation, the rats were anesthetized with halothane in oxygen. The brains were perfusion-fixed by an injection of 4% formaldehyde (buffered to pH 7.4) into the ascending aorta, preceded by a 30-sec rinse with saline [8]. Both solutions were prewarmed to 37°C and infused at a pressure of 135 mmHg. The brains were removed on the following day and stored in cold fixative. They were cut coronally into about 3 mm thick slices and dehydrated in graded strengths of ethanol over 2 days. The slices were embedded in paraffin and subserially sectioned at 6 μm on a rotary microtome (Rechert-Jung 820-II) and stained with celestine blue-acid fuchsin. With this staining the necrotic neurons appear bright red under the light microscope. Quantification of neuronal damage in the parietal cortex and in the dorsal hippocampus was done in the brain slices taken approximately 3.8 mm caudal to the bregma at a magnification of 400x. Ischemic neuronal damage was graded on a scale of 0–3: 0, normal; 1, slight damage (neuronal damage less than 10%); 2, moderate damage (neuronal damage 10%–50%); and 3, severe damage (neuronal damage more than 50%) [9].

Changes in physiological variables with time were examined by Student's t-test for paired data. The changes in behavioral and memory parameters were examined by one-way analysis of variance (ANOVA) for repeated measures, and when indicated, ANOVA was followed by Fischer's least significant difference test. Physiological variables, behavioral and memory parameters between the groups were compared using either ANOVA or Student's t-test for unpaired data. Histopathologic scores were compared among the groups using the Kruskal-Wallis test, followed by the Mann-Whitney test with a Bonferroni correction for multiple comparisons. $P < 0.05$ was considered significant.

Results

Physiological variables in the rats subjected to the assessment of ambulatory behavior or memory function are shown in Tables 1 and 2, respectively. There

Table 1. Physiological variables in the rats subjected to ambulatory behavior test

n	Sham-operated 8	10-min ischemia 9		15-min ischemia 6	
		Pre	Post	Pre	Post
B W (g)	318 ± 7	303 ± 9	$305 \pm 18^\S$	295 ± 3	$260 \pm 13^\S$
MAP (mmHg)	125 ± 6	$109 \pm 3^\#$	$111 \pm 2^\#$	$112 \pm 6^\#$	$112 \pm 4^\#$
B T (°C)	37.1 ± 0.1	37.7 ± 0.3	37.4 ± 0.2	36.8 ± 0.1	36.7 ± 0.0
PaO$_2$ (mmHg)	123 ± 9	110 ± 8	$131 \pm 8^*$	126 ± 6	116 ± 6
PaCO$_2$ (mmHg)	35 ± 1	38 ± 2	35 ± 1	37 ± 1	38 ± 1
pH$_a$	7.39 ± 0.02	7.34 ± 0.01	$7.40 \pm 0.01^*$	7.38 ± 0.02	7.35 ± 0.02
H b (g/dl)	18.6 ± 0.3	17.7 ± 0.8	17.4 ± 0.5	19.6 ± 0.7	18.8 ± 0.6
B G (mg/dl)	85 ± 10	84 ± 14	$106 \pm 19^*$	98 ± 15	116 ± 24

Values are mean ± SEM. §, body weight 7 days following ischemia; #, sigificantly different from the sham-operated group ($P < 0.05$); *, significantly different from the pre-ischemic values ($P < 0.05$)

Table 2. Physiological variables in the rats subjected to memory function test

n	Sham-operated 6	10-min ischemia 9	
		Pre	Post
B W (g)	300 ± 16	308 ± 10	$227 \pm 15^{\S*}$
MAP (mmHg)	119 ± 7	124 ± 3	121 ± 4
B T (°C)	36.9 ± 0.1	36.9 ± 0.1	36.8 ± 0.7
PaO$_2$ (mmHg)	131 ± 12	121 ± 7	113 ± 5
PaCO$_2$ (mmHg)	39 ± 0.4	37 ± 0.4	$39 \pm 0.7^*$
pH$_a$	7.38 ± 0.01	7.35 ± 0.02	7.39 ± 0.01
H b (g/dl)	19.8 ± 0.6	19.2 ± 0.4	$18.5 \pm 0.5^*$
B G (mg/dl)	105 ± 17	87 ± 6	$104 \pm 8^*$

Values are mean ± SEM. §, body weight 7 days following ischemia; *, significantly different from the pre-ischemic values ($P < 0.05$)

Takefumi Sakabe et al.

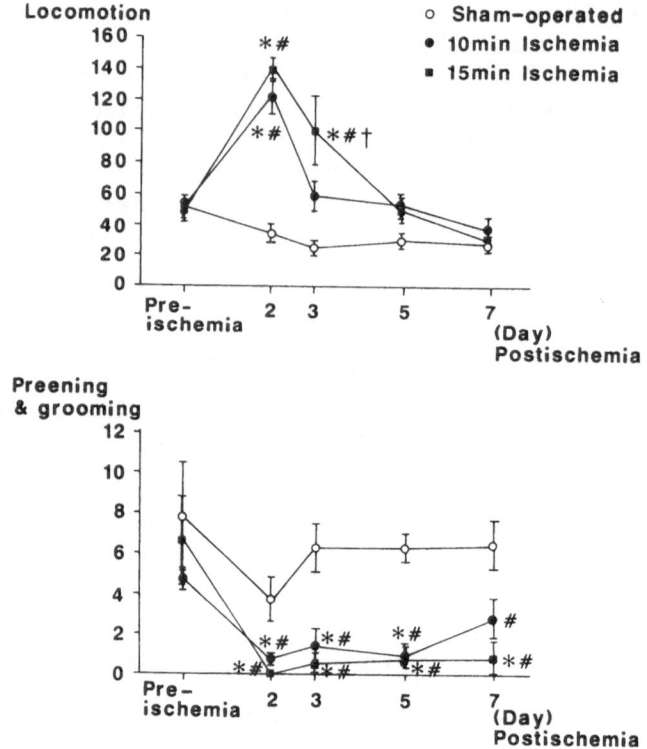

Fig. 1. Ambulatory behavior assessed by open field method. Open circles, closed circles and closed squares indicate data of rats assigned for the sham-operated, 10-min ischemia, and 15-min ischemia groups, respectively. Ordinate indicates the number of behavioral activity observed during a 3 min-examination period. In the ischemia rats significant increase in locomotion was observed at 2–3 days after ischemia, while preening and grooming were decreased significantly throughout a 7-day postischemic observation period. *, Significantly different from the pre-ischemic values; #, Significantly different from the values of the sham-operated group; +, Significantly different from the values of the 10-min ischemia group

was no difference in physiological variables, including MAP, BT, PaO_2, $PaCO_2$, pHa, Hb, and BG, between the ischemia and sham-operated groups except for a higher MAP in the sham-operated group which had been subjected to the assessment of ambulatory behavior. However, the mean preischemic MAPs in all groups were within physiological ranges (109–125 mmHg). The mean values of postischemic PaO_2, pHa, and BG in the 10 min ischemia group—although significantly different when compared to the preischemic values—were also within normal physiologic ranges.

The results of the assessment of ambulatory behavior including locomotion, preening, and grooming are shown in Fig. 1. The rats subjected to either 10-min or 15-min ischemia showed a marked increase in locomotion for the 2–3 days

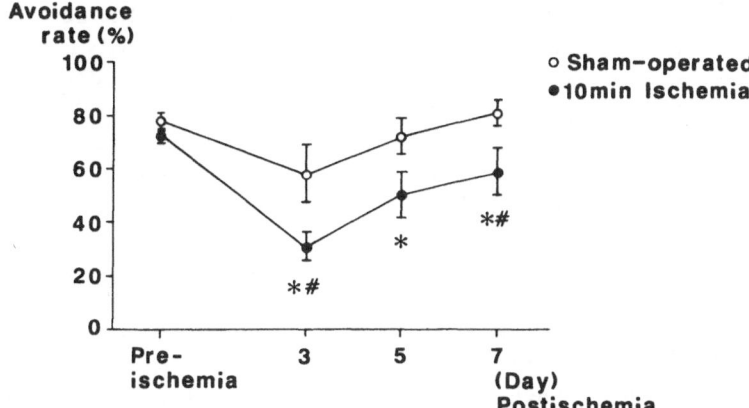

Fig. 2. Memory function assessed by the active avoidance method. Open circles and closed circles indicate the data (expressed as an avoidance rate) of the sham-operated and 10-min ischemia groups, respectively. Avoidance rate was significantly decreased at 3, 5, and 7 days after ischemia in the rats subjected to 10-min ischemia. Symbols for statistics are as in Fig. 1.

following ischemia. Locomotion, however, returned to the preischemic level between 4 and 7 days after ischemia. Preening and grooming were decreased throughout the 7 day postischemic observation period. In the sham-operated group, no increase in locomotion was observed, and preening and grooming were unchanged.

The results of the assessment of memory function (as evaluated by active avoidance) are shown in Fig. 2. In the rats subjected to a 10-min ischemia, active avoidance response was impaired at 3, 5, and 7 days after ischemia, the mean avoidance rate being 31%, 50%, and 59%, respectively. In the sham-operated group, the active avoidance response was impaired only at 3 days after ischemia, the mean avoidance rate being 58%; however, at 5 and 7 days after ischemia no impairement was observed. The mean avoidance rates at 3 and 7 days postischemia in the ischemia group were significantly lower than those in the sham-operated group.

All rats rendered ischemic for 10 min and 15 min showed severe ischemic damage in CA1 and CA4 sectors of the hippocampus, with a mean score of 2.3–3. Ischemic damage in the CA3 and dentate gyrus was rated as slight to moderate, with a mean score of 0.6–2. In the cortex ischemic neuronal damage was rated as slight, with a mean score of 1–1.6. In the sham-operated group, as expected, there was no apparent ischemic neuronal damage in any area.

Discussion

The present study reveals that increased locomotion and impairment of memory function develop after transient forebrain ischemia in rats. We are unaware of

any other reports which demonstrate increased locomotion after cerebral ischemia. It has been suggested that an increase in locomotor activity is initiated by
an activation of the central dopaminergic system, which includes the nucleus
accumbens and tuberculum olfactorium (mesolimbic projections) [10,11].
Marked reduction in dopamine concentration in the nucleus accumbens, tuberculum olfactorium, and hypothalamus has also been reported following ischemia
[12]. The decrease in dopamine concentration is assumed to be a reflection of
enhanced activity of the dopaminergic systems [13]. Therefore, our finding of
increased locomotion following ischemia may indicate that the dopaminergic
system is either activated or disinhibited following cerebral ischemia. However,
an involvement of other systems such as the cholinergic, serotoninergic, or
GABAergic systems cannot be excluded [14]. Preening and grooming are considered to be emotional behaviors and the disturbances in these patterns observed
in the present study may indicate that damage to the limbic system (which is
known to be closely related to emotion) is involved, at least in part.

Memory function after cerebral ischemia has been studied by several investigators using the radial maze task method which reflects spatial memory [1–4].
Their results indicate that a significant impairment of memory function, especially in working performance, may occur following transient forebrain ischemia. In
contrast, reference performance was not significantly impaired and even improved over trials[2]. The present study assessed memory function using the active
conditioned avoidance method, focusing especially on the retention of memory
after ischemia in the pretrained rats. This method assesses operant behavior
(conditioned avoidance) and this type of learning is known to be the prototype
of an intentional "forward looking" behavior [15]. Our results demonstrate that
retention of memory is impaired in rats subjected to 10 min of ischemia.
Although the mechanism for learning avoidance and learning mazes may not be
identical, the learning ability for both tasks has been reported to be correlated
[15]. Therefore, it can be said that transient forebrain ischemia, even if its duration is as short as the 10 min limit used in the present study, produces memory
impairment.

It has been reported that rats with selective damage to the CA1 region of the
hippocampus produced by aspiration were impaired in learning to perform a
complex spatial task, but that their performance was not affected if the task was
learned preoperatively. Fimbrial and complete hippocampal lesions impaired
both postoperative acquisition and performance of the preoperatively learned
task [16]. Damage to the CA3 region of the hippocampus produced by a local
injection of kainic acid has also been reported to impair spatial memory in the
rat [17]. In an ischemia model, Volpe et al. [1] demonstrated impairement of
memory in association with severe neuronal damage in the CA1 region. All rats
subjected to ischemia in the present study, regardless of the ischemic period (10
or 15 min), showed marked neuronal damage in both the CA1 and CA4 of the

---

[2]In behavioral analysis of maze performance, evaluation of two types of performance is
possible: one is reference performance (entering a baited arm) and working performance
(not entering a baited arm after the food is taken)

hippocampus, while damage in the CA3 and dentate gyrus was slight to moderate, and damage in the cortex was slight. These histopathological findings are comparable to those reported previously in this model [18]. Our findings that memory dysfunction was associated with severe CA1 (and CA4) damage with less damage in CA3 (and dentate gyrus) in all ischemic rats are in good agreement with those of Volpe et al. [1] although the ischemia model and the method for the assessment of memory are different. Thus, hippocampal damage—especially that in the CA1 region—seems to contribute to the memory dysfunction occuring after forebrain ischemia. However, the contribution to memory impairement of damage in other brain structures should be further evaluated because neuronal damage in the present model is not restricted to the hippocampus.

It is not clear whether there is any correlation between increased locomotion and the impairment of memory function observed in this study. Handelmann and Olton [17] reported that kainic acid injection into the hippocampal CA3 sector simultaneously induced increased locomotion and impairment of memory function. However, even after memory function recovered, increased locomotion was still present. From these results, they suggested that the neuroanatomical mechanisms responsible for memory function and locomotor activity are independent. Our results showed that memory impairment was present when increased locomotor activity had subsided. We have no satisfactory explanation for the sequential difference of both phenomena between their study and ours. It may be due to the differences in the models and of the methods for the determination of the memory function.

Our preliminary data suggest that the [^{3}H] L-glutamate binding sites as well as [^{3}H]-Quinuclidinyl benzilate binding sites (acetylcholine binding sites) were markedly decreased in the hippocampus in the rats. Recent evidence indicates a close relationship between these neurotransmitters and memory function [19,20]. Although the data are preliminary, this finding suggests that the memory disturbances observed in this study are related to the disintegration of glutamate and acetylcholine receptors. The neurochemical and histopathological bases for the impairment of behavior and memory function after ischemia, especially for the long-term functional changes, must be further studied.

Summary. An increase in locomotor activity and impairment of retention of memory occurred during the first 3 days after forebrain ischemia was produced by bilateral carotid artery occlusion and induced hypotension. The memory impairment lasted at least one week despite recovery of locomotor activity. Damage in the vulnerable neurons, particularly in the CA1 sector of the hippocampus, may be the cause of the memory impairment following transient forebrain ischemia.

References

1. Volpe BT, Pulsinelli WA, Tribuna J, Davis HP (1984) Behavioral performance of rats following transient forebrain ischemia. Stroke 15:558–562

2. Davis HP, Tribuna J, Pulsinelli WA, Volpe BT (1986) Reference and working memory of rats following hippocampal damages induced by transient forebrain ischemia. Physiol Behav 37:387–392
3. Davis HP, Baranowski JR, Pulsinelli WA, Volpe BT (1987) Retention of reference memory following ischemic hippocampal damage. Physiol Behav 39:783–786
4. Volpe BT, Waczek B, Davis HP (1988) Modified T-maze training demonstrates dissociated memory loss in rats with ischemic hippocampal injury. Behav Brain Res 27:259–268
5. Smith M-L, Bendek G, Dahlgren N, Rosén I, Wieloch T, Siesjö BK (1984) Models for studying long term recovery following forebrain ischemia in the rats: II. A two-vessel occlusion model. Acta Neurol Scand 69:385–401
6. Hall CS (1934) Emotional behavior in the rat: I. Defecation and urination as measures of individual differences in emotionality. J Comp Psychol 18:385–403
7. Collins RL (1964) Inheritance of avoidance conditioning in mice: A diallel study. Science 143:1188–1191
8. Auer RN, Olsson Y, Siesjö BK (1984) Hypoglycemic brain damage: Correlation of density of brain damage with the EEG isoelectric time. Diabetes 33:1090–1098
9. Schmidt-Kastner R, Hossmann K-A (1988) Distribution of ischemic neuronal damage in the dorsal hippocampus of rat. Acta Neuropathol 76:411–421
10. Pijnenburg AJJ, Honig WMM, Van Rossum JM (1976) Effects of chemical stimulation of the mesolimbic dopamine system upon locomotor activity. Eur J Pharmacol 35:45–58
11. Kelly PH, Seviour PW, Iversen SD (1975) Amphetamine and apomorphine responses in the rat following 6-OHDA lesions of the nucleus accumbens septi and corpus striatum. Brain Res 94:507–522
12. Zervas NT, Hori H, Negora M, Wurtman RJ, Larin F, Lavyne MH (1974) Reduction in brain dopamine following experimental cerebral ischemia. Nature 247:283–284
13. Thierry AM, Tassin JP, Blanc G, Glowinski J (1976) Selective activation of the meso-cortical DA system by stress. Nature 263:242–243
14. Jones DL, Mogenson GJ, Wu M (1981) Injections of dopaminergic, cholinergic, serotoninergic and gabaergic drugs into the nucleus accumbens: Effects on locomotor activity in the rat. Neuropharmacology 20:29–37
15. Bovet D, Bovet-Nitti F, Oliverio A (1969) Genetic aspects of learning and memory in mice. Science 163:139–149
16. Jarrard LE (1978) Selective hippocampal lesions: Differential effects on performance by rats of a spatial task with preoperative versus postopertaive training. J Comp Physiol Psych 92:1119–1127
17. Handelmann GE, Olton DS (1981) Spatial memory following damage to hippocampal CA3 pyramidal cells with kainic acid: Impairement and recovery with preoperative training. Brain Res 217:41–58
18. Smith M-L, Auer RN, Siesjö BK (1984) The density and distribution of ischemic brain injury in the rat following 2–10 min of forebrain ischemia. Acta Neuropathol (Berl) 64:319–332
19. Rauca CH, Kammerer E, Matthies H (1980) Choline uptake and permanent memory storage. Pharmacol Biochem Behav 13:21–25
20. Lynch G, Baudry M (1984) The biochemistry of memory: A new and specific hypothesis. Science 224:1057–1063

9

Neurobehavioral Changes Following Cerebral Ischemia and Treatment by a TRH Derivative

Akira Tamura, Makoto Hirakawa, and Keiji Sano[1]

Introduction

Multiple symptoms including reduced spontaneous activity, hemiparesis, and cognitive disturbance are frequently observed in the chronic phase of cerebral vascular diseases in humans. For investigating pathogenesis and drug therapy in the chronic phase of cerebral vascular diseases, it is important to observe the long-term functional consequences in a model of focal cerebral ischemia. Pathological and functional changes including neurological deficits, cognitive disturbances, brain edema, decreased cerebral blood flow and metabolism, and catecholamine levels after focal cerebral ischemia have been well documented in animals. However, most of these changes were observed only in the acute phase and few reports are available concerning the long-term functional consequences of focal cerebral ischemia [1].

The aim of this study is to describe the behavioral changes after focal cerebral ischemia using a middle cerebral artery (MCA)-occlusion model in the rat.

Surgical Preparation

Rats were anesthetized with 2% halothane inhalation and the proximal portion of the left MCA was permanently occluded by a microsurgical technique which was modified from our original method for the purpose of chronic phase experiments [2–4]. After the temporalis muscle was simply retracted via a trans retro orbital approach (without removal of the temporalis muscle and zygomatic arch), the animals underwent a left subtemporal small craniectomy. The stem of the MCA was electrocauterized just medial to the olfactory tract, and was cut to

[1]Department of Neurosurgery, Teikyo University School of Medicine, 2-11-1 Kaga, Itabashi-ku, Tokyo, 173 Japan

ensure the completeness of the vascular occlusion. A sham operation was done in the same manner, but the artery was left intact.

Behavioral Changes After Focal Cerebral Ischemia

Neurological Deficits

Since neurological deficits, such as altered spontaneous activity and hemiparesis, can interfere with cognitive tasks in animals [4], the changes in neurological deficits were examined. These deficits were evaluated by determining the degree of spontaneous activity and the degree of hemiparesis. The spontaneous activity was classified into 3 grades by the following criteria: grade 1, inactive; grade 2, mildly inactive (sometimes sneaking about); and grade 3, actively sneaking about.

Three features were independently observed for the assessment of hemiparesis: walking ability, paresis of the hind legs, and muscle power of the forelimbs. The walking ability and the paresis of the hind legs were devided into three grades: grade 1, severely disabled; grade 2, mildly disabled; and grade 3, normal. The muscle power of the forelimbs were classified as weak (grade 1) or normal (grade 2). Assessments of the hemiparesis were obtained by a summed score of the three items, which ranges from 3–8.

Figure 1 shows the changes of the spontaneous activity in the normal, sham-operated, and MCA-occluded animals. No significant change was observed between the groups. The incidence of hemiparesis in each group is shown in the lower part of Fig. 1. Significant hemiparesis was observed only in the MCA-occluded animals during the first 4 weeks after the occlusion. Maximal hemiparesis was observed at day 3 after occlusion after which it gradually improved.

Passive Avoidance Learning

The training was carried out according to the one-trial step-through procedure [4,5]. The rat was placed in an illuminated safe compartment (40 × 25 × 25 cm) with a hole in the wall, through which the rat could enter into a dark compartment (20 × 15 × 25 cm) which had a grid on the floor. Once the four paws were on the grid, a scrambled foot-shock (60V, 50Hz) was delivered to the grid. The rat could escape from the shock only by stepping back into the safe, illuminated side. In the test trial for evaluating the acquisition and retention abilities, the rat was again placed in the safe compartment and the latency responce to enter into the dark compartment was measured. The latency of animals which did not move into the dark compartment during the observation period of 600 sec was assumed to be 600 sec.

When the training was performed at day 3 after surgery, the latency of step-through in the MCA-occluded group was significantly shortened compared to that observed in the sham-operated group (Fig. 2). When the rats were retrained 8 weeks after MCA occlusion, there was also a significant shortened latency of the rat's entering into the dark compartment observed in the MCA-occluded

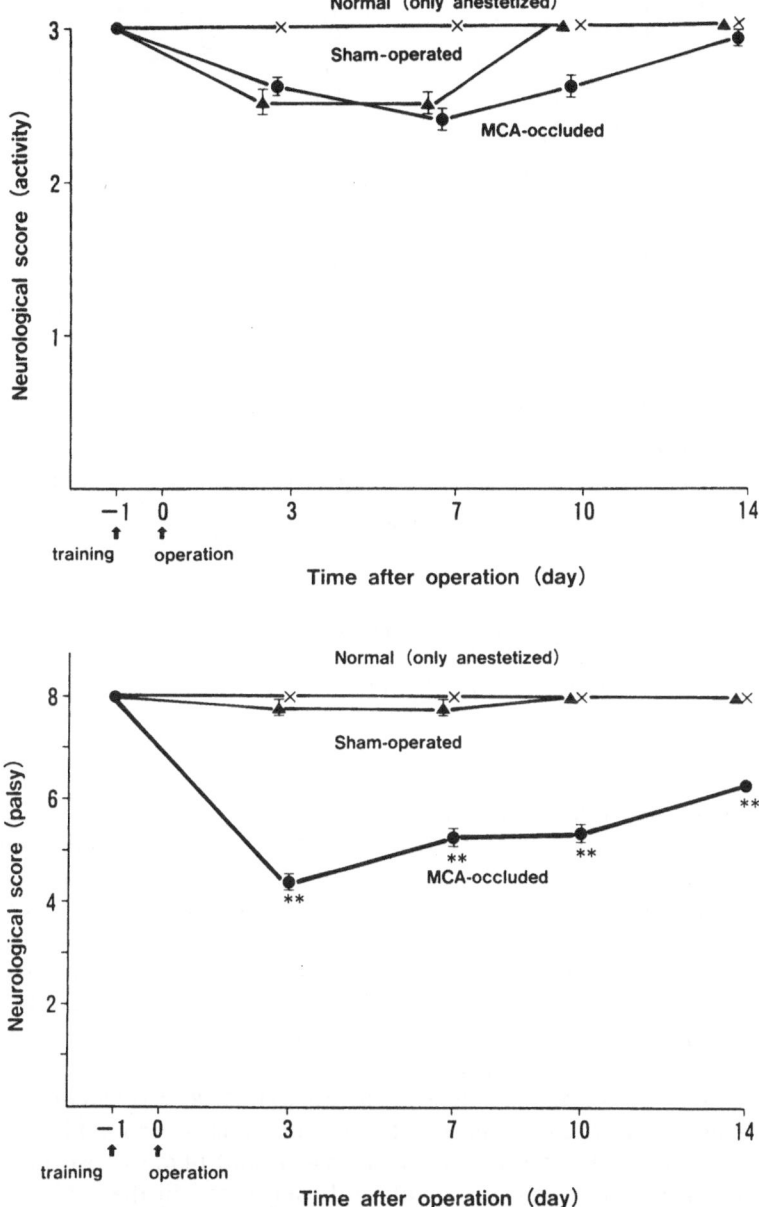

Fig. 1. Changes in spontaneous activity in normal, sham-operated and MCA-occluded rats (*upper*) and incidence of hemiparesis in these three groups (*lower*). Each point represents the mean ± SEN. Significantly different from sham-operated values: **$P < 0.01$

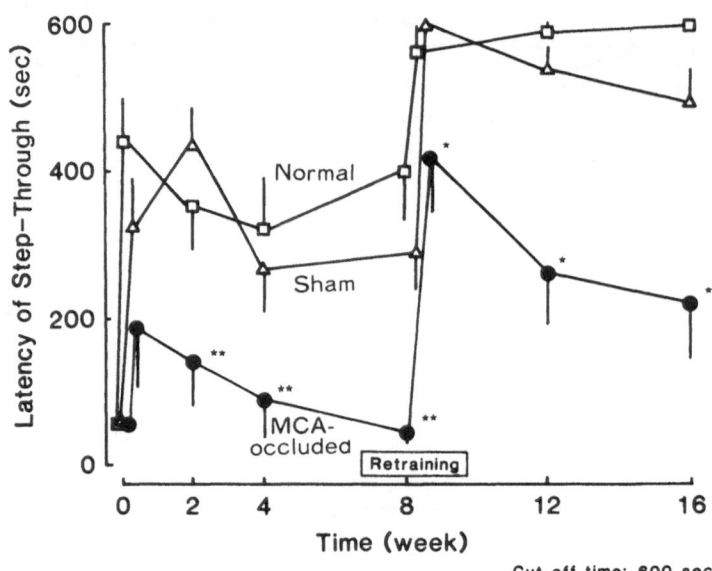

Fig. 2. Latency of step-through in passive avoidance task in normal, sham-operated and MCA-occluded rats. The rats were trained at 3 days and 8 weeks (re-training) after the operation. Each point represents the mean ± SEM. Significantly different from sham-operated values: *$P<0.05$, **$P<0.01$ (Mann-Whitney's U-test)

group compared to that in the sham-operated group (Fig. 2). These findings indicate that MCA occlusion severely disturbed the acquisition and retention of the passive avoidance task. It is well known that changes in spontaneous movements influence cognitive tasks in animals. In this study, no difference was observed among these three groups in the spontaneous movements. Therefore, the shortening of latency in the MCA-occluded group did not result from changes in the activity but rather was due to the disturbance of cognitive ability itself. In this post-operative training study, primarily the acqustion of the passive avoidance task was examined.

The pre-operative training study was performed in order to evaluate the retention of memory after MCA occlusion. The training was carried out in 51 rats the day before surgery. The test trials were given 4, 7, and 14 days post-operatively. The latency of step-through was significantly shortened in the MCA-occluded group when the training was carried out pre-operatively (Fig. 3). It was clearly demonstrated that the retention of the passive avoidance task was severely disturbed in the MCA-occluded group.

Active Avoidance Learning

The training was carried out according to the discrete lever-press avoidance procedure [6,7]. The temporal factors of the discrete avoidance schedule were an

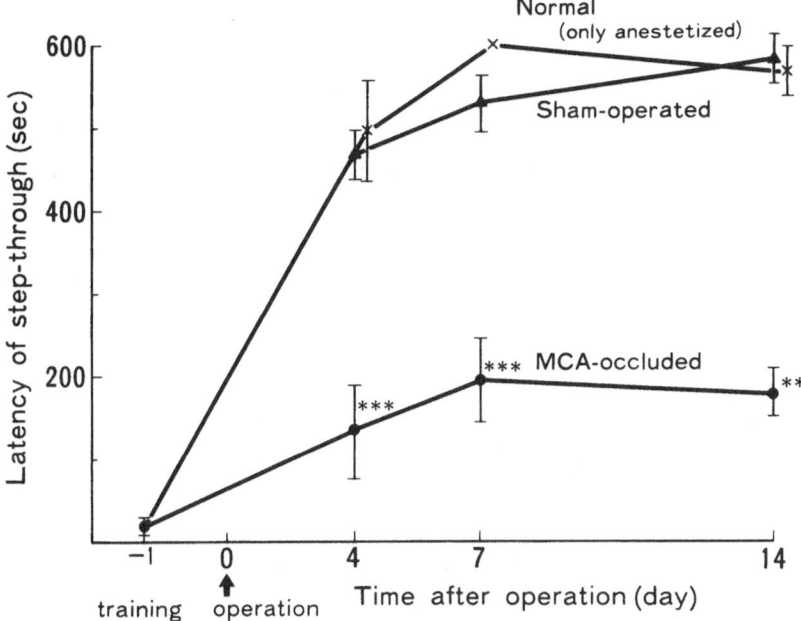

Fig. 3. Latency of step-through in passive avoidance task in normal, sham-operated and MCA-occluded rats. The training was performed the day before surgery. Each point represents the mean ± SEM. Significantly different from sham-operated values: ** $P<0.01$, *** $P<0.001$ (Mann-Whitney's U-test)

intertrial interval of 25 sec and a warning duration of 5 sec. Visual and and auditory stimuli, lighting of a 30 w pilot lamp and, a pure tone of 600 Hz at 60 db were used as the warning signals. The shock was given in the form of an electric current of 70V, 3mA, 50Hz for 5 sec during the training session. A lever-pressing action within the warning duration made the warning signs stop and the trial program return to the starting point of the intertrial interval of 25 sec. Each avoidance session (consisting of 30 min training per day) was held every other day during the training period. The index of the discrete avoidance behavior was indicated as the avoidance rate (number of correct avoidance responses/number of trials). After 14 sessions of routine training procedure, the animals which achieved above a critical level of avoidance rate (60% in the overall last three sessions) were selected as good performing animals. These were randomly assigned into three groups: normal control with anesthesia, sham-operation, and MCA-occlusion.

The mean values of the avoidance rates before surgery in the normal, sham-operation, and MCA-occlusion group were 71.0%, 77.1%, and 74.3%, respectively. The avoidance rate in the MCA-occluded group was significantly decreased in each point during the 14 days after occlusion (Fig. 4). This finding suggests that the retention of the active avoidance task was significantly disturbed after MCA occlusion.

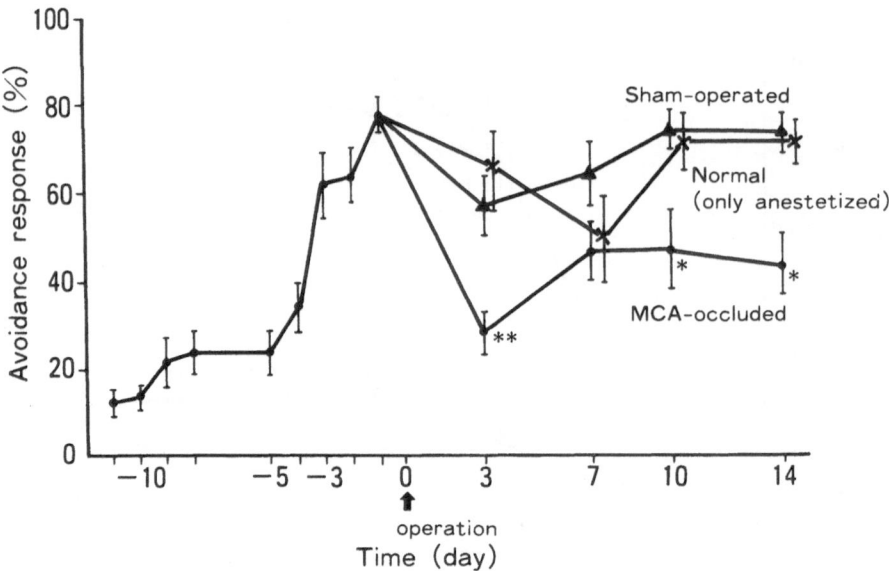

Fig. 4. Pre-operative acquisition and post-operative retention of descrete lever-press avoidance response (%) in normal, sham-operated and MCA-occluded rats. Each point represents the mean ± SEM. Significantly different from sham-operated values: $*P<0.05$, $**P<0.01$

On the other hand, the active avoidance responses may be affected by neurological deficits, such as hemiparesis. In this study, slight but significant hemiparesis was still shown in the MCA-occluded animals at day 14 after the occlusion. Therefore, further examination should be done.

Effects of a TRH Derivative on Behavioral Changes

It is reported that the thyrotropin-releasing hormone (TRH) possesses facilitatory effects on impaired motor function and learned behavior after anoxia or ischemia [8,9]. However, TRH is known to be rapidly metabolized in vivo. Therefore, the effect on behavioral changes of a new long-acting TRH derivative, YM-14673 [10], were studied using focal cerebral ischemia model as described above.

Under halothane anesthesia, the left MCA was occluded. The administration of YM-14673 was started immediately after the surgical operation and was repeated once a day for 3 weeks. In a total of 31 rats, either the drug (0.03, 0.1, or 0.3 mg/Kg) or saline was administered by intraperitoneal route. During the 3 weeks following MCA occlusion, behavioral changes including neurological deficits and disturbance of learned behavior were observed in a random manner 30 min after the drug injection. One, 2 and 3 weeks after MCA occlusion, neurological

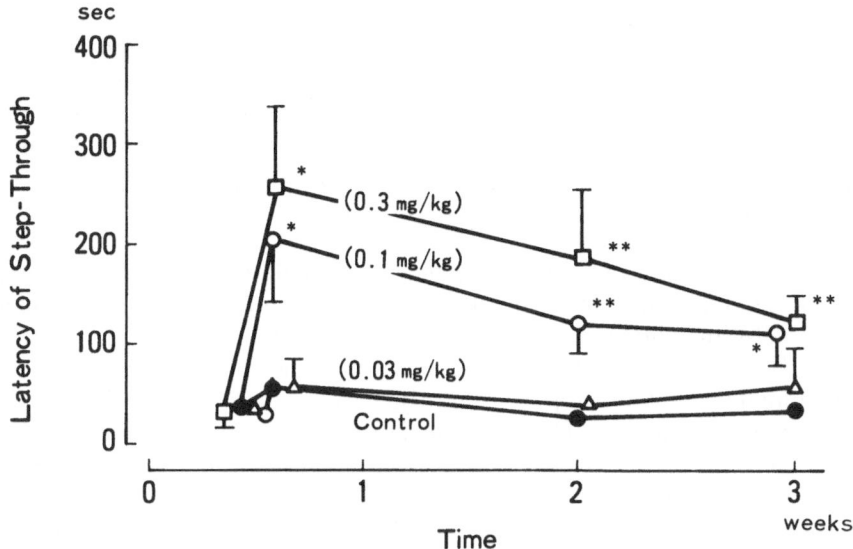

Fig. 5. Effects of YM-14673 on latency of step-through in passive avoidance task in MCA-occluded rats. The drug was administered intraperitoneally daily for 3 weeks. Each point represents the mean ± SEM. Significantly different from saline-treated control values: * $P < 0.05$, ** $P < 0.01$ (Mann-Whitney's U-test)

deficits were evaluated by measuring the degrees of hemiparesis and abnormal posture. After training by one-trial passive avoidance task at day 3 after operation, test trials were given 4 days and 2 and 3 weeks after the operation for evaluating acquisition and retention abilities. Three weeks after the surgical procedure, rats were perfusion fixed and a neuropathological study was carried out.

Neurological deficits were clearly observed 1 week after occlusion after which they gradually recovered. YM-14673 significantly accelerated the recovery of the neurological deficits compared to the saline control 2 and 3 weeks after occlusion.

The latency of step-through in the one-trial passive avoidance task in each group is shown in Fig. 5. It has been already shown in our previous study that the latency of step-through in the MCA-occluded group was significantly shortened compared to sham-operated and non-operated groups. YM-14673 significantly prolonged the shortened latency compared to the saline control 4 days and 2 and 3 weeks after MCA occlusion.

In the MCA-occluded model in the rat, ischemic damage is commonly observed in the cerebral cortex of the frontal, sensorimotor, and auditory areas as well as in the lateral part of the caudate nucleus [2,11]. In this study, the values of infarction ratio to the total area of both hemispheres in five coronal sections are $10.4 \pm 1.1\%$ (Mean ± SE) in the saline group, 13.0 ± 2.0 in 0.03 mg/kg of the YM-14673 group, 14.5 ± 3.1 in 0.1 mg/kg pf the YM-14673 group, and 12.7 ± 0.5 in 0.3 mg/kg of YM-14673 group. There was no significant difference between these groups. The infarction ratio in the saline-treated group was relatively smaller than that reported in our previous studies [11]. In the present study, neuropathological examinations were carried out 3 weeks after MCA occlusion. Therefore, some of the differences in the infarct ratio may be due to the development of atrophy during the 3 weeks after the occlusion.

In the preliminary study, YM-14673 did not affect spontaneous movement. Therefore, it is clearly shown that YM-14673 significantly accelerated the recovery of the neurological deficits and ameliorated cognitive disturbance, although the drug did not show any influence on ischemic infarct itself. A probable mechanism for the ameliorating effects of YM-14673 on neurological deficits and disturbed learned behavior can be proposed as follows: in this focal cerebral ischemia model in rats, hemiparesis may be induced mainly by a disturbance of pyramidal motor function. Thus, the ameliorating effects of this drug on hemiparesis may be due to its facilitatory effects on the pyramidal system. Involvement of the central monoaminergic and cholinergic systems is well known in regulating learned behavior. Since TRH possesses facilitatory effects on the central monoaminergic and cholinergic systems, ameliorationg effects of YM-14673 on the disturbance of learned behavior may be attributable to its facilitatory effects on these systems. In addition, it is reported that TRH exerts a trophic influence on long-term cultures of fetal rat motor neurons. Therefore, synaptic plasticity in the presence of damage observed in MCA-occluded rats may be promoted by the TRH analogue.

In conclusion, behavioral changes in the chronic phase of MCA-occluded rats were ameliorated by the new long-acting TRH derivative, YM-14673, without effect on ischemic brain damage. Furthermore, the MCA-occluded model in the rat is useful for quantitatively measuring functional changes in the chronic phase of focal cerebral ischemia.

Summary. We investigated behavioral changes after occlusion of the left middle cerebral artery (MCA) in rats. Incidence of neurological deficits evaluated by the degree of spontaneous activity and the degree of hemiparesis was observed during the first 4 weeks after the occlusion. Learning behavior in onetrial passive avoidance task or discrete lever-press active avoidance task was disturbed for the entire 2 weeks' periods when rats were trained before MCA occlusion. Behavioral changes during the chronic phase of MCA-occluded rats were ameliorated by a new thyrotropin-releasing hormone derivative, YM-14673, without any effect on ischemic brain damage.

Acknowledgements. The excellent technical assistance of Miss N Tomukai is greatfully acknowledged. The helpful comments of Dr. T. Kirino and Dr. O. Gotoh are particularly appreciated. This work was supported in part by Grant-in-Aid for Scientific Research (No. 63440054, 63570687) from the Japanese Ministry of Education, Science, and Culture and by a Research Grant for Cardiovascular Diseases (63C-5) from the Ministry of Health and Welfare.

References

1. Robinson RG (1979) Differential behavioral and biochemical effects of right and left hemispheric cerebral infarction in the rat. Science 205:707–710
2. Tamura A, Graham DI, McCulloch J, Teasdale GM (1981) Focal cerebral ischaemia in the rat: I. Description of technique and early neuropathological consequences following middle cerebral artery occlusion. J Cereb Blood Flow Metab 1:53–60
3. Tamura A, Gotoh O, Sano K (1986) Focal cerebral infarction in the rat: I. Operative technique and physiological monitoring for chronic model. Brain and Nerve (Tokyo) 38:747–751.
4. Yamamoto M, Tamura A, Kirino T, Shimizu M, Sano K (1988) Behavioral changes after focal cerebral ischemia by left middle cerebral artery occlusion in rats. Brain Res 452:323–328
5. Jarvik ME, Kopp A (1967) An improved one-trial passive avoidance learning situation. Psychol Rep 21:221–224
6. Kuribara H, Tadokoro S (1984) Conditioned lever-press avoidance response in mice: Acquisition processes and effects of diazepam. Psychopharmacology (Berlin) 82:36–40
7. Kuribara H, Tadokoro S (1985) Effects of psychoactive drugs on conditioned avoidance response in Mongolian gerbils (*Moriones unguiculatus*): Comparison with Wister rats and dd mice. Pharmacol Biochem Behav 23:1013–1018
8. Yamazaki N, Shintani M, Saji Y, Nagawa Y (1985) Effect of TRH and its analog DN-1417 on anoxia-induced amnesia in mice. Yakubutsu Seishin Kodo 5:1–9
9. Yasuhara N, Naito H (1983) Effects of TRH-T and DN-1417 on the central nervous system: An electrophysiological study of arousal reaction and evoked muscular discharges. Int J Neurosci 21:197–224
10. Yamamoto M, Shimizu M (1987) Facilitative effects of a new TRH analogue, YM-14673 on the central nervous system. Naunyn Schmiedebergs Arch Pharmacol 336:561–565
11. Tamura A, Gotoh O, Sano K, Nagashima T, Matsutani M, Orii H, Graham DI (1986) Focal cerebral infarction in the rat: 2. Neuropathological study and local cerebral blood flow pattern. Brain and Nerve (Tokyo) 38:859–863

10

Regeneration of Cerebral Microvessels After Cold Injury: Biological Significance of Edema Fluid

Tetsuji Orita, Takafumi Nishizaki, Tatsuo Akimura, Toshifumi Kamiryo, Kunihiko Harada, Haruhide Ito, and Hideo Aoki[1]

Introduction

Brain edema is a frequent and serious problem in man. Many investigators have performed morphological studies on various aspects of brain edema, including blood-brain barrier (BBB) damage and vascular permeability, for which they have often employed the cold-lesion model of vasogenic edema. In this cold-injury model, it has been clearly established that edema fluid passes into the brain tissue and accumulates within the abnormally distended extracellular space [1]. Many neurons as well as all other cellular elements are separated from each other by the hematogenous edematous fluid, and eventually begin to float freely in the fluid. However, it seems that little attention has been paid to the biological significance of the edema fluid. Mitosis of astrocytes is observed in edematous tissue. By cinematography, it has also been noted that many cells showing ruffle movement swim about feverishly in the edematous tissue [2]. Obviously, many cells in the edematous tissue will take part in lesion repair, under conditions where the vascular network in the edematous tissue is sparse [3]. However, it is still unknown from where and how many cells receive the energy for lesion re-pair. In the present study, by revealing sequential changes in endothelial cell kinetics following different types of cold injury using immunohistochemical tech-niques, we intended to verify if edematous fluid is an important factor for lesion repair. In addition, it was considered important to observe if transferrin recep-tors (Tf-R) are present when transferrin in edematous fluid is needed. Transfer-rin is the major serum iron-transport protein and, along with iron, plays an im-portant role in cell growth and metabolism. Key reactions in energy metabolism and DNA synthesis are catalyzed by iron-containing enzymes. The first step in the delivery of iron to cells by transferrin involves its binding to a specific cell-

[1]Department of Neurosurgery, Yamaguchi University, School of Medicine, 1144 Kogushi, Ube, 755 Japan

surface receptor [4]. However, synthesis of transferrin does not occur in the brain [5]. In the normal brain, transferrin receptors are expressed on brain capillaries, facilitating transport of transferrin into brain tissues [6].

Materials and Methods

Wistar rats were anesthetized by intraperitoneal injections of pentobarbital sodium (40 mg/kg). The skull over the right frontoparietal cortex was exposed, and a focal cold injury was produced by contact with a metal plate cooled with a mixture of dry ice and acetone. It allowed the severity and extent of the lesion to be constantly changed without having to consider factors such as infection when interpreting the experimental results. There were two experimental groups. In group I, the metal plate was 5.5 mm in diameter and the duration of contact was 60 sec, while in group II, the metal plate was 3 mm in diameter and the duration of contact was 20 sec. The animals were examined at 1, 2, 3, 4, 5, 7, and 14 days post-injury, respectively.

Rats were injected intravenously with 2% Evans blue (2 ml/kg) and 10 mg/kg of bromodeoxyuridine (BrdU, Sigma Chemical Company, St. Louis, MO) 1 hour prior to sacrifice. BrdU is one of the halopyrimidines and, like thymidine, is incorporated into cellular nuclei at the time of mitotic DNA synthesis [7]. It is rapidly degraded by the liver (more than 90% of a single bolus BrdU is debrominated within 20 min) [8]. Fourteen out of 56 rats were not injected with Evans blue for immunofluorescent staining. The coronal sections through the lesions were prepared for immunohistochemical staining using cryostat.

The following three monoclonal antibodies and antisera were used: anti-BrdU antibody (Beckton Dickinson Monoclonal Center Inc., Mountain View, CA) diluted 1:100; anti-rat Tf-R (Serotec Ltd., Blackthorn, Bicester, England) diluted 1:200, and OX-43 (Serotec) diluted 1:200. OX-43 exists in rat endothelium outside the brain capillary endothelium [9]. Two polyclonal antibodies were used: anti-glial fibrillary acidic protein antibody (GFAP, DAKO Corporation, CA) and anti-factor VIII-related antigen antibody (VIII-Ag, DAKO Corporation).

The labeled streptavidin-biotin technique was employed for staining of BrdU, as described elsewhere [3,10]. In brief, frozen sections were fixed in ether and sequentially incubated with 4-N-HCl, the primary antibody, biotinylated anti-mouse IgG, and peroxidase-labeled streptavidine, supplied in a Stravigen B-SA kit (Biogenex Laboratories, Dublin, CA). When using polyclonal primary antibodies, the peroxidase-anti-peroxidase technique was applied. Peroxidase binding sites were detected in 20 mg of diaminobenzidine tetrahydrochloride and 100 ml 0.005 M Tris buffer containing 0.0005% H_2O_2.

For double-staining with monoclonal antibodies (Tf-R, OX-43) and VIII-Ag, ether-fixed sections were incubated with monoclonal antibodies, biotinylated anti-mouse IgG, and fluorescein (FITC)-conjugated streptavidin (Amersham International plc, Amersham, UK) as the first step, and the antiserum and

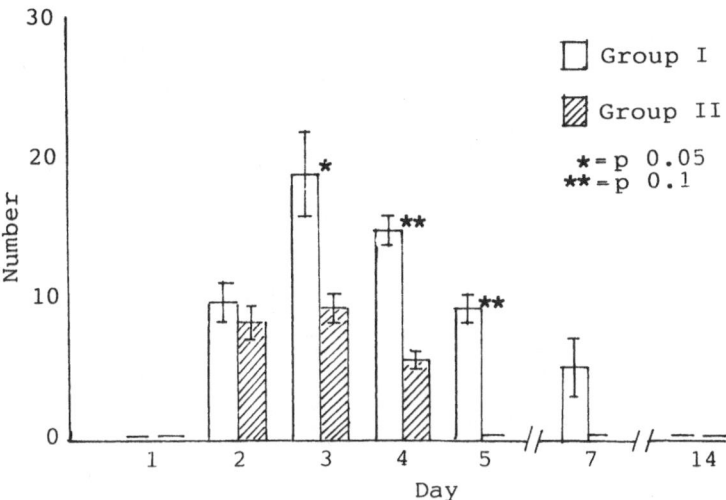

Fig. 1. Number of bromodeoxyuridine-positive endothelial cells in the lesion. Values are means ± SEM from three animals on each day after injury. *$P < 0.05$, **< 0.1 by Student's *t*-test (From [10] with permission)

rhodamine-conjugated swine anti-rabbit IgG as the second step. FITC and rhodamine were analysed using a Nikon fluorescence microscope.

Results

The extent of each edematous lesion was estimated from the intensity and extent of the Evans blue (EB)-stained area in the coronal brain slices. The lesions were at maximal extent on day 1, when the diameter was about 4 mm in group I and 2 mm in group II. The EB-stained area disappeared at day 14 post-injury in group I and after day 5 in group II. Immunohistochemical sections prepared with BrdU demonstrated immunoreaction products exclusively in the cellular nuclei. Several BrdU-positive astrocytes were observed in the edematous tissue. In both groups, the earliest evidence of the presence of BrdU-positive endothelial cells (BrdU+end) was observed at day 2 post-injury, and the number of these cells increased most markedly at day 3. They could not be detected at day 14 in group I and at day 5 in group II. There were significant differences between the number of BrdU+end in both groups from 3–7 days post-injury (Fig. 1). The earlist evidence for the presence of all BrdU-positive cells except BrdU+end was observed at day 1 and none could be detected at day 14 in either group. There were also significant differences between the cell numbers in both groups from 2–7 days post-injury (Fig. 2). Repair of the cerebral endothelium was one day behind that of other cells. In both groups, BrdU+end were always seen within the EB-stained area (Fig. 3). At day 2, most of them were not in contact with

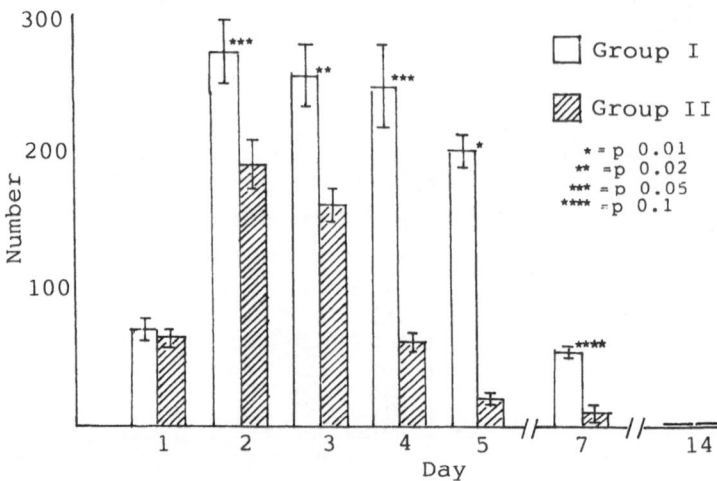

Fig. 2. Number of bromodeoxyuridine-positive cells except bromodeoxyuridine-positive endothelial cells in the lesions. Values are means ± SEM from three animals on each day after injury. $*P < 0.01$, $**P < 0.02$, $***P < 0.05$, $****P < 0.1$ by Student's t-test (From [10] with permission)

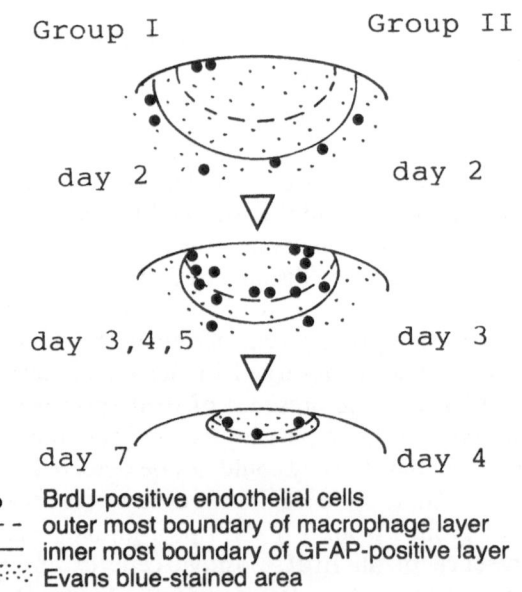

● BrdU-positive endothelial cells
- - - outer most boundary of macrophage layer
——— inner most boundary of GFAP-positive layer
⋯⋯⋯ Evans blue-stained area

Fig. 3. Illustrations of distribution of bromodeoxyuridine-positive endothelial cells in the lesions (From [10] with permission)

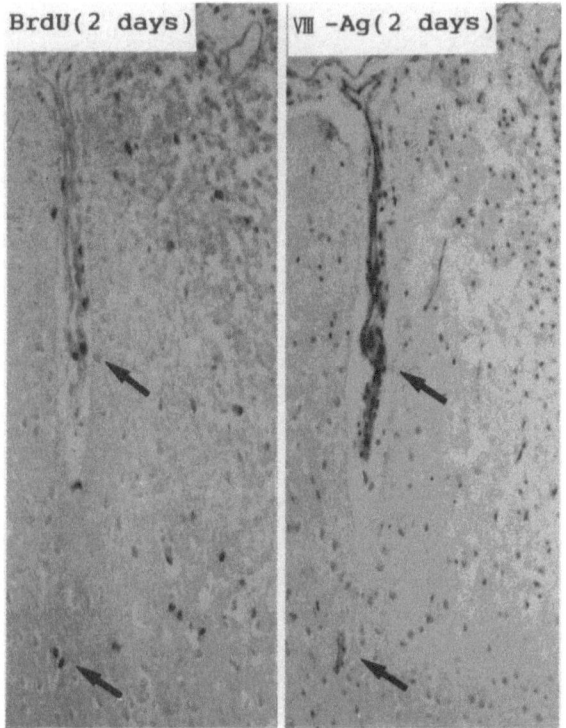

Fig. 4. In group I, *left* and *right* are serial sections on the second day. Immunostaining with mouse monoclonal antibody for BrdU. × 200. BrdU-positive endothelial cells (*arrows*) are not in contact with macrophages (*left*). Immunostaining with rabbit polyclonal antibody for VIII-Ag. × 200. There are positive endothelial cells (*arrows*) in venules and capillaries (*right*)

macrophages and were observed at the periphery of the layer of GFAP-positive cells (Fig. 4). After day 3, BrdU+end were seen in the macrophage layer (Fig. 5), regenerating from the edge toward the center of the lesion.

In normal rat brain, the capillary endothelial cells were labelled with Tf-R, whereas endothelial cells in large vessels were not. Many Tf-R-positive cells were observed in the injured brain (Fig. 6). The distribution of Tf-R-positive cells was slightly less than those of the BrdU-positive cells. The transition pattern of the numbers of Tf-R-positive cells was the same as for BrdU-positive cells. From 2–5 days after injury in group II, tortuous capillaries were labelled with Tf-R. Tf-R-positive endothelial cells were not seen in the large vessels. Many Tf-R-positive cells in the center of the lesion did not stain with anti-VIII-Ag and anti-GFAP antibodies. Their number was increased markedly at 2 or 3 days and thereafter decreased gradually. At the edge of the lesion, several Tf-R-positive cells were stained with anti-GFAP antiserum 2 or 3 days postinjury. In normal and injured brain, OX-43 staining of capillary endothelial cells was not observed. However, endothelial cells in major vessels were OX-43-positive.

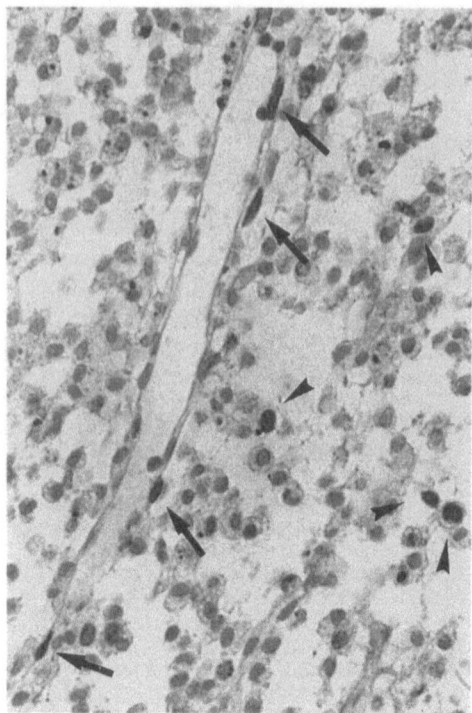

Fig. 5. Immunostaining with mouse monoclonal antibody for BrdU. × 400. In group I, positive endothelial cells (*arrows*) lie among many macrophages on the third day. BrdU-positive macrophages (*arrowheads*) are seen

Discussion

There have been a few studies of cerebral endothelial cell kinetics after brain injury, using [³H] thymidine [11,12]. However, one of the major disadvantages of autoradiography is its cost and time-consuming nature. An alternative approach is to use immunohistochemical staining of BrdU, one of the halopyrimidines, which, like thymidine, is incorporated into cellular nuclei at the time of mitotic DNA synthesis [7]. Although this new method using BrdU and anti-BrdU monoclonal antibodies has recently been employed for the study of tumor cell kinetics [13,14], there have been only two previous studies using this method for elucidating the mechanism of tissue repair [3,10].

The earlist evidence for the presence of BrdU+end was observed at 2 days post-injury, the injured endothelial cells regenerating from the edge toward the center of the lesion in both groups. However, there has been a report describing the incorporation of [³H] thymidine into endothelial cells at the edge of the lesion beginning at day 3 of post-injury [11]. In this report, [³H] thymidine was injected 24 h prior to sacrifice. As [³H] thymidine is rapidly incorporated into cellular nuclei, we consider that the latter result at day 3 to be equivalent to the appearance of BrdU+end on the second day seen in our study. The difference in the process of endothelial regeneration was initiated on day 3 and there was no difference at day 2, while the difference in the process of regeneration of BrdU-

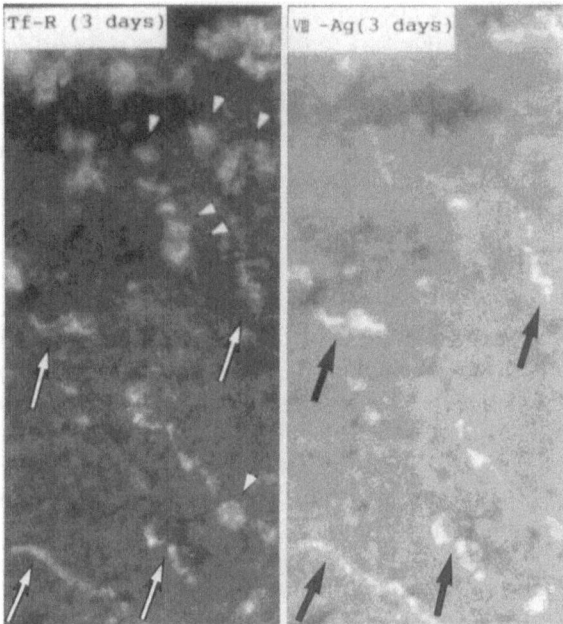

Fig. 6. Double immunofluorescent staining for Tf-R (*left*) and VIII-Ag (*right*). Many Tf-R-positive cells (*arrowheads*) except regenerated endothelial cells (*arrows*) are observed in the lesion. 3 days post-injury (× 200)

positive cells (except BrdU+end) was apparent from day 2 and there was no difference on day 1. Although these findings are very interesting, we are as yet unable to clarify the reason. Nevertheless, the distribution and transition pattern of the numbers of regenerated endothelial cells were the same in both groups.

Since macrophages are very closely associated with necrotic and regenerative vessels, and appear just before the onset of endothelial DNA synthesis, these cells are considered to stimulate cerebral endothelial regeneration following cold injury [11,15]. There is also a report stating that depletion of macrophages by irradiation significantly decreases the [³H] thymidine labeling index of the cerebral endothelium at 3, 4, and 5 days post-injury in comparison with that in non-irradiated animals [12]. The signals for cerebral endothelial regeneration following cold injury are undoubtly complex and may differ from those for endothelium in other organs. Since BrdU+end were always found in EB-stained areas, and the time of their disappearance coincided with the disappearance of EB-stained areas, we suggest that edematous fluid may be an important factor in endothelial regeneration. In other words, a substance may be present in edematous fluid, or this fluid may be an important medium for transmission of a stimulator substance from macrophages. Ikuta et al. [2] first drew attention to the biological significance of edematous fluid as a source of energy for macrophages and astrocytes in lesion repair.

Tf-R is a transmembrane glycoprotein that mediates the cellular uptake iron. The transferrin and iron take part in brain energy metabolism enzyme function [16]. In the normal brain, Tf-R are expressed in the capillary endothelium and the developing neuron [6]. Many Tf-R-positive cells were also present in the lesions we produced. These cells consisted of regenerated endothelial cells, reactive astrocytes [17], and other cells which seemed to be macrophages from their distribution and morphology. The distribution and number of Tf-R-positive cells were less than those of the BrdU-positive cells. Considering these findings, we suggest that part of the expression of transferrin receptors in the lesion is not involved in cell proliferation, although proliferation cells express much larger numbers of Tf-R than do resting cells [18]. Strong reactions with non-dividing cells have been clearly established on the basis of labelling of hepatocytes, Kupffer cells, and contractile heart and muscle cells [4]. Transferrin in edematous fluid would bind Tf-R and be taken into the cells, which would consequently receive energy for activation of lesion repair under conditions where the vascular network is sparse [3]. In the fetal brain, transferrin, albumin, and α-fetoprotein are localized by immunohistochemical methods in neuroblasts, the wide extracellular space of the cortical plate, the migrating zone, and the paraventricular matrix layer [2,19]. Since the network of cerebral vessels is sparse in the early developmental stages [20], migrating neuroblasts may receive the energy for their movements from the extracellular fluid of the fetal brain. Newly formed vessels in the injured brain resemble the vascular architecture of the fetal and newborn rat cortex, revealed by three-dimensional observations of the microvascular architecture [3]. The vascular pattern in a large proportion of gliomas is also based on the mode of vascularization observed in the normal developing brain [21], just as the blood vessels in metastatic brain tumors are known to imitate the structure of the vasculature specific for that organ [22]. The regenerated processes of edematous lesions may be a revival of the mechanisms of the normal development in the fetal brain. However, it is still unknown what are the further characteristics of the regenerated endothelium. In the tortuous capillaries which are considered to be regenerated vessels [3], Tf-R which are expressed in normal brain capillary endothelium [6], were positive, while OX-43 which exist in rat endothelium outside the brain capillary [9], were negative on the days examined. From this immunohistochemical study, it appears that even from the early stages of regeneration the characteristics of the endothelium may differ from those of the vessels in other organs and may have the nature of normal brain endothelium. Additionally regenerated endothelial cells may regulate the transport of transferrin from blood vessels into tissue spaces around neuronal and glial cells.

Summary. The earliest evidence for the presence of BrdU-positive endothelial cells (BrdU+end) was observed at day 2 after injury, with the injured endothelial cells regenerating from the edge toward the center of the lesion in both groups. The distribution and transition patterns of the number of BrdU+end were also the same. BrdU+end were always found in EB-stained areas. We therefore considered that edematous fluid could act as an important factor for lesion repair, even though edema itself is definitely a serious pathological condition. The presence of transferrin receptors was also demonstrated in regenerated endothelial cells, reactive astrocytes, and other cells, and probably in macrophages. Transferrin and iron play an important role in cell growth and metabolism. The receptors may bind transferrin present in edematous fluid to provide an energy source for lesion repair. Therefore, edematous fluid can effect lesion repair, and the term "blood-brain barrier (BBB) damage" can be reconsidered to be "physiological opening of the BBB."

References

1. Hirano A, Zimmerman HM, Levine S (1967) Fine structure of cerebral fluid accumulation. In: Klatzo I, Seitelberger F (eds) Proceedings of the symposium on brain edema, Vienna 1965, Springer, New York, pp 569–589
2. Ikuta F, Yoshida Y, Ohama K, Oyanagi S, Takeda S, Yamazaki K, Watabe K (1983) Revised pathophysiology on BBB damage. The edema as an ingeniously provided condition for cell motality and lesion repair. Acta Neuropathol [suppl] (Berl) 8:103–110
3. Orita T, Nishizaki T, Kamiryo T, Harada K, Aoki H (1988) Cerebral microvascular architecture following experimental cold injury. J Neurosurg 68:608–612
4. Bomford AB, Munro HN (1985) Transferrin and its receptor: Their roles in cell function. Hepatology 5:870–875
5. Bockmeer MF, Morgan EH (1977) Identification of transferrin receptors in reticulocytes. Biochim Biophys Acta 468:437–450
6. Jefferies WA, Brandon MR, Hunt SV, Williams AF, Gatter KC, Mason DY (1984) Transferrin receptor on endothelium of brain capillaries. Nature 312:162–163
7. Goz B (1977) The effects of incorporation of 5-halogenated deoxyuridines into the DNA of eukaryotic cells. Pharmacol Rev 29:249–272
8. Kriss JP, Révész L (1962) The distribution and fate of bromodeoxyuridine and bromodeoxycytidine in the mouse and rat. Cancer Res 22:254–265
9. Robinson AP, White TM, Mason DW (1986) MRC OX-43. A monoclonal antibody which reacts with all vascular endothelium in the rat except that of brain capillaries. Immunology 57:231–237
10. Orita T, Akimura T, Kamiryo T, Nishizaki T, Furutani K, Harada K, Ikeyama Y, Aoki H (1989) Cerebral endothelial regeneration following experimental brain injury. Variation in the regeneration process according to the severity of injury. Acta Neuropathol (Berl) 77:397–401
11. Cancilla PA, DeBault LE (1980) Freeze injury and repair of cerebral microvessels. Adv Exp Med Biol 131:257–269
12. Beck DW, Hart MN, Cancilla PA (1983) The role of macrophage in microvascular regeneration following brain injury. J Neuropathol Exp Neurol 42:601–614
13. Nishizaki T, Orita T, Saiki M, Furutani Y, Aoki H (1988) Cell kinetics studies of human brain tumors by in vitro labeling using anti-BUdR monoclonal antibody. J Neurosurg 69:371–374
14. Nishizaki T, Orita T, Furutani Y, Ikeyama Y, Aoki H (1989) Comparison of Ki-67 monoclonal antibody labeling and in vitro labeling using anti-BUdR monoclonal antibody in human brain tumors. J Neurosurg 70:379–384
15. Cancilla PA, Formes BS, Kahn LE, Debault LE (1979) Regeneration of cerebral

microvessels. A morphologic and histochemistry study after local freeze-injury. Lab Invest 40:74–82

16. Angelova-Gateva P (1980) Iron transferrin receptors in rat and human cerebrum. Agressologie 21:27–30
17. Orita T, Akimura T, Nishizaki T, Kamiryo T, Ikeyama Y, Aoki H, Ito H (1990) Transferrin receptors in injured brain. Acta Neuropathol (Berl) 79:686–688
18. Faulk WP, Hsi BL, Stevens PJ (1980) Transferrin and transferrin receptors in carcinoma of the breast. Lancet II:390–392
19. Toran-Allerand CD (1980) Coexistence of α-fetoprotein, albumin and transferrin immunoreactivity in neurons of the developing mouse brain. Nature 286:733–735
20. Yoshida Y, Ikuta F (1984) Three-dimensional architecture of cerebral microvessels with a scanning electron microscope. A cerebrovascular casting method for fetal and adult rats. J Cereb Blood Flow Metab 4:290–296
21. Orita T, Nishizaki T, Kamiryo T, Aoki H, Harada K, Okamura T (1988) The microvascular architecture of human malignant glioma. A scanning electron microscopic study of a vascular cast. Acta Neuropathol (Berl) 76:270–274
22. Hirano A, Zimmerman HM (1972) Fenestrated blood vessels in a metastatic renal carcinoma in the brain. Lab Invest 26:465–468

11

The Effects of Sodium Bicarbonate on Cerebrospinal Fluid Acid-Base Disturbances in Total Cerebral Ischemia

Akitsugu Kohama, Nobukatsu Takasu, Shinichi Ishimatsu, Akiyuki Maenosono, and Kouichiro Suzuki[1]

Introduction

If normal circulation or ventilation is stopped, a living body progressively falls into hypoxemia; continuation of this condition will worsen tissue and organ damage. In order to ameliorate this condition, cardiopulmonary cerebral resuscitation (CPCR) is usually performed. In CPCR, cerebral resuscitation is critical because of the difficulty in accomplishing it and because, if it does not succeed, the human being ceases to exist as a human being.

Since we are particularly interested in the resuscitation of non-hospitalized victims of cardiac arrest, such as incidents of near-drowning, asphyxia, and myocardial infarction, we have made a great deal of effort to achieve the recovery of cerebral function in these patients.

Cerebrospinal fluid (CSF) maintains the interior milieu of the brain, and regulation of the acid-base balance in CSF seems to be an indispensable factor for maintaining normal cerebral function. Thus, to achieve successful cerebral resuscitation, the investigation of CSF acid-base disturbances during and after cerebral ischemia is important [1,2]. Therefore, we performed the following experiments on dogs to learn what acid-base disturbances occur in CSF and to determine the effects of sodium bicarbonate ($NaHCO_3$) on these disturbances.

Methods

Mongrel dogs (10–23 kg in weight) were divided in three groups: in group I ($n = 7$) 10–12 min total cerebral ischemia (TCI) was produced by the aortic occlusion balloon (AOB) catheter method. Arterial blood and CSF pH, PCO_2, and HCO_3^-, and brain pH and PCO_2 were measured before, during, and after

[1] Department of Emergency and Critical Care Medicine, Kawasaki Medical School, 577 Matsushima, Kurashiki, 701-01 Japan

TCI. The observation time was 120 min. In group II ($n = 7$) two mEq/kg of $NaHCO_3$ was administered immediately after recirculation of TCI. The parameters for the acid-base disturbances were the same as in group I. In group III ($n = 7$) 10 mEq/kg of $NaHCO_3$ was administrated immediately after recirculation of TCI. The parameters for the acid-base disturbances were the same as in the other groups. Blood pressure (BP), heart rate (HR), intracranial pressure (ICP), and serum Na and K were measured at the same time.

Atropine sulfate (0.5 mg) and ketamine hydrochloride (10 mg/kg) were injected intramuscularly, and pancronium bromide (4 mg) was injected intravenously 10 min later to achieve muscle relaxation. Next, endotracheal intubation was performed and the tube was connected to the ventilator (AIKA R-60). Ventilation was adjusted to maintain arterial PO_2 and PCO_2 at about 100 and 30 mmHg, respectively. Nitrous oxide 25% in oxygen were used for anesthesia during the experiments. Both brachial arteries and the right femoral artery were used for the measurement of blood pressure and sampling of arterial blood. The right femoral vein was catheterized for infusion of sodium lactated Ringer solution 5–10 ml/kg/hr, and the left femoral artery was used for the AOB catheter which would produce the TCI. A Fogarty balloon catheter (8F in size) used to reduce the venous return to the heart was inserted into the left femoral vein. ECG and BP were measured by a polygraph recorder (RM-85 Nihon Electric Co.).

TCI was produced for 10–12 min by the AOB catheter method (Fig. 1) [3]. Initially, the balloon of the Fogarty catheter, which was inserted by way of the left femoral vein to the inferior vena cava, was inflated to stop blood flow in the inferior vena cava. Then, the AOB catheter balloon, located at the aortic arch, was inflated to stop total systemic circulation. The success of TCI was confirmed by the disappearance of the pressure waves of both brachial arteries. The TCI was discontinued 10–12 min later by deflation of both balloons.

Arterial pH, PCO_2 and HCO_3^- were measured by a blood gas analyzer (Radiometer Co.). Brain and CSF pH and PCO_2 were measured continuously for 120 min by pH and PCO_2 sensors based on the pH-ISFET (ion selective field effect transistor (Kuraray Co.) [4]. The 90% response time of PCO_2 and pH sensors were 30–90 and 2 s, respectively. The CSF HCO_3^- was calculated using a carbonic acid pH of 6.13 and a CO_2 solubility factor of 0.0312 [5]. pH/PCO_2 monitor instruments (KR-500 Kuraray Co.) and multipenrecorders were used to record these parameters.

The method used to measure brain and CSF pH and PCO_2 with pH-ISFET sensors was as follows: after experimental preparations were completed, the dog was placed in the prone position with its head fixed on a stand. For the measurement of brain pH and PCO_2, a burr hole was made in the left lateral bone, and pH and PCO_2 sensors were inserted through the dura mater into brain tissue to a depth of about 1.0 cm. For the CSF pH and PCO_2, a skin incision of 5–7 cm was made in the occipitocervical area, and the dura mater was exposed by laminectomy. Thereafter, a 1–2 mm longitudinal incision was made into the dura mater, and pH and PCO_2 sensors were implanted into the cisterna magna. Intracranial pressure under the dura mater was measured by a catheter method.

Fig. 1. A-O Balloon Catheter method producing total cerebral ischemia (TCI)

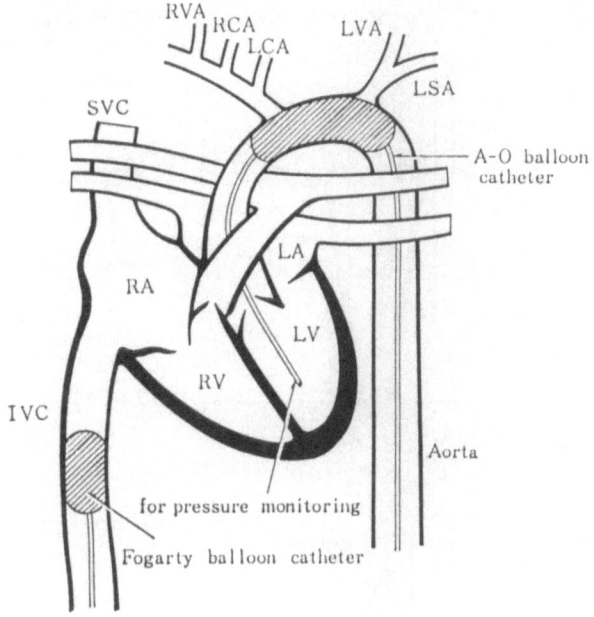

Serum Na and K were measured by the ion selective electrode method (NOVA-1 NOVA Co.).

The experimental results were presented as mean ± SE. The paired Student's *t*-test was used to evaluate possible statistical differences. A P-valve less than 0.05 was considered statistically significant.

Results

The results of brain and CSF acid-base disturbances were recorded continuously during the experiments, with the exception of arterial pH and PCO_2. To calculate the mean value and standard error, however, the results were evaluated intermittently.

Blood Pressure and Heart Rate

Changes in BP and HR in each group are shown in Fig.2. In group I (control) and II (2 mEq/kg of NaHCO), BP recovered within 3 min after recirculation of TCI. In group III (10 mEq/kg of $NaHCO_3$), BP showed a tendency to return to the control level after recirculation of TCI, but decreased below the control level for 30 min by the administration of 10 mEq/kg of $NaHCO_3$ ($P<0.05$). HR increased in each group during TCI ($P<0.05$), but returned to the control level after recirculation of TCI with exception of group III. In group III, HR decreased temporally by the administration of 10 mEq/kg of $NaHCO_3$ ($P<0.05$).

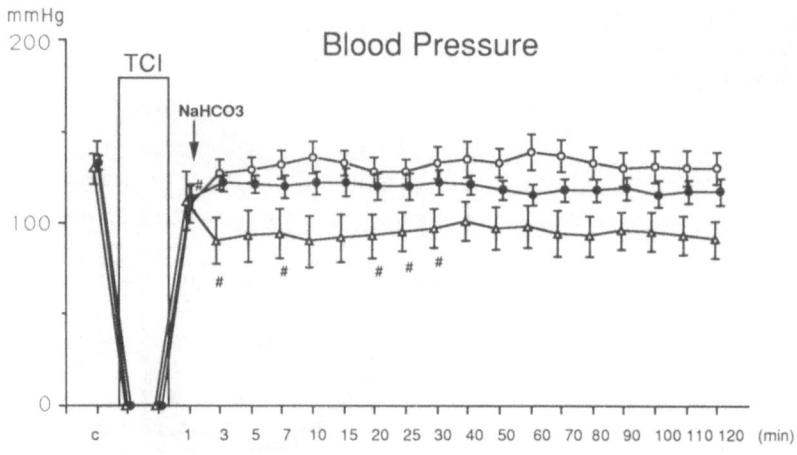

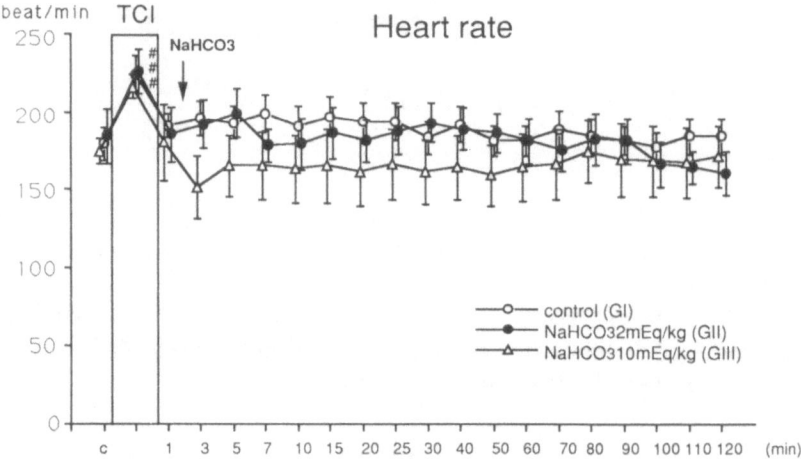

Fig. 2. Changes in blood pressure (BP) and heart rate (HR) before, during, and after recirculation of TCI. Each point is the mean ± SE for seven dogs. #, significantly different from control (c); $P < 0.05$

Changes in Arterial pH, PCO₂ and HCO₃⁻

Arterial pH and HCO_3^- decreased and PCO_2 increased after recirculation of TCI ($P < 0.05$). Although pH and HCO_3^- showed a tendency to return to the control level, they did not return to the control level during the time of observation ($P < 0.05$). PCO_2 returned to the control level within 10 min after recirculation of TCI (Fig. 3).

In group II, pH returned to the control level and HCO_3^- and PCO_2 were increased by the administration of 2 mEq/kg of $NaHCO_3$ ($P < 0.05$). Then, HCO_3^- and PCO_2 decreased gradually to the control level by the end of the experiments. In group III, the administration of 10 mEq/kg of $NaHCO_3$ elevated

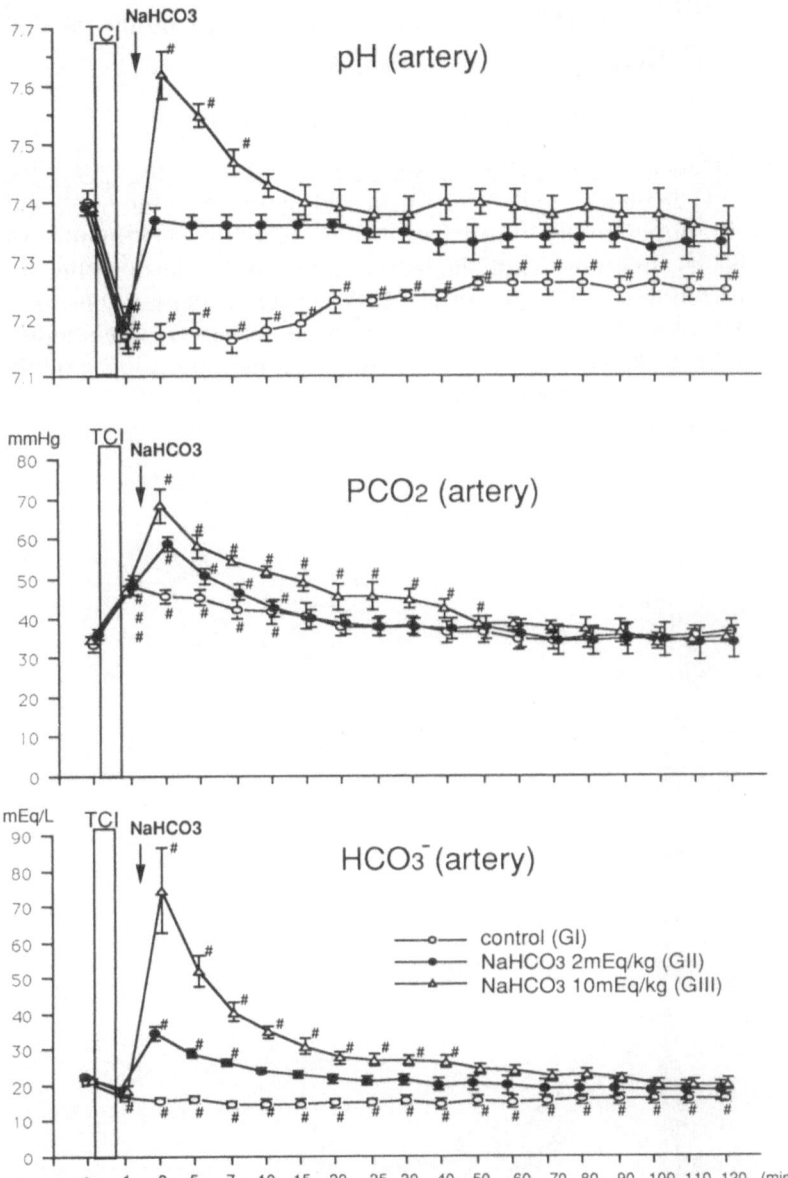

Fig. 3. Changes of pH, PCO_2 and HCO_3^- in arterial blood before, during, and after recirculation of TCI. Each point is the mean $\pm$ SE for seven dogs. #, significantly different from control (c); $P < 0.05$

arterial pH, PCO_2, and HCO_3^- to above the control level during the time of the experiments ($P<0.05$).

Changes in Brain pH and PCO_2

Brain pH decreased and PCO_2 increased as a result of the TCI ($P<0.05$). However, pH increased and PCO_2 decreased gradually after recirculation of TCI. PCO_2 returned to the control level by the end of the experiments, while pH did not (Fig. 4). In group II, pH increased to the control level within 30 min by the administration of 2 mEq/kg of $NaHCO_3$. PCO_2 increased, but returned to the control level within 30 min of recirculation. In group III, the administration of 10 mEq/kg of $NaHCO_3$ elevated brain pH continuously ($P<0.05$). Brain PCO_2 increased, but showed a tendency to decrease to the control level during the time of observation.

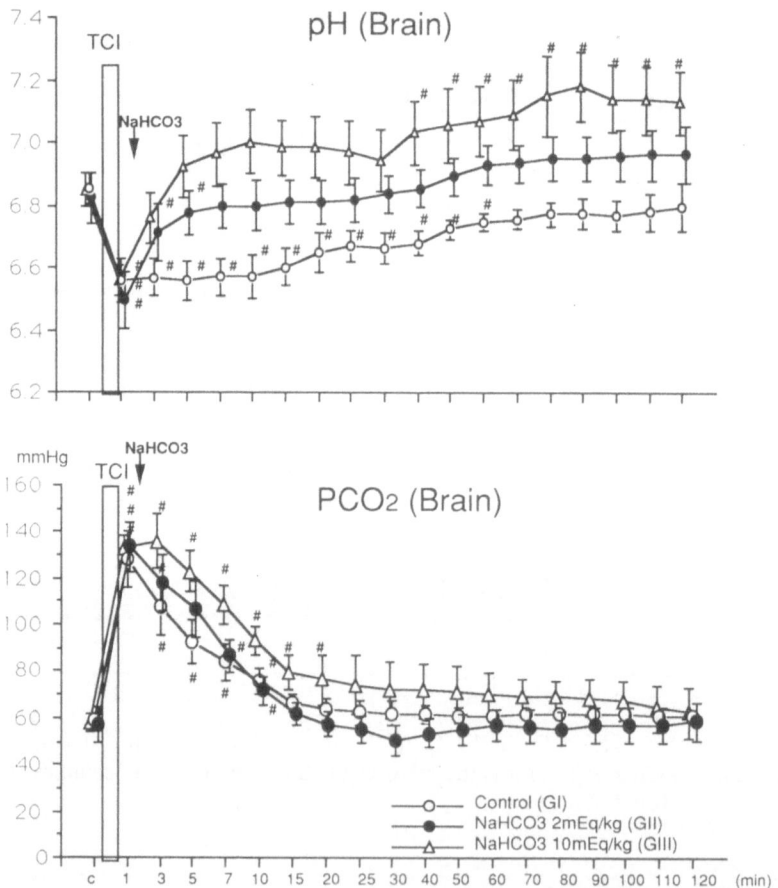

Fig. 4. Changes in pH and PCO_2 in brain before, during, and after recirculation of TCI. Each point point is the mean ± SE for seven dogs. #, significantly different from control (c); $P<0.05$

Changes in CSF pH, PCO$_2$ and HCO$_3^-$

TCI caused decreases in CSF pH and HCO$_3^-$ and an increase in PCO$_2$ ($P<0.05$). PCO$_2$ showed a decreased tendency to the control level, but pH and HCO$_3^-$ did not show an increased tendency to the control level within 120 min of observation ($P<0.05$) (Fig. 5). In group II, CSF pH and PCO$_2$ increased following the administration of 2 mEq/kg of NaHCO$_3$. pH returned to the control level within 15 min of recirculation. PCO$_2$ showed a tendency to decrease to the control level during the time of observation. CSF HCO$_3^-$ increased with the administration of 2 mEq/kg of NaHCO$_3$, and returned to the control level after 5 min of recirculation.

In group III, the administration of 10 mEq/kg of NaHCO$_3$ increased CSF pH, HCO$_3^-$ and PCO$_2$ excessively ($P<0.05$). PCO$_2$ decreased gradually to the control level during the time of the experiments, but did not return to the control level until the end of the experiments. pH and HCO$_3^-$ continued increasing until the end of the experiments ($P<0.05$).

Changes in Intracranial Pressure

ICP increased temporarily after recirculation of TCI, but returned to the control level after 10–20 min of recirculation.

Changes in Serum Na and K

Serum Na concentration did not show any remarkable changes in groups I and II after recirculation of TCI. In group III, the administration of 10 mEq/kg of NaHCO$_3$ increased serum Na concentration after recirculation of TCI ($P<0.05$). Serum K concentration increased slightly in groups I and II after recirculation of TCI. In group III, serum K concentration decreased sightly during the time of observation. However, these changes were within normal range (Fig. 6).

Discussion

Cardiac and respiratory arrests or disturbances cause hypoxemia and accelerate the production of lactate. CO$_2$ also accumulates in disturbances of circulation and ventilation. Accumulations of lactate and CO$_2$ (metabolic and respiratory acidosis) remarkably decrease the systemic blood pH. These changes also occur in the brain and disturb the cerebral function. Therefore, to achieve successful brain resuscitation, clarification of acid-base disturbances in the brain after the recirculation of TCI would be important. These disturbances must be elucidated by investigating changes in the acid-base balance in blood and CSF, since blood supplies oxygen and energy to the brain and removes waste products from it and CSF circulates through the intraventricular and subarachnoid spaces to maintain homeostasis and convert hormones and chemical transmitters in the brain.

From these points of view, we studied the changes in arterial, brain and CSF

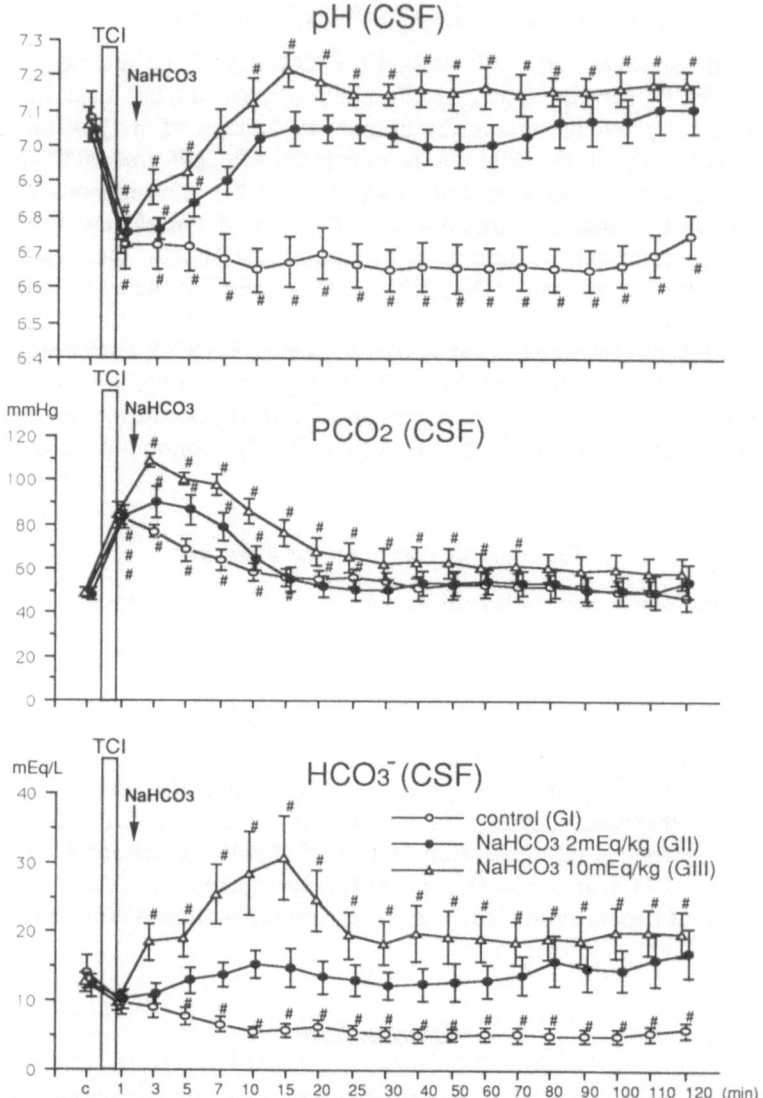

Fig. 5. Changes in pH, PCO_2 and HCO_3^- in CSF before, during, and after recirculation of TCI. Each point is the mean ± SE for seven dogs. #, significantly different from control (c); $P < 0.05$

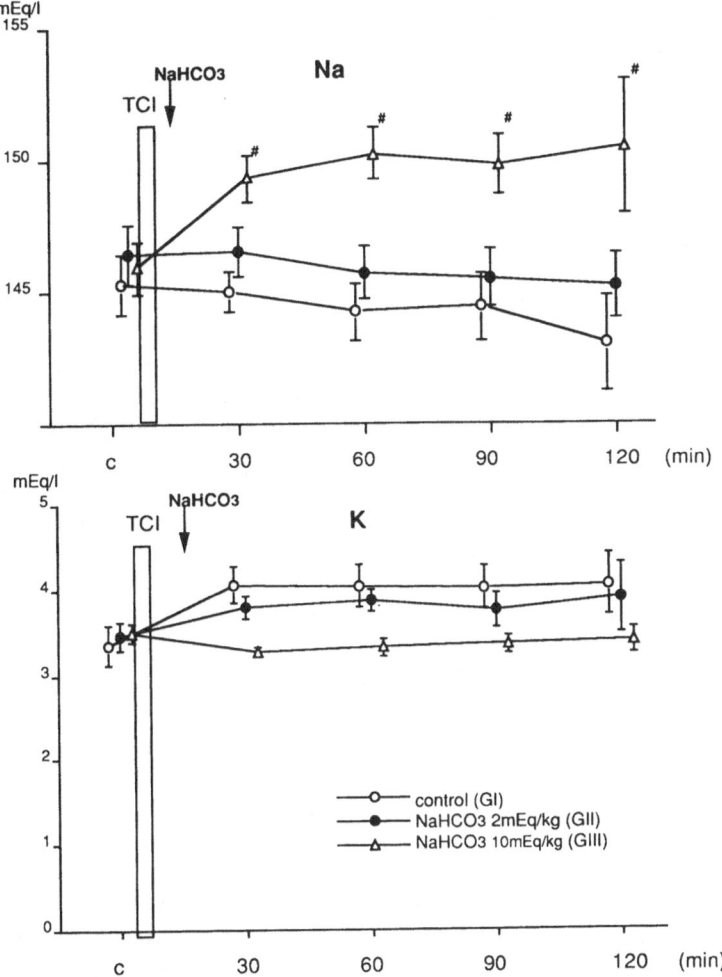

Fig. 6. Changes in serum Na and K before, during, and after recirculation of TCI. Each point is the mean ± SE for seven dogs. #, significantly different from control (c); $P < 0.05$

acid-base disturbances after recirculation of TCI in dogs. In these experiments, arterial pH and HCO_3^- decreased and PCO_2 increased after arrest of the systemic circulation. These changes showed a tendency to return to the control level after 10–20 min of recirculation, but pH and HCO_3^- did not return to within the control level during the time of observation. In these experiments, the changes in HCO_3^- were more remarkable than those in PCO_2. This fact suggests that the metabolic factor (HCO_3^-) has a greater effect on the decreased pH in TCI than the respiratory factor (PCO_2).

In the acid-base disturbances of CSF, pH and HCO_3^- never returned to the control level and PCO_2 showed a tendency to decrease to the control level, but did not return to the control level within the period of observation. These results indicate that in the CSF, in contrast to the blood and brain, metabolic and respiratory factors affect the decrease in pH. There are several posslble explanations for the delay in recovery in CSF acid-base disturbances. First, that the circulation of CSF in the dogs was very slow compared with that of blood in the artery. Judging from the production rate (0.047 ml/kg) and total volume (10–13 ml) of CSF in dogs (12–17 kg in weight) [6], the turnover time of CSF is about 3.5–4.6 h. These facts explain why the recovery of pH and HCO_3^- in the CSF is delayed. Secondly, that the normal circulation of CSF was disturbed by the cerebral ischemia. Two factors seem to be involved in circulation disturbances in CSF, the disturbance of its production and its absorption. Production disturbances in CSF in a state of hypoxia are considered to be caused by: (1) CSF secretion disturbances in the choroid plexus, (2) metabolic disturbances of the choroid plexus, or (3) blood flow disturbances in the choroid plexus. The causes of absorption disturbances in CSF are considered to be the same as those for disturbances in the arachnoid villi. The increased ICP and decreased cerebral blood flow after recirculation of TCI also affect circulation disturbances in the CSF. A third explanation is that the buffering capacity of CSF was insufficient to correct the disturbances. The mechanism of acid-base regulation in CSF is still unclear. Usually, it is considered that changes in lactate and PCO_2 in CSF depend on the metabolism and circulation in the brain [7,8]. HCO_3^- seems to be regulated by active transport [9]. It is generally accepted that lactate and HCO_3^- in blood never enter into the CSF due to the blood brain barrier (BBB) [10]. In the state of cerebral hypoxia, however, the normal functioning of the BBB is disturbed, and lactate and HCO_3^- seem to pass freely back and forth in the blood and CSF [11]. Lastly, cerebral edema, elevated ICP, and decreased regional cerebral blood flow may interfere with the correction of CSF acid-base disturbances.

There have been many studies of the relationships between acid-base disturbance and cerebral function. Posner et al. [12,13] and Ohman and Kozak [14] suggested that acidosis and decreased pH in the CSF can result in unconsciousness. The relationship between decreased pH and cerebral dysfunction is still unclear, but normal circulation and metabolism in the CSF are disturbed by abnormal pH and these disturbances will impair cerebral function. Therefore, correction of CSF acid-base disturbances seems to be important for successful cerebral resuscitation. Siesjö [15]

$NaHCO_3$ has been used commonly as an alkalizing agent in cases of metabolic(lactic) acidosis [16]. We administered 2 and 10 mEq/kg of $NaHCO_3$ for decreased pH in arterial blood, the brain, and CSF after recirculation of TCI. Two mEq/kg of $NaHCO_3$ effectively prevented any decrease in pH. The administration of 10 mEq/kg of $NaHCO_3$ increased arterial, brain, and CSF pH excessively. These results suggested that the administration of 2 mEq/kg of $NaHCO_3$ may be a suitable dose for the correction of metabolic acidosis in these experiments. The administration of $NaHCO_3$ increased arterial, brain and CSF PCO_2. In the 2 mEq/kg of $NaHCO_3^-$ administered group (group II), arterial, brain, and CSF PCO_2 decreased to the control level within 10–25 min after recirculation of TCI, but in the 10 mEq/kg of $NaHCO_3$-administered group (group III), CSF PCO_2 did not decrease to the control level until after 70 min of recirculation.

Recently, there have been many reports disapproving of the use of $NaHCO_3$ in such cases because: (1) it has not improved the survival rate and success of defibrillation in experimental dogs [17], (2) it has exacerbated metabolic acidosis [18], (3) it has lowered the intracellular pH (paradoxical acidosis) and decreased cardiac and cerebral function by the production of CO_2 [13,19,20], (4) it has produced hypernatremia and hyperosmolarity [20,21], (5) it has generated arrhythmia [22], disturbed the separation of HbO_2 [23], and decreased the cerebral blood flow as a results of excessive alkalemia, (6) it has increased the level of lactate by acceleration of glycolysis [24], and it has worsened central venous acidosis [25].

The results of our study appear to differ somewhat from those of other investigators. The following factors were considered as the reasons for amelioration: (1) we produced cerebral ischemia using the AOB catheter method, which maintained good hemodynamics during recirculation, (2) the dogs were well ventilated, (3) the pH and PCO_2 sensors picked up exact changes in the pH and PCO_2 in CSF, and (4) $HCO_3{}^-$ passed through the BBB in the state of hypoxia.

There have been some reports which support the administration of $NaHCO_3$ in a state of acidemia. Sessler et al. [26] administered 10 mEq/kg $NaHCO_3$ for lactic acidosis caused by hypoxia, and noted the effects of $NaHCO_3$ on the decreased pH of blood and cerebral cells. PCO_2 increased by only 10 mmHg after the administration of 10 mEq/kg $NaHCO_3$ and paradoxical acidosis, which occurs following the administration of $NaHCO_3$, was not observed. From the results of their experiments, they concluded that: (1) the elevated PCO_2 may work as a good factor because it increased cardiac output and decreased intracerebral lactate by the dilatation of cerebral vessels, (2) the administration of $NaHCO_3$ may increase serum osmolarity but helps prevent cerebral edema, and (3) in the state of hypoxia, lactate passes through the BBB because of dysfunction of the barrier.

Wikland et al. [27] administered an average of 3.4 mEq/kg of $NaHCO_3$ to piglets. Although the CSF PCO_2 was elevated, the prevention of a fall in CSF pH was observed. Based on these results, they concluded that $NaHCO_3$ passes through the BBB during CPCR. They also maintained that since the elevation of PCO_2 by the administration of $NaHCO_3$ was related to a decrease in cardiac output and ventilation, reinforcement of cardiac output and ventilation would lower the PCO_2 level in CSF.

These reports suggest that if a suitable dose of $NaHCO_3$ were administered, and circulation and ventilation were well maintained, the administration of $NaHCO_3$ would prevent the fall in CSF pH after recirculation of TCI, and the production of CO_2 would not be a problem in the resuscitation of the brain. As many investigators have noted, however, $NaHCO_3$ may not be the best alkalizing agent for the correction of metabolic (lactic) acidosis because of CO_2 production. Bishop [20] emphasized the need for buffer agents, such as tromethamine, monohydrated sodium carbonate etc., which ameliorate metabolic acidosis without production of CO_2. Wiklund et al. [27] recommended the use of tris-buffer, which does not produce CO_2, for the correction of metabolic acidosis. Recently, Filley and Kindig [28] and Bersin and Arieff [29] have reported the use of a new alkalizing agent called "Carbicarb". They indicated that this buffer corrected the lowered pH in extracellular fluid and never produced the intracellular acidosis which is induced by PCO_2.

Summary. Arterial, brain, and CSF acid-base disturbances in dogs were studied before, during and after TCI. pH decreased and PCO_2 increased during and after TCI. The changes in the arterial blood and brain, showed a tendency to recover to the control level during the time of observation. In CSF, PCO_2 showed a tendency to recover to the control level, but pH and HCO_3^- did not show a tendency to recover to the control level during the time of observation. These results suggest the existence of metabolic acidosis in CSF after recirculation of TCI. The intravenous administration of 2 mEq/kg of $NaHCO_3$ to correct the metabolic acidosis in CSF improved the decreased pH in CSF. The administration of 10 mEq/kg of $NaHCO_3$ increased pH in CSF excessively. These results suggest that if circulation and ventilation are well maintained and a suitable dose of $NaHCO_3$ is administered, the administration of $NaHCO_3$ will ameliorate the decreased pH in CSF after recirculation of TCI.

References

1. Siesjö BK, Kjällzuist Ä, Zwetnow N (1968) The CSF lactate/pyruvate ratio in cerebral hypoxia. Life Sci 7:45–52
2. Metzel E, Zimmermann WE (1971) Changes of oxygen pressure, acid-base balance, metabolites and electrolytes in cerebrospinal fluid and blood after cerebral injury. Acta Neurochi (wien) 25:177–188
3. Tabuse H, Fukuda A (1981) Cerebral pathophysiological changes after cardiopulmonary resuscitation (in Japanese). Jpn J Acute Med 5:317–323
4. Kohama A, Nakamura Y, Nakamura M, Yano M, Shibatani K (1984) Continuous monitoring of arterial and tissue PCO_2 Crit Care Med 12:940–942
5. Mitchell RA, Herbert DA, Carman CT (1965) Acid-base constants and temperature coefficients for cerebrospinal fluid. J Appl Physiol 20:27–30
6. Bering EA, Sato O (1963) Hydrocephalus: Changes in formation and absorption of cerebrospinal fluid within the cerebral ventricles. J Neurosurg 20:1050–1055
7. Valenca LM, Shannon DC, Kazemi H (1971) Clearance of lactate from the cerebrospinal fluid. Neurology 21:615–620
8. Kazemi H, Johnson DC (1986) Regulation of cerebrospinal fluid acid-base balance. Physiol Rev 66:953–1037
9. Mima T, Takakura K (1987) Cerebrospinal fluid circulation and electrolytes transport (in Japanese). Jpn J Clin Med 45:263–272
10. Posner JB, Plum F (1967) Independence of blood and cerebrospinal fluid lactate. Arch Neurol 16:492–496
11. Jaraheri S, Clendening A, Papadakis N, Brody JS (1984) pH Changes on the surface of brain and in cisternel fluid in dogs in cardiac arrest. Stroke 15:553–557
12. Posner JB, Swanson AG, Plum F (1965) Acid-base balance in cerebrospinal fluid. Arch Neurol 12:479–496

13. Posner JB, Plum F (1967) Spinal fluid pH and neurologic symptoms in systemic acidosis. N Engl J Med 277:605–613
14. Ohman JL, Kozak GP (1971) The cerebrospinal fluid in diabetic ketoacidosis. New Engl J Med 284:283–290
15. Siesjö BK (1984) Administration of base via the CSF route: A clinically useful treatment of cerebral acidosis? Intensive Crit Care Digest 3:5–9
16. Stewart JSS (1964) Management of cardiac arrest, with special reference to metabolic acidosis. Br Med J 1:476–479
17. Minuck M, Sharma GP (1977) Comparison of THAM and sodium bicarbonate in resuscitation of the heart after ventricular fibrillation in dogs. Anesth Analg 56:38–45
18. Graff HW, Leach W, Arieff Al (1985) Evidence for a detrimental effect of bicarbonate therapy in hypoxic lactic acidosis. Science 227:754–757
19. Berenyi KJ, Wolk M, Killip T (1975) Cerebrospinal fluid acidosis complicating therapy of experimental cardiopulmonary arrest. Circulation 52:319–324
20. Bishop RL, Weisfeldt ML (1976) Sodium bicarbonate administration during cardiac arrest, effect on arterial pH, PCO_2 and osmolality. JAMA 235:506–509
21. Niemann JT, Rosborough JP (1984) Effects of acidemia and sodium bicarbonate therapy in advanced cardiac life support. Ann Emerg Med 13:781–784
22. Lawson NW, Butler GH III, Roy CT (1973) Alkalosis and cardiac arrythmia. Anesth Analg 52:951–965
23. Bellingham AJ, Detter JC, Lenfant C (1971) Regulatory mechanisms of hemoglobin-oxygen affinity in acidosis and alkalosis. J Clin Invest 50:700–706
24. Relman AS (1972) Metabolic consequences of acid-base disorders. Kidney Int 1:347–358
25. Weil MH, Rackow EC, Trevino R, Grundler W, Falk JL, Griffel Ml (1986) Difference in acid-base state between venous and arterial blood during cardio-pulmonary resuscitation. N Engl J Med 315:153–156
26. Sessler D, Mills P, Gregory G, Litt L, James T (1987) Effects of bicarbonate on arterial and brain intracellular pH in neonatal rabbits recovering from hypoxic lactic acidosis. J Pediatr 111:817–823
27. Wiklund LW, Soderberg D, Henneberg S, Rubertsson S, Stjernström H, Groth T (1986) Kinetics of carbon dioxide during cardiopulmonary resuscitation. Crit Care Med 8:1015–1022
28. Filley GF, Kindig NB (1984) Carbicarb, an alkalinizing ion-generating agent of possible clinical usefulness. Trans Am Clin Climatol Assoc 96:141
29. Bersin RM, Arieff AL (1988) Improved hemodynamic function during hypoxia with Carbicarb, a new agent for the management of acidosis. Circulation 77:227–233

12

Effects of Dichloroacetate on Survival Rate, Brain ATP, Lactate, and Water Content Following Cerebral Ischemia in Spontaneously Hypertensive Rats

Toshizoh Ishikawa, Toshiko Ueda, Takefumi Sakabe,
Tsuyoshi Maekawa, and Hiroshi Takeshita[1]

Introduction

Previous reports have demonstrated that brain acidosis associated with the accumulation of lactate may aggravate ischemic-hypoxic brain damage [1,2]. Therefore, measures preventing or attenuating lactate accumulation may be protective against the ischemic brain damage. Dichloroacetate (DCA) is known to activate the pyruvate dehydrogenase (PDH) complex in various tissues in vitro by inhibiting PDH kinase [3] and to reduce serum levels of lactate and pyruvate by its action on PDH [4]. A recent in vivo investigation revealed that DCA significantly lowered the lactate and glucose concentrations of the brain [5] and that it has therapeutic potential in brain ischemia. Subsequently, Biros and colleages [6–8] have demonstrated that the increase in brain lactate content following incomplete ischemia was reduced by both pre- and post-treatment with DCA. However, the effect of DCA on the neurological outcome of animals subjected to brain ischemia has not been investigated.

Therefore, we examined the effect of DCA on survival rate after bilateral carotid artery occlusion (BCAO) in spontaneously hypertensive rats (SHR), and subsequently attempted to relate survival rate with brain adenosine triphosphate (ATP), lactate, and water contents.

Materials and Methods

Animals

Male SHR, 12–16 weeks old and weighing 250–350g, were used. They were fasted overnight preceding the experiment but had free access to tap water.

[1] Department of Anesthesiology-Resuscitology, Yamaguchi University, School of Medicine, 1144 Kogushi, Ube, 755 Japan

Experimental Groups

Two studies were performed, one with permanent BCAO and the other with 3 h BCAO. In the permanent BCAO study, the rats were anesthetized with enflurane (2%) in oxygen, a catheter was inserted into the left femoral vein for drug injection, and both carotid arteries were exposed. After stabilization for 30 min, both carotid arteries were occluded with surgical threads. Immediately after occlusion one group of animals ($\tilde{n} = 18$) received DCA 125 mg/kg (DCA group) intravenously, and the other group ($\tilde{n} = 21$) received saline (untreated group). Thereafter, anesthesia was discontinued and the rats were returned to the cages. Neurologic deficits were assessed by McGraw's stroke index score [9] every 4 h and the survival rate at 24 h was calculated. In the 3 h BCAO study, the rats were anesthetized with enflurane (2%) in oxygen and mechanically ventilated through a tracheostomy tube. Muscle relaxation was produced by d-tubocurarine, 1.5 mg/kg initially, and followed by 0.5 mg/kg as needed. Catheters were inserted into the femoral artery for blood pressure monitoring and blood gas analysis and into the femoral vein for drug injection. PaO_2, $PaCO_2$, pH, and rectal temperature were maintained at 149 ± 52 mmHg, 39 ± 4 mmHg, 7.40 ± 0.04, and $37.0 \pm 0.5°C$, respectively. The rats were divided into three groups: control (sham-operation only), untreated (saline given after BCAO), and DCA (125mg/kg DCA given after BCAO). The rats in the control group were decapitated 3 h after the stabilization without BCAO. The rats with DCA and untreated groups were decapitated at the end of 3 h BCAO. The brains were quickly frozen in liquid nitrogen and stored in a freezer ($- 40°C$).

Analysis of Brain ATP, Lactate, and Water Contents

Two regions of the brain, forebrain (frontal cortex, hippocampus, caudate-putamen) and midbrain (thalamus, central gray, hypothalamus) were dissected and weighed in a glove box ($- 20°C$). For the determinations of brain ATP (luciferine-luciferase method) and lactate (enzymatic method), tissue samples were extracted with 0.1 M perchrolic acid. Brain water content was determined by the dry-wet method.

Statistics

Survival rate at 24 h after permanent BCAO between the groups was tested by Fisher's exact test. Brain ATP, lactate, and water content among the groups were tested by one-way analysis of variance with critical difference testing. $P < 0.05$ was considered significant.

Results

Permanent BCAO Study

After permanent BCAO, most of the untreated rats showed decreased locomotor activity followed by repeated paw lifting within 3–4 h. This corresponded to a stroke index score of 3–10 as shown in Fig. 1. Within 8 h most of these rats

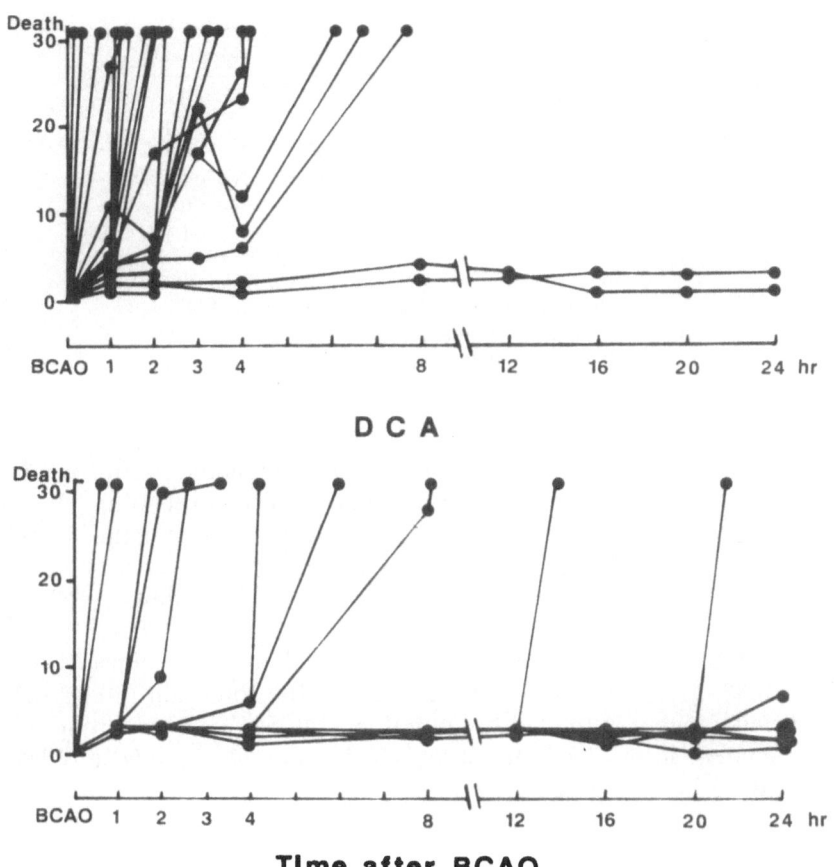

Fig. 1. Neurologic deficits (McGraw's stroke index score) after bilateral carotid artery occlusion (BCAO). The development of neurologic deficits was delayed in the DCA group compared with those in the untreated group in which most of the rats initially showed a decrease in locomotor activity and then showed repeated paw lifting 3–4 h after BCAO

developed seizures associated with respiratory failure and death. In the DCA group, the development of these neurologic deficits was delayed when compared to the untreated group. The survival rate at 24 h of 44% in the DCA group was significantly higher than that in the untreated group (10%).

Three-hour BCAO Study

Table 1 shows the brain ATP and lactate contents in the forebrain and midbrain at the end of 3 h of BCAO. In both the untreated and DCA groups, the ATP

Table 1. Brain ATP and lactate contents at the end of 3 h bilateral carotid artery occlusion

		Forebrain	Midbrain
Control	ATP	4.0 ± 0.7	3.7 ± 0.9
	Lactate	1.8 ± 0.3	2.3 ± 0.1
Untreated	ATP	$0.4 \pm 0.1^*$	$0.4 \pm 0.1^*$
	Lactate	$20.0 \pm 3.9^*$	$16.9 \pm 2.4^*$
DCA	ATP	$0.9 \pm 0.2^*$	$0.9 \pm 0.2^*\#$
	Lactate	$15.4 \pm 3.2^*$	$10.3 \pm 1.1^*\#$

Values (μ mol/wet g) indicate mean $\pm$ SE. *, significantly different from control ($P < 0.05$); #, significantly different between the groups of 3 h BCAO ($P < 0.05$)

content was significantly lower by 76–90% than that in the control group and lactate content was 5- to 10-fold higher than that in the control group. However, the decrease in ATP content and the increase in lactate content in the midbrain were significantly less in the DCA group than in the untreated group. Figure 2 shows the brain water content at the end of 3 h of BCAO. Following BCAO, water contents in the forebrain ($81.4 \pm 0.3\%$) and midbrain ($80.9 \pm 0.5\%$) in the untreated group were higher than those in the control group (78.9 ± 0.5 and $79.0 \pm 0.5\%$, respectively). Water content in the forebrain ($80.0 \pm 0.5\%$) was significantly higher in the DCA group than that in the control group. However, water content in the midbrain in the DCA group was significantly lower than that in the untreated group.

Discussion

The present study demonstrates that DCA, when given immediately after incomplete ischemia, improves the survival rate at 24 h and attenuates the increase in lactate and water contents, and the decrease in ATP occurring in the midbrain following 3 h of ischemia in SHR.

The neurologic deficits after BCAO observed in the untreated group were compatible with those reported by Sadoshima et al. [10]. The cause of death seems to be supratentorial brain edema. Before discussing our results, some comments on brain blood flow after ischemia in the present model may be necessary. We previously reported that regional blood flow in the forebrain and midbrain following 3 h of BCAO decreased to about 20% and 40% of preischemic values, respectively [11]. Thus, the model used in this study may provide a different degree of brain ischemia, in the forebrain and midbrain. Because of this the results of the present study are reported for the forebrain and midbrain separately. The sampling time (following a 3 h-ischemic period) was selected because in this model, lactic acidosis in the whole brain develops within the first hour [12]

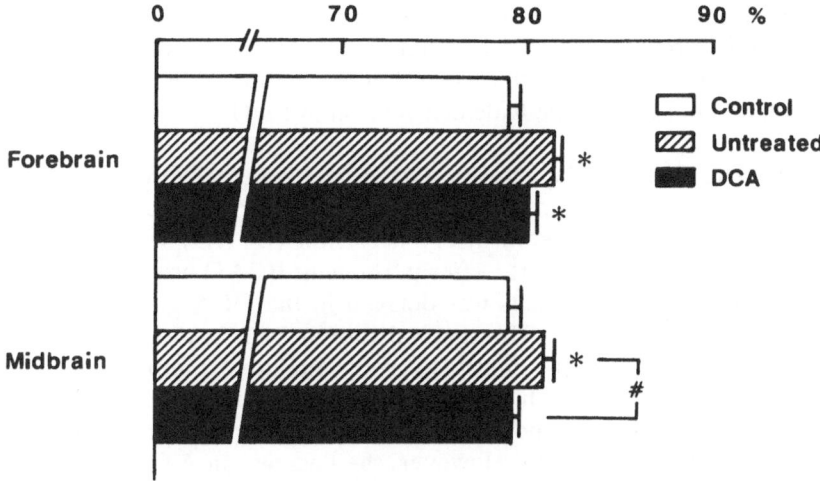

Fig. 2. Brain water contents at the end of 3 h BCAO. In the midbrain, water content in the DCA group was significantly lower than in the untreated group. *, significantly different compared to the control group ($P < 0.05$); #, significantly different between the untreated and DCA groups ($P < 0.05$)

and results in severe cell damage at 3 h [13]. The dose of 125 mg/kg dichloroacetate was chosen because this amount is known to activate the PDH complex and reduce brain lactate content in normal rats [5].

Most of the untreated rats died within 8 h after BCAO in the present study. In contrast, the development of neurologic deficits was significantly delayed in the DCA group, resulting in a higher survival rate in the 24 h time period studied. It is reasonable to believe that the improvement of survival rate was due to DCA's metabolic effect on the brain because this substance principally acts as an enzyme inhibitor which can cross the blood-brain barrier [5]. We therefore examined the metabolic effects of DCA, especially focusing on the brain ATP and lactate contents at 3 h after BCAO. It is well known that severe lactic acidosis causes edema due to swelling of dendrites and glial cells [14]. Our results demonstrated that lactate accumulation and ATP reduction were attenuated, and the increase in water content was also reduced with DCA. However, these favorable effects of DCA were only significant in the midbrain. The most likely explanations for the regional differences are that any drug reaching the forebrain where the ischemia is most severe may not be sufficient to reduce the lactate accumulation, or that the ischemia in the forebrain may be so severe that any effect of DCA cannot be undetected. From these results, it is suggested that the DCA's effects depend on the severity of ischemia. However, the possibility that DCA has differing effects on the different brain tissues can not be excluded.

As stated before, acidosis aggravates neurologic damages following brain ischemia-hypoxia [1,2]. Furthermore, Rehncrona et al. [15] clearly showed that excessive accumulation of lactate caused deterioration of brain energy state.

Therefore, a therapy which reduces acidosis induced by an accumulation of lactate may ameliorate neurological damages. DCA may be a suitable drug to achieve this goal, but its side-effects on other organs and/or tissues must be carefully evaluated because of the wide distribution of PDH.

Summary. The effects of the activation of the pyruvate dehydrogenase complex by dichloroacetate (DCA) on survival rate, brain ATP, lactate, and water contents following bilateral carotid artery occlusion (BCAO) were examined in spontaneously hypertensive rats. After permanent BCAO was achieved the development of neurologic deficits was delayed in the DCA group compared to that in the untreated group. The survival rate at 24 h in the DCA group was significantly higher than in the untreated group. In both the untreated and DCA-treated groups subjected to 3 h BCAO, the ATP content was significantly lower than in the control group, and lactate content was higher than in the control group at the end of 3 h BCAO. However, the decrease in ATP content and the increase in lactate content in the midbrain were significantly less in the DCA group than in the untreated group. Water content in the midbrain in the DCA group was significantly lower than in the untreated group. These results suggest that the improvement of survival rate with DCA may be related, at least in part, to a reduction in lactate accumulation and water content of the ischemic brain.

References

1. Myers RE (1979) Lactic acid accumulation as a cause of brain edema and cerebral necrosis resulting from oxygen deprivation. In: Korobkin R, Guilleminault G (eds) Advances in Perinatal Neuroloyy, Spectrum, New York, pp 85–114
2. Siesjö BK (1981) Cell damage in the brain: A speculative synthesis. J Cereb Blood Flow Metab 1:155–185
3. Whitehouse S, Randle PJ (1973) Activation of pyruvate dehydrogenase in perfused rat heart by dichloroacetate. Biochem J 134:651–653
4. Evans OB, Stacpoole PW (1982) Prolonged hypolactatemia and increased total pyruvate dehydrogenase activity by dichloroacetate. Biochem Pharmacol 31:1295–1300
5. Kuroda Y, Toshima K, Watanabe T, Kobashi H, Ito M, Takeda E, Miyao M (1984) Effects of dichloroacetate on pyruvate metabolism in rat brain in vivo. Pediatr Res 18:936–938
6. Biros MH, Dimlich RVW, Barsan WG (1986) Postinsult treatment of ischemia-induced cerebral lactic acidosis in the rat. Ann Emerg Med 15:397–304
7. Biros MH, Dimlich RVW (1987) Brain lactate during partial global ischemia and reperfusion: Effect of pretreatment with dichloroacetate in a rat model. Am J Emerg Med 5:271–277
8. Kaplan J, Dimlich RVW, Biros MH (1987) Dichloroacetate treatment of ischemic cerebral lactic acidosis in the fed rat. Ann Emerg Med 16:298–304
9. McGraw CP (1977) Experimental cerebral infarction. Effects of pentobarbital in Mongolian gerbils. Arch Neurol 34:334–336
10. Sadoshima S, Nakatomi Y, Fujii K, Oobashi H, Ishitsuka T, Ogata J, Fujishima M (1988) Mortality and histological findings of the brain during and after cerebral ischemia in male and female spontaneously hypertensive rats. Brain Res 454:238–243

11. Ishikawa T, Sano T, Kuroda Y, Soejima Y, Maekawa T, Sakabe T, Takeshita H (1987) Effects of buflomedil on survival rate, local cerebral blood flow and glucose utilization after cerebral ischemia in spontaneously hypertensive rats (abstract in English). Folia Pharmacol Jpn 90:303–312
12. Fujishima M, Omae T (1976) Cerebral lactate, pyruvate and ATP concentrations, and arterial acid-base balance at various time intervals following bilateral carotid artery occlusion in normotensive and spontaneously hypertensive rats. Acta Neurol Scand 54:13–21
13. Ogata J, Fujishima M, Tamaki K, Nakatomi Y, Omae T (1977) An ultrastructural study of developing cerebral infarction following bilateral carotid artery occlusion in spontaneously hypertensive rats. Acta Neuropathol (Berlin) 40:171–177
14. Siesjö BK (1985) Acid-base homeostasis in the brain: Physiology, chemistry, and neurochemical pathology. In: Kogure K, Hossmann K-A, Welsh FA (eds) Progress in brain research. Elsevier, Amsterdam, pp 121–153
15. Rehncrona S, Rosen I, Siesjö BK (1981) Brain lactic acidosis and ischemic cell damage : 1. Biochemistry and neurophysiology. J Cereb Blood Flow Metab 1:297–311

13

Effects of Heparin-Urokinase, Diazepam, or Nimodipine on Brain Damage Induced by Complete Global Brain Ischemia

Hidenori Hashimoto, Masanori Kondo, Yoshimasa Takeda,
Masaki Sato, Shino Oka, Takayuki Okamoto, Masahiro Ohkawa,
Yutaka Yaida, Yutaka Shimoda, Toshiko Ikeda,
Hidehiko Yatsuzuka, and Futami Kosaka[1]

Introduction

The outcome of brain insults induced by ischemia is influenced by two major factors, recirculation disturbance and metabolic derangement. Besides the reactive hyperemia, two types of recirculation disturbance can be distinguished; the no-reflow phenomenon [1] and delayed post-ischemic hypoperfusion [2–5], those have been shown to add a secondary ischemic insult to the tissue. Excessive release of excitatory neurotransmitters (e.g., glutamate, aspartate) during ischemia has been shown to play an important role in metabolic derangement, resulting in selective hyperexcitability after ischemia which leads to postsynaptic ionic influxes (e.g., sodium, calcium), causing neuronal damage [6–8].

In this study, we used heparin-urokinase (H-U) to improve disturbed recirculation, nimodipine to reduce postsynaptic calcium influx in addition to improvement of disturbed recirculation, and diazepam to lessen hyperexcitability caused by excessive release of excitatory neurotransmitters. We administered these drugs in the post-ischemic period to make this study relevant to the human post-cardiac-arrest setting. We designed this study to determine whether or not these drugs ameliorate brain damage induced by complete global brain ischemia in dogs.

Materials and Methods

Fifty-eight fasting anesthetized adult mongrel dogs weighing 7–10 kg were studied. Dogs were divided into two groups for the acute and chronic studies. In the acute study, 26 dogs were used to study the effects of.H-U, diazepam, or nimodipine on cerebral blood flow (CBF), electroencephalogram (EEG), intracranial

[1] Department of Anesthesiology and Resuscitology Okayama University Medical School,
2-5-1 Shikata-cho, Okayama, 700 Japan

Table 1. Drugs administered after the restoration of circulation to the brain

Group	Drug	Initial administration	Continuous infusion
H-U	Heparin	100 I.U./kg	0.25 I.U./kg per min
	Urokinase	3000 I.U./kg	8.3 I.U./kg per min for 6 h
		15 min after ischemia	
D	Diazepam	0.5 mg/kg	2.5 μg/kg per min
		15 min after ischemia	* acute study for 6 h
			* chronic study for 24 h
N	Nimodipine	10 μg/kg	1.0 μg/kg per min for 6 h
		15 min after ischemia	
C	None		

pressure (ICP), and other hemodynamic parameters for the first 7 h after ischemia. In the chronic study, 32 dogs were evaluated for neurologic outcome up to seven days after ischemia.

In each study, dogs were divided into four groups according to the drugs administered after ischemia: H-U-treated group (group H-U), diazepam-treated group (group D), nimodipine-treated group (group N), and control group (group C). Group H-U received heparin 100 I.U./kg and urokinase 3000 I.U./kg, i.v. 15 min after ischemia followed by a continuous infusion of heparin 0.25 I.U./kg/min and urokinase 8.3 I.U./kg/min for 6 h. Group D received diazepam 0.5 mg/kg, i.v. 15 min after ischemia followed by a continuous infusion of 2.5 μg/kg per min for 6 h in the acute study or for 24 h in the chronic study. Group N received nimodipine 10 μg/kg, i.v., 15 min after ischemia followed by a continuous infusion of 1.0 μg/kg per min for 6 h. Group C did not receive any of these drugs (Table 1).

CBF was measured using a hydrogen clearance method. EEG was monitored with bipolar parietal and occipital leads. ICP was measured with a subdurally placed transducer. Hemodynamic parameters were measured with a Swan-Ganz catheter inserted via the femoral vein. In the chronic study, dogs were under intensive care with controlled ventilation for the first 24 h after ischemia and then weaned successfully. Thereafter the neurologic outcome was evaluated.

In the acute study, two dogs in group H-U which suffered excessive hemorrhage were excluded. Additionally, in the chronic study, eight dogs—two dogs in each group who had died within 24 h after ischemia—were excluded.

Anesthesia was induced with ketamine 20 mg/kg and atropine sulfate 0.05 mg/kg, i.m., and maintained with 1.5% halothane in oxygen. Ventilation was controlled to keep $PaCO_2$ 30–35 mmHg. Body temperature was measured with an esophageal thermometer and was maintained at 37–38°C. Anesthesia was discontinued at least 20 min before the ischemic insult in order to avoid the influence of the anesthesia itself.

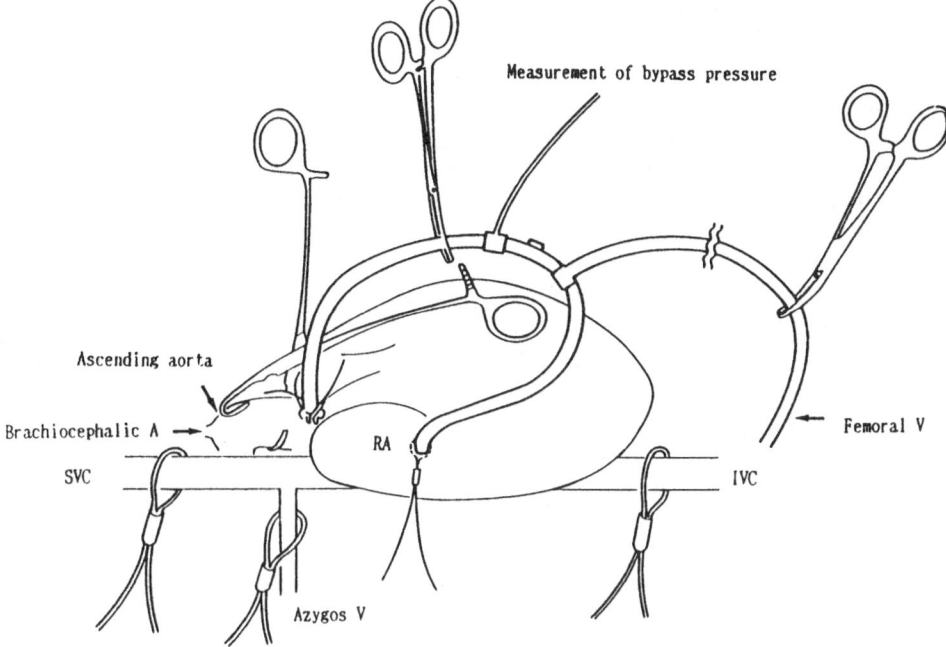

Fig. 1. Complete global brain ischemia for 18 min was produced by clamping ascending aorta with aorto-atrial and aorto-femoral bypass formation. Clamping of the vessels was done in the sequence of IVC, ascending aorta, SVC, and azygous vein. (Ao, aorta; RA, right atrium; IVC, inferior vena cava; SVC, superior vena cava)

Induction of Ischemia

All surgical procedures were done by aseptic technique. An 18 min period of complete global brain ischemia was produced by clamping the ascending aorta with aorto-atrial and aorto-femoral bypass formation (Fig. 1). Umbilical tapes were placed around the superior vena cava (SVC), inferior vena cava (IVC), and azygous vein. The aortic tip of the bypass tubing was inserted into the ascending aorta, the atrial end was inserted into the right auricle, and the femoral end was inserted into the femoral vein. Clamping of the vessels was achieved in the sequence of IVC, ascending aorta, SVC, and azygous vein in order to avoid overfilling the heart and congesting the brain. The bypass tubing was opened as soon as the aorta was clamped. Without the aorto-femoral bypass, pulmonary edema often followed ischemia as a result of high pressure in the pulmonary circulation. To avoid this problem, we added an aorto-femoral bypass to the aorto-atrial bypass circuit. Whenever pulmonary arterial pressure increased, this bypass was temporarily opened to reduce pulmonary arterial pressure. Additionally, whenever intra-bypass pressure decreased, IVC was temporarily opened to gain proper the filling volume. With this method, we could maintain pulmonary arterial pressure below 40 mmHg in order to avoid pulmonary edema after ischemia. Intra-bypass pressure was maintained from 60–80 mmHg. In this preparation, it is not necessary to use any catecholamines during and after bypass due to good preservation of myocardial function. We could almost ignore extra-cerebral complications which have often influenced post-ischemic brain damage. Functions of the brain stem were well preserved to survive without respiratory and hemodynamic support. On the other hand, neurologic functions were critically deteriorated. We could therefore evaluate the effect of the drug regimens on neurologic outcome more precisely.

For statistical comparison of the neurologic outcomes between dogs in each group (group H-U, group D, and group N) and control group (group C), the Wilcoxon's U test was used. For all other comparisons between dogs in each group, unpaired Student's t-test was used. P values less than 0.05 were considered statistically significant. All data were presented as mean ± SE.

Results

Just after aortic clamping, the mean arterial pressure (MAP) fell to near zero and its waveform disappeared instantaneously. EEG became isoelectric within 26 sec after the initiation of ischemia. Following the release of aortic clamping, MAP returned to the pre-ischemic value within 1 min and remained over the pre-ischemic value for several minutes. ICP increased up to 300%–400% of the pre-ischemic value immediately after the initiation of recirculation and then decreased to the pre-ischemic value within 20–40 min. There was no significant difference in the timing of EEG return among the groups. A burst-suppression pattern appeared within 40–90 min after the initiation of recirculation. However, a significant decrease in the frequency of burst-suppression pattern and in the amplitude of the EEG were observed in group D.

Figure 2 shows the changes in heart rate (HR), MAP, and cardiac index (CI) for the first 7 h after ischemia in all groups. There were no significant differences among the groups except for group N. In group N, HR and MAP were lower, although CI was well preserved compared with that in other groups.

Figure 3 shows the changes in percent of CBF in all groups. In group C, during the first 20–40 min following reperfusion, there was a reactive hyperemia followed by a relentless fall off to a blood flow of about 40%–50% of the pre-ischemic value. This delayed post-ischemic hypoperfusion had been observed throughout the study. However, in groups H-U and group N, CBF was significantly higher than that in group C, whereas in group D, CBF was not improved compared with group C.

The neurologic outcome was evaluated according to the neurologic deficit score (NDS) established by Safar et al. [9]. The NDSs in all groups are summarized in Fig. 4. In group C, the NDS was approximately 40–50, and the typical one was opisthotonic, responding only to noxious stimuli. All dogs received intravenous hyperalimentation (IVH) throughout the study. In the same group, one of six dogs died due to aspiration pneumonia on day 5 after ischemia and

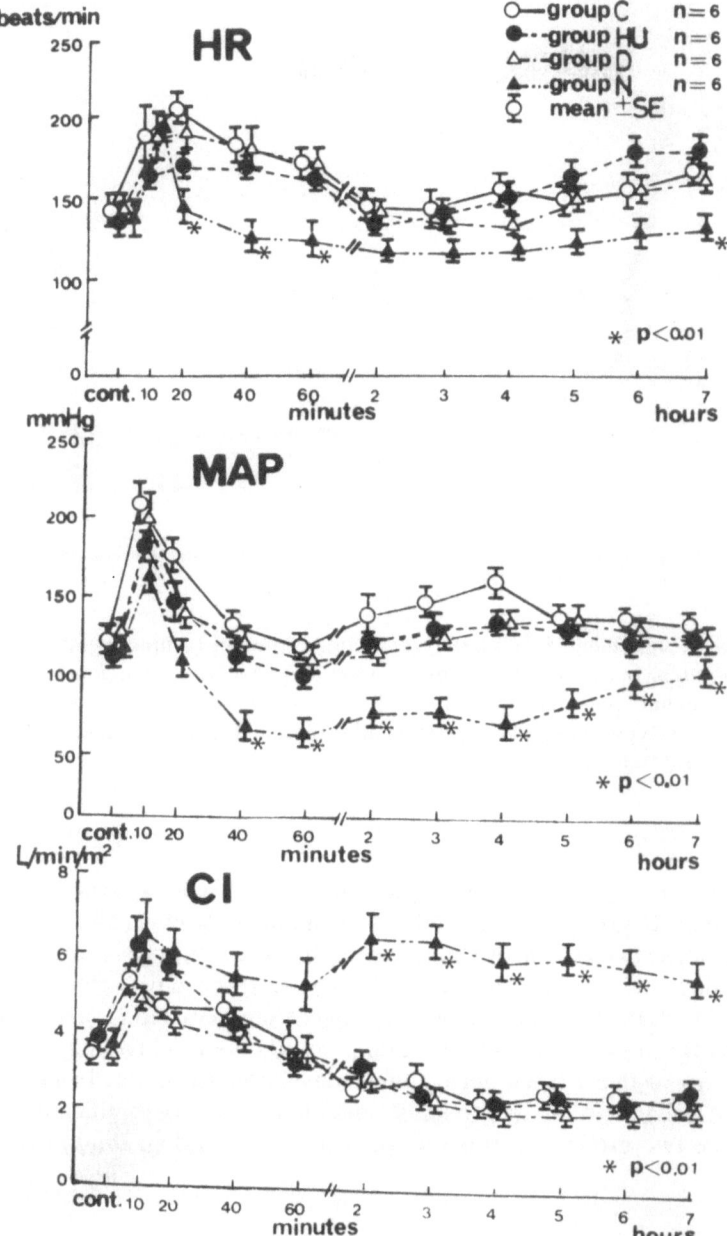

Fig. 2. Changes in heart rate (HR), mean arterial pressure (MAP), and cardiac index (CI) after ischemia for the first 7 h in all groups. Each value represents mean ± SE. A statistically significant decrease in HR was found during the period starting from 20 min through 60 min and 7 h after ischemia. Significant decrease in MAP and increase in CI were also observed during the period from 40 min through 7 h and from 2 hours through 7 h after ischemia, respectively. The *asterisks* indicate values that are significantly different from control at $P < 0.01$

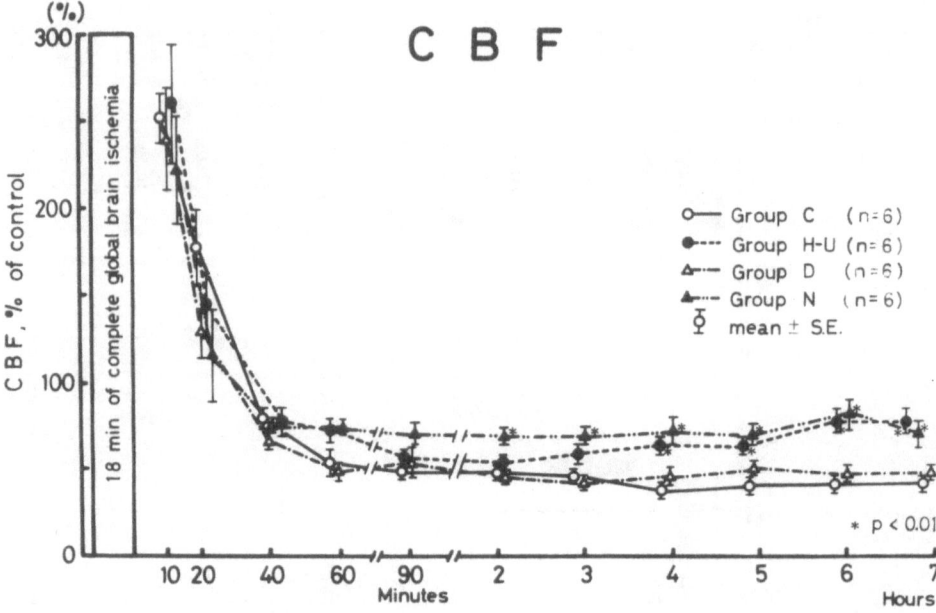

Fig. 3. CBF values (mean ± SE) before and after 18 min of complete global brain ischemia in all groups. In groups H-U (*closed circle*) and N (*closed triangle*), CBF remained
significantly higher than that in groups C (*open circle*) and D (*open triangle*) after the
initial period of hyperemia. The *asterisks* indicate values that are significantly different
from control at $P < 0.01$

two of remaining five died due to septicemia caused by long term IVH on day 6
after ischemia. In group H-U, the NDS was significantly lower than that in group
C, and all dogs survived throughout the study: three out of six dogs could take
food two days after ischemia and could get up three days after ischemia. In
group D, the NDS was also significantly lower than in group C, and all survived
throughout the study: three out of six dogs could take food two days after ischemia, and two of these three get up three days after ischemia. In group N, the
NDS was similar to that in group C, and there was no significant difference
between the two groups. In this group, two out of six dogs died at day 2 after
ischemia.

Discussion

The no-reflow phenomenon [1] and the post-ischemic hypoperfusion [3–6] are
major complications of post-ischemic recirculation [1,2]. The no-reflow phenomenon is the combined result of many factors, such as increased viscosity of
blood [10], disseminated intravascular coagulopathy [11], and others [12–15].
The delayed post-ischemic hypoperfusion is mainly due to an increase in calcium

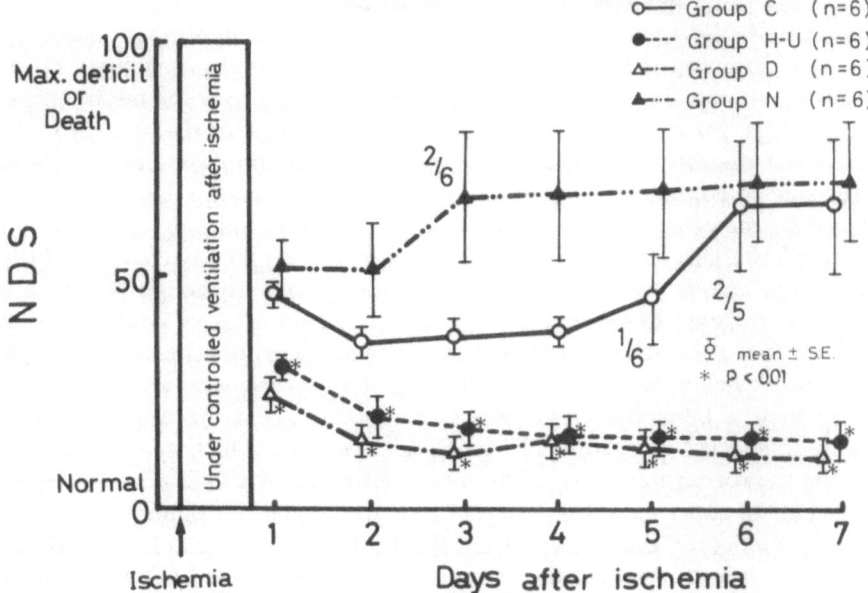

Fig. 4. NDS values (mean ± SE) up to 7 days after 18 min of complete global brain ischemia. In group C (*open circle*), one dog died (1/6) on day 5, and two died (2/5) on day 6 after ischemia. In group N (*closed triangle*), two dogs died (2/6) on day 3 after ischemia. In groups H-U (*closed circle*) and D (*open triangle*), neurologic outcome was significantly improved compared with those in the other two groups. The *asterisks* indicate values that are significantly different from control at $P < 0.01$

influx into the vascular smooth muscle [16] and is independent of the no-reflow phenomenon [6].

Heparin acts as an anticoagulant in the presence of antithrombin III. Urokinase has been known to be a thrombolytic agent due to activation of plasminogen. In this study, H-U improved both the post-ischemic CBF and neurologic outcome. This result suggests that H-U improved both the post-ischemic CBF and neurologic outcome due to amelioration of the no-reflow phenomenon, which may play an important role in the brain damage induced by ischemia and may have some effect in the delayed post-ischemic hypoperfusion. This hypothesis needs support from histopathological studies.

Nimodipine, a potent calcium entry-blocking agent, has been shown to improve post-ischemic hypoperfusion in dogs [17] and cats [18] when administered pre-ischemia, and in dogs when administered post-ischemia [19,20]. However it has not been shown whether post-ischemic nimodipine improves neurologic outcome. In previous studies, improvement of the neurologic outcome had been demonstrated in dogs when administered pre-ischemia [17]. Similar results had been obtained in a primate model when administered post-ischemia [21]. Contrarily, in a recent study, failure of nimodipine to prevent ischemic neuronal damage has been demonstrated in a rat model [22]. In this study, nimodipine administered post-ischemia also failed to improve the neurologic outcome in spite of improvement in the post-ischemic CBF. Several possibilities must be considered to explain why the nimodipine failed to improve the neurologic outcome in dogs. First, during severe ischemia such as that of the present study, a massive and possibly irreversible influx of calcium had occurred, and it might have be too late to administer nimodipine in order to reduce the deleterious effects induced by the calcium influx. Second, the dose of nimodipine administered in this study (10 μg/kg, followed by 1.0 μg/kg per min) might have been too little to achieve full effects on cerebral calcium binding sites, although the dose was enough to decrease MAP significantly. Third, the duration of administration in this study (6 h after ischemia) might have been too short to achieve enough effect on cerebral calcium binding sites or the delayed post-ischemic hypoperfusion.

Diazepam, a benzodiazepine (BDZ) derivative, seems to exert its action by binding to specific receptors [23,24], BDZ receptors, which appear to interact with the neurotransmission mediated by γ-aminobutyric acid (GABA). Diazepam seems to exert inhibitory effects on neuronal cells by binding BDZ receptors and mimicking the effects of GABA on Cl$^-$ ionophore [25]. GABAergic neurons are not equally sensitive to ischemia throughout the brain [26], and once GABAergic neurons are damaged profoundly, a denervation supersensitivity phenomenon [27] seems to occur. These data suggest that diazepam may sufficiently offer its inhibitory effect on the neuronal cells , independent of the number of GABAergic neurons which have escaped ischemic damage. Recently, excitotoxins which aggravate the brain damage induced by ischemia have been studied in delayed neuronal death. Excitatory amino acids, such as glutamate and aspartate, are the major substances among this group. To avoid this hyperexcitability, a number of drugs, especially anti-glutaminergics, have been investigated, and the effects of these drugs on neuronal cells which had suffered ischemic insult have been studied [28–30]. In this study, we used diazepam to lessen hyperexcitability following ischemia, and improvement of the neurologic outcome was obtained. This suggests that diazepam might improve the neurologic outcome due to its inhibitory effect on neuronal cells, and post-ischemic hyperexcitability may play an important role in brain damage induced by ischemia. Additional experiments are needed to evaluate this hypothesis.

Summary. The efficacy of heparin-urokinase, diazepam, and nimodipine in preventing ischemic brain damage was evaluated in 58 dogs subjected to 18 min of brain ischemia by a method of clamping the ascending aorta with aorto-right atrial and aorto-femoral vein bypassses. Dogs were divided into three groups according to the drugs, administered 15 min in bolus after restoration of circulation to the brain followed by a continuous infusion for 6 h: group H-U (heparin 100 I.U./kg and urokinase 3000 I.U./kg, followed by heparin 0.25 I.U./kg per min and urokinase 8.3 I.U./kg per min); group D (diazepam 0.5 mg/kg, followed by 2.5 μg/kg per min); group N (nimodipine 10 μg/kg, followed by 1.0 μg/kg per min), and the control group C (none). Post-ischemic cerebral blood flow and hemodynamic parameters were measured for 7 h after ischemia in 26 dogs. The neurologic outcome was evaluated for seven days after ischemia in 32 dogs. H-U improved both the post-ischemic CBF and neurologic outcome. Diazepam improved the neurologic outcome without any effect on the post-ischemic CBF. Nimodipine failed to improve the neurologic outcome in spite of improvement of the post-ischemic CBF. Therefore, this study revealed a beneficial effect of both H-U and diazepam, but not of nimodipine, on the neurologic outome when administered in the post-ischemic period.

References

1. Ames A III, Wright RL, Kowada M, Thurston JM, Majno G (1968) Cerebral ischemia: II. The no-reflow phenomenon. Am J Pathol 52:437–453
2. Hossmann KA, Lechtape-Gruter H, Hossmann V (1973) The role of cerebral blood flow for the recovery of the brain after prolonged ischemia. J Neurol 204:281–299
3. Nemoto EM, Snyder JV, Carroll RG, Morita H (1975) Global ischemia in dogs: Cerebrovascular CO_2 reactivity and autoregulation. Stroke 6:425–431
4. Siesjö BK (1978) Brain energy metabolism. Wiley, New York
5. Miller CL, Lampard DG, Alexander K, Brown WA (1980) Local cerebral blood flow following transient cerebral ischemia: I. Onset of impaired reperfusion within the first hour following global ischemia. Stroke 11:534–541
6. Kirino T (1982) Delayed neuronal death in the gerbil hippocampus following ischemia. Brain Res 239:57–69
7. Pulsinelli WA, Brierley JB, Plum F (1982) Temporal profile of neuronal damage in a model of transient forebrain ischemia. Ann Neurol 11:491–498
8. Suzuki R, Yamaguchi T, Li CL, Klatzo I (1983) The effects of 5 minute ischemia in Mongolian gerbils: II. Changes of spontaneous neuronal activity in cerebral cortex and CA1 sector of hippocampus. Acta Neuropathol (Berl) 60:217–222.
9. Safar P, Stezoski W, Nemoto EM (1976) Amelioration of brain damage after 12 minutes' cardiac arrest in dogs. Arch Neurol 33:91–95
10. Fischer EG (1973) Impaired perfusion following cerebrovascular stasis. Arch Neurol 29:361–36
11. Hossman KA, Hossman V (1977) Coagulopathy following experimental cerebral ischemia. Stroke 8:249–254
12. Cantu RC, Ames A III, DiGiacinto G, Dixon J (1969) Hypotension: A major factor limiting recovery from cerebral ischemia. J Surg Res 9:525–529
13. Arsenio-Nunes ML, Hossmann KA, Farkas-Bargeton E (1973) Ultrastructural and histochemical investigation of the cerebral cortex of cat during and after complete ischaemia. Acta Neuropathol (Berl) 26:329–344
14. Zimmermann V, Hossmann V, Hossmann K-A (1975) Intracranial pressure after prolonged cerebral ischemia. In: Lundberg N, Ponten U, Brock M (eds) Intracranial pressure II. Springer, Berlin, Heidelberg, New York, pp 177–182
15. Dietrich WD, Busto R, Ginsberg MD (1984) Cerebral endothelial microvilli: Formation following global forebrain ischemia. J Neuropathol Exp Neurol 43:72–83
16. Hoffmeister F, Kazda S, Krause HP (1979) Influence of nimodipine (BAY e 9736) on the post-ischemic changes of brain function. Acta Neurol Scand 60 (Suppl 72):358–359

17. Steen PA, Newberg LA, Milde JM, Michenfelder JD (1983) Nimodipine improves cerebral blood flow and neurologic recovery after complete cerebral ischemia in the dog. J Cereb Blood Flow Metab 3:38–43
18. Kazda S, Garthoff B, Krause HP, Schlosmann K (1982) Cerebrovascular effects of the calcium antagonistic dihydropyridine derivative nimodipine in animal experiments. Arzneimittelforschung 32:331–338
19. Steen PA, Newberg LA, Milde JH, Michenfelder JD (1984) Cerebral blood flow and neurologic outcome when nimodipine is given after complete cerebral ischemia in the dog. J Cereb Blood Flow Metab 4:82–87
20. Milde LN, Milde JH, Michenfelder JD (1986) Delayed treatment with nimodipine improves cerebral blood flow after complete cerebral ischemia in the dog. J Cereb Blood Flow Metab 6:332–337
21. Steen PA, Gisvold SE, Milde JH, Newberg LA, Scheithauer BW, Lanier WL, Michenfelder JD (1985) Nimodipine improves outcome when given after complete cerebral ischemia in primates. Anesthesiology 62:406–414
22. Vibulsresth S, Dietrich WD, Busto R, Ginsberg MD (1987) Failure of nimodipine to prevent ischemic neuronal damage in rats. Stroke 18:210–216
23. Braestrup C, Squires RF (1978) Pharmacological characterization of benzodiazepine receptors in the brain. Eur J Pharmacol 78:263–270
24. Young WS III, Niehoff D, Kuhar MJ, Lippa AS (1981) Multiple benzodiazepine receptor localization by light microscopic radiohistochemistry. J Pharmacol Exp Ther 216:425–430
25. Bowery NG, Price GW, Hudson AL, Hill DR, Wilkin GP, Turnbull MJ (1984) GABA receptor multiplicity, visualization of different receptor types in the mammalian CNS. Neuropharmacol 23:219–231
26. Francis A, Pulsinelli W (1982) The response of GABAergic and cholinergic neurons to transient cerebral ischemia. Brain Res 243:271–278
27. Kuriyama K, Kurihara E, Ito Y, Yoneda Y (1980) Increase in striatal [^{3}H] muscimol binding following intrastriatal injection of kainic acid: A denervation supersensitivity phenomenon. J Neurochem 35:343–348
28. Hallmayer J, Hossmann KA, Mies G (1985) Low dose of barbiturates for prevention of hippocampal lesions after brief ischemic episodes. Acta Neuropath (Berl) 68:27–31
29. Kirino T, Tamura A, Sano K (1986) A reversible type of neuronal injury following ischemia in the gerbil hippocampus. Stroke 17:455–459
30. Wong EHF, Kemp JA, Priestley T, Knight AR, Woodruff GN, Iversen LL (1986) The anticonvulsant MK-801 is a potent N-methyl-D-aspartate antagonist. Proc Natl Acad Sci USA 83:7104–7108

14
Cerebral Protection With Barbiturates, Nizofenone and Steroids

Shinya Manaka[1], Takaaki Kirino[2], and Mamoru Sasaki[3]

Introduction

Steroids and barbiturates have been considered to be very reasonable drugs for cerebral protection from ischemic or traumatic insult. However, controversies on the effectiveness of these drugs have been growing [1–3]. On one hand there are experimental studies which strongly support the utility of these drugs for cerebral trauma or ischemia, but recent clinical studies based on randomized well-controlled trials have failed to show any advantages for these drugs [4–8]. The initial clinical enthusiasm for the widespread application of barbiturates to protects the brain from head injury, circulatory arrest, asphyxia, or ischemia has almost disappeared. Clinical and experimental data showed that complete global ischemic injury has not been favorably altered by barbiturate coma therapy, while barbiturates do appear to protect the brain damage from focal or incomplete ischemia/hypoxia [2,3]. If barbiturates are given before permanent occlusion, the size of the ensuing infarct is significantly reduced [9]. Barbiturate protection from cerebral ischemia/hypoxia is, at least partially, the result of a depressed metabolic demand for oxygen. Astrup demonstrated that the metabolic depression resulted from inhibition of neurotransmission [10].

Barbiturates cause vasoconstriction that results in the rapid fall of intracranial pressure (ICP) [11]. This effect of barbiturates on ICP is established and has been used clinically, particularly in head-injured patients. Among the many arguments concerning barbiturates and steroids, three topical themes are focused upon and reviewed in this paper: first, prevention of "delayed neuronal death" with barbiturates; second, the promise of a cerebral protector,

Department of Neurosurgery,
[1] Ichihara Hospital, Teikyo University, School of Medicine, 3426-3 Anesaki, Ichihara, 299-01 Japan
[2] Teikyo University School of Medicine, 2-11-1 Kaga Itabashi-ku, Tokyo, 173 Japan
[3] Teraoka Memorial Hospital, Shinichi-machi, Hiroshima, 729-31, Japan

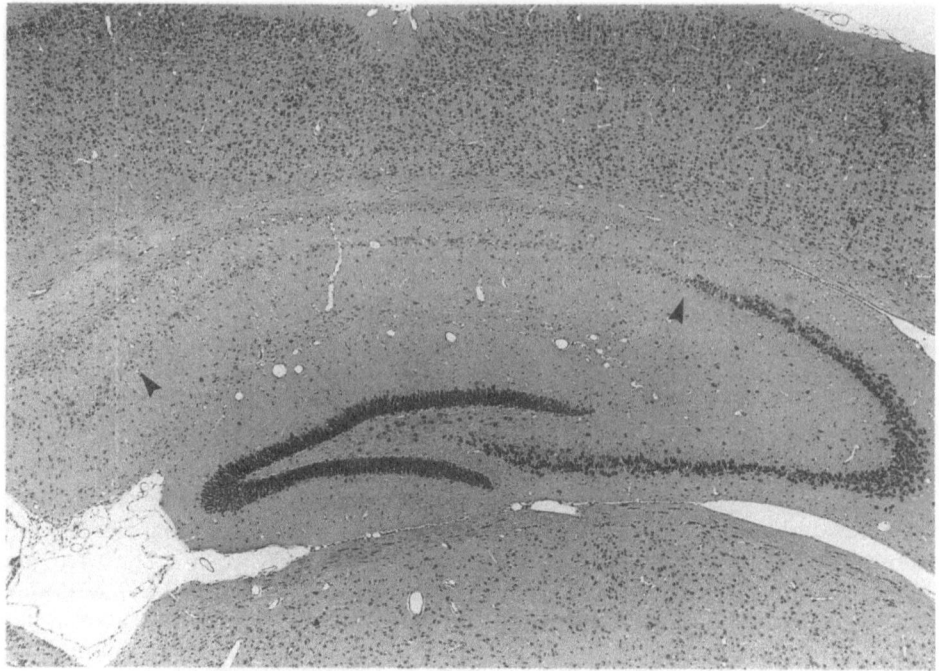

Fig. 1. The dorsal hippocampus of the gerbil. Seven days after 5 min of ischemia, the neurons in the CA1 sector (between the *arrowheads*) were extensively destroyed. Hematoxylin and eosin x20

nizofenone, in place of barbiturates; and third, optimal timing of steroid administration for an effect upon cerebral injury.

Pharmacological Prevention of Delayed Neuronal Death

The Mongolian gerbil is known to develop delayed neuronal death in the hippocampus following brief cerebral ischemia [12]. The effect of pentobarbital on this slow process of neuronal damage was investigated [13]. Adult Mongolian gerbils weighing 60–80 gm were used. The carotid arteries on both sides were occluded for 5 min. Seven days after brief ischemia, the neurons in the CA1 sector were extensively destroyed (Fig. 1 between the *arrowheads*). In 8 unoperated normal gerbils, the average neuronal cell density of the CA1 sector was more than 200/mm. The neuronal density of the abnormal, ischemic CA1 sector was divided into five grades; the neuronal density of grade 0, more than 160/mm; grade 1, from 12–160/mm; grade 2, from 80–120/mm; grade 3, from 40–80/mm; and grade 4, under 40/mm. Neuronal density over 160/mm is assumed to be within normal limits.

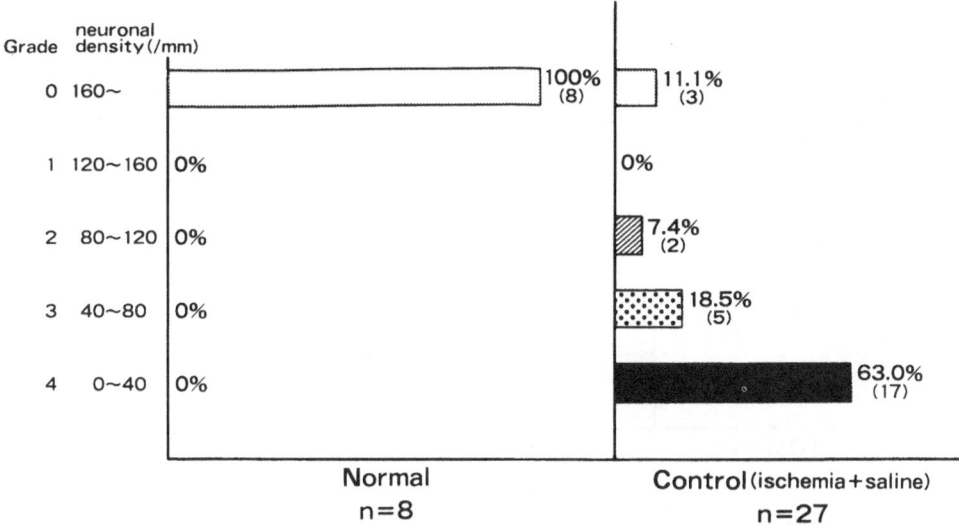

Fig. 2. Neuronal density (per mm) in the CA1 sector of unoperated normal gerbils (*left*) and that of control group subjected to 5 min of ischemia and saline injection (*right*)

The Effect of Pentobarbital on Delayed Neuronal Death

The neuronal density of all normal animals belonged to the group of grade 0, whereas in the untreated ischemic group the neuronal loss in the CA1 sector was remarkable. Only 11% of gerbils belonged to grade 0, whereas 63% fell into grade 4 (Fig. 2). Various doses of pentobarbital were given immediately after ischemia and neuronal density was measured. In animals treated with pentobarbital of 20mg/kg and 40mg/kg, the loss of the neuronal density was remarkably prevented, whereas in a dose of 10 mg/kg, the protective effect was not significant (Fig. 3). When pentobarbital of 40mg/kg was given after a 1 or 2 h delay following ischemia, there was no significant improvement of the neuronal density in the CA1 subfield.

Akiten and Schiff demonstrated that in hippocampal tissue slices in vitro exposed to hypoxia, pentobarbital significantly increased the survival of CA1 pyramidal cells [14]. These results indicate that the neuronal damage in the process of "delayed neuronal death" is reversible if treatment is started at a sufficiently early period following ischemia. Neurons in CA1 can remain morphologically normal for longer than 24 h, however the brain-protecting effect of pentobarbital is dominant at only the initial stage of ischemia and rapidly diminishes after some min.

The Effect of Nizofenone and Diazepam

The protective effect on delayed neuronal death of nizofenone, a recently developed brain protector, and of diazepam was also investigated [15]. The results

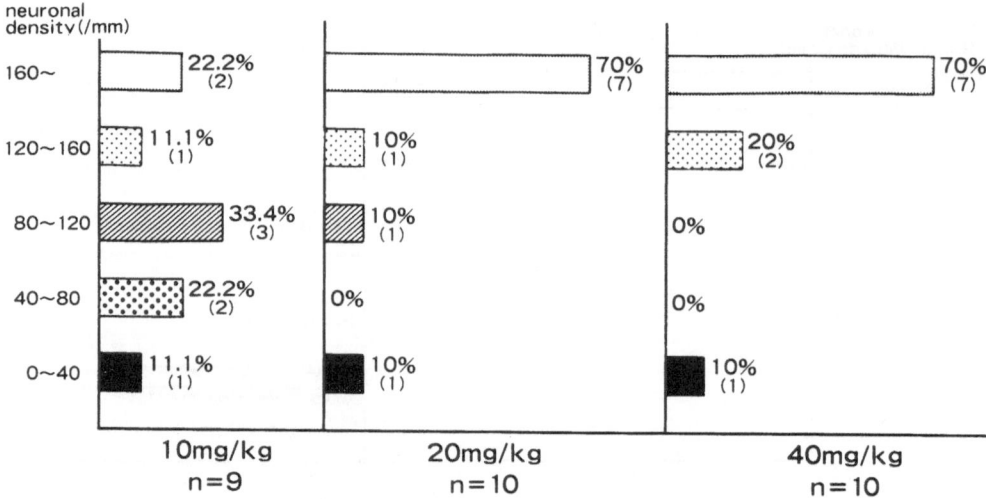

Fig. 3. Neural density in the CA1 sector in pentobarbital-treated group. In 20mg/kg and 40mg/kg groups, an improvement of the neuronal density is definite

are shown in Fig. 4. The neural density in the CA1 subfield in the nizofenone-treated group was maintained in the 12.5 mg/kg and 25 mg/kg groups. The protective effect of early nizofenone is comparable with that of pentobarbital. The lower part of Fig. 4 shows the results of early treatment with diazepam. An obvious increase of density appears in the 10 mg/kg and 20 mg/kg group.

Discussion

The process of delayed hippocampal neuronal death is considered to be affected via an excitotoxic mechanism. Neurons recovering from ischemia are exhausted by excessive excitatory neurotransmission. Glutamate is postulated as an important excitotoxin. Benveniste et al. demonstrated that following 10 min ischemia. concentration of L-glutamate as well as L-aspartate in the hippocampal extracellular space was remarkably elevated for 20–25 min after ischemia [16]. Excessive excitation of CA1 neurons induces an intracellular over-loading of calcium in CA1 pyramidal cells. Excessive calcium entry causes proteolysis or lipolysis, followed by neuronal death through a membrane-mediated phenomena. Neuronal survival obtained by pentobarbital is probably produced by inhibition of excessive neuronal excitation.

Cerebral Protection With Nizofenone

Nizofenone derived from imidazole appears promising as a brain protector. The effects of this drug are reviewed briefly in this paper.

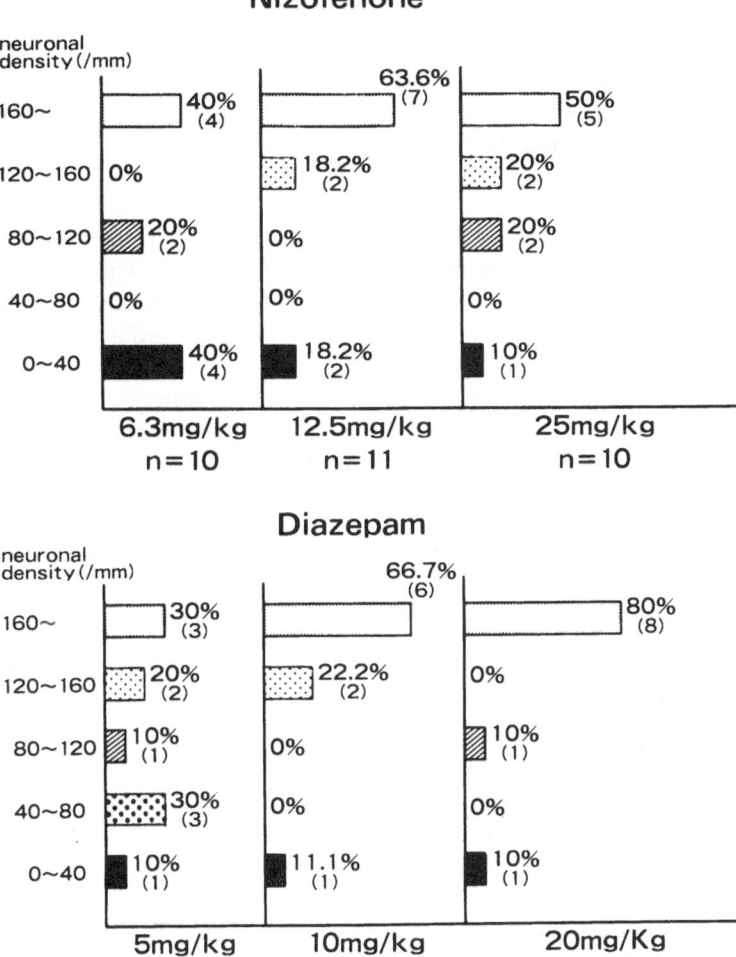

Fig. 4. Neuronal density in the CA1 sector in nizofenone-treated group (*upper*) and that in diazepam treated group (*lower*)

Experimental Data

Nizofenone significantly prolonged the survival time of mice and rats under hypoxic conditions and reduced KCN fatality in rats. This drug had significantly prolonged the time until the flattening of the cortical EEG in rats after cessation of artificial respiration, and prompted the re-appearance of EEG activity. These data suggested that nizofenone has a strong anti-hypoxic action. Regional cerebral ischemia was produced in cats by permanent occlusion of the middle cerebral artery (MCA). A significant decrease of infarction rate was found in nizofenone-treated groups [17].

Clinical Trials

Early clinical trials demonstrated that nizofenone had brain protective effects. In order to confirm these findings, a preliminary multi-center double-blind study was performed by Saito and his colleagues [18]. The results of this study indicated that nizofenone might be useful in the therapy of delayed ischemic neurological deficits following subarachnoid hemorrhage (SAH).

A second multi-center controlled double-blind clinical study was performed by Ohta et al. [19]. This study was carried out in 30 institutes on 266 patients with SAH who were treated within 2 weeks of the attack. The purpose of this study was to investigate the ischemic deficit following cerebral vasospasm in the acute stage following SAH. The test drug was administered as well as conventional treatment. Of the 208 patients studied, 102 were treated with nizofenone 30 mg/day for 2 weeks and 106 with placebo. The percentage of cases in which therapy was judged as "effective" in the cases treated with nizofenone was 31%, whereas that with placebo was 18% ($P < 0.05$). Percentage of good recovery at 1 month in nizofenone-treated group was 73%, whereas that in placebo group was 56% ($P < 0.05$). The effect of nizofenone was notable in patients with delayed ischemic symptoms, moderately severe preoperative deficits (Hunt and Hess grade II or III), and diffuse high-density areas in pre- and postoperative CT scans. No significant side effects were observed. Further, nizofenone has also been used in patients with severe head injury with beneficial effects [20].

Discussion

Although nizofenone is similar to barbiturates with regard to brain protective action, there are some differences between them. First, nizofenone has no anesthetic action. The decrease in cerebral metabolic rate for oxygen induced by nizofenone was milder than that induced by pentobarbital in experiments on dogs using the venous outflow method to measure CBF and $CMRO_2$ [21]. The distribution of cerebral blood flow (CBF) was monitored for 1 week after MCA occlusion in cats. In the animals treated with pentobarbital, CBF in the ischemic area was more decreased, whereas in animals treated with nizofenone CBF was not further reduced. Nizofenone prevents vasogenic edema and scavenges free radical, whereas barbiturates have little or no effect. Thus, both experimental and clinical results suggest that nizofenone is a promising drug, functioning as a cerebral protector in place of barbiturates.

Optimal Timing of Steroids to Influence the Effect of Head Injury Outcome

Although steroids are routinely used for metastatic brain tumors with favorable results, there is little evidence that they are effective in the treatment of acute stroks and severe head injuries. Nevertheless, steroids have been used to treat the ischemic or injured brain in hope that they will reduce edema and ameliorate

Fig. 5. Schematic illustration of experimental head injury model of mice

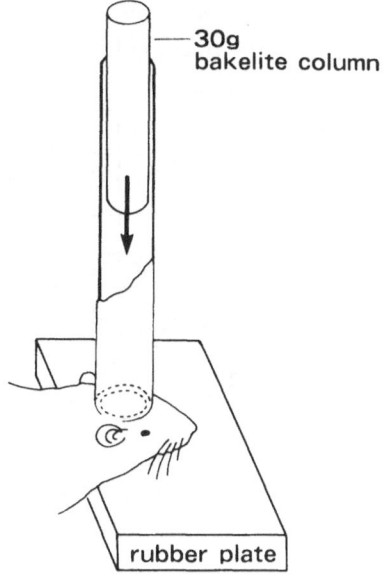

microcirculation. We have studied the timing of administration on the effect of dexamethasone using experimental head-injured mice [22].

Effect of Dexamethasone on an Experimental Head Injury Model

Awake male mice of the dd-strain were restrained and subjected to head injury using a bakelite rod weighing 30gm dropped from a height of 18 cm to the vertex (Fig. 5). This impact resulted in immediate loss of consciousness and was fatal in about 40%. Convulsion was provoked in approximately 30% of the animals. The duration of impaired consciousness was estimated from the interval between the time of impact and the appearance of righting reflex (RR time) or of spontaneous movement (SM time).

Four mg/kg of dexamethasone phosphate were given intraperitoneally 0.5–24 h before the head injury. In animals pretreated by dexamethasone 4–6 h before the impact, the mortality was significantly improved, whereas all groups treated a half hour or later than 12 h before impact failed to improve (Fig. 6). Dexamethasone did not affect the incidence of convulsions. Dexamethasone given 4–12 h before impact shortened the duration of disturbed consciousness, whereas in the other groups there was no effect. These results suggest that dexamethasone requires an optimal incubation period before the effect is manifest.

In order to investigate the participation of protein synthesis in the protective effect of steroids, 0.5 mg/kg of actinomycin D, an inhibitor of messenger RNA synthesis, was given intravenously 1 h prior to the administration of dexamethasone. The mortality-reducing effect of steroids given 4, 6, or 8 h before impact

Shinya Manaka et al.

Mortality

	Death (N) (+)	(−)	Mortality (‰)		
Control	50	70	42%		
0.5hours	23	51	31%		
4	17	55	24% p=0.017		
6	11	32	26% p=0.092		
12	27	35	44%		
18	25	28	47%		
24	19	27	41%		

Convulsion

	Convulsion (N) (+)	(−)	Rate of Convulsion (‰)		
Control	85	35	71%		
0.5hours	42	32	57%		
4	47	25	65%		
6	25	18	58%		
12	46	16	74%		
18	44	9	83%		
24	33	13	72%		

Fig. 6. The effects of dexamethasone on mortality and incidence of convulsion on the experimental head injury model

	Death (N) (+)	(−)	Mortality (%)		
Control	42	40	51%		
Actinomycin-D 0.5mg/kg	36	52	41%		
DXM 4h. before HI	19	25	43%		
DXM 6h. before HI	11	17	39%		
DXM 12h. before HI	23	18	56%		

DXM : Dexamethasone
HI : Head injury

Fig. 7. Influence of actinomycin D before dexamethasone treatment on mortality. Actinomycin D (0.5mg/kg) was administered i.v. 1 h before dexamethasone (4mg/kg) treatment

was almost reversed (Fig. 7). Disturbance of consciousness was not affected by steroids when the animal was pretreated with actinomycin D.

The relationship between dose and time lag to the effect of dexamethasone was studied. The dose of dexamethasone was varied from 2–8 mg/kg. If 8 mg/kg were given 30 min before impact, the mortality was significantly low, whereas in the animal treated with 6mg/kg, the mortality was low but not significanty (Fig. 8). There is tendency towards the condition that the larger the dose, the shorter was the incubation period of the effect. These results indicate that steroids need

		Death (N)		Mortality		
		(+)	(−)	(‰)	50	100
	Control	19	28	40%		
30 min.	Dexamethasone 6.0mg/kg	6	20	23%		
30 min.	8.0mg/kg	2	23	8%	P=0.009	
2 hours	4.0mg/kg	7	22	24%		
2 hours	6.0mg/kg	4	28	13%	P=0.015	
4 hours	2.0mg/kg	13	18	42%		

Fig. 8. Relationship between dose and time lag of the effects of dexamethasone on mortality. Dexamethasone was given i.p. 0.5, 2, or 4 h prior to head injury

a time lag before they can affect brain injury. Therefore steroids administered at any time after injury may have little merit.

Discussion

A critical factor for the success of a pharmacological treatment in the acute therapy of cerebral injury must be its time of initiation and dose [23]. In experimental studies, steroids have been administered within 1 h following the injury, whereas in clinical studies steroids have been given later than 1 h. By 6 h after the insults, neuronal/axonal deterioration had occurred with accompanying edema, marked ischemia, and generally advanced structural degeneration. Any pharmacological agent as late as 12–24 h after the injury would not reverse the damage or alter the ultimate outcome [23].

In a recent report by Jane and colleagues, a significant improvement in outcome in patients with severe head injuries was observed in those aged 16 years or younger who received an initial 15–30 mg/kg dose of methylprednisolone at the scene of the accident [24]. Giannotta et al. demonstrated that head-injured patients who received the high dose of steroids experience a mortality of 39% as compared to a 52% mortality in the low dose and placebo groups ($P < 0.05$). Mortality difference were most marked in patients less than 40 years old [25]. Braughler and Hall recommended that an initial 30 mg/kg intravenous dose of methylprednisolone, or alternatively a 6–15 mg/kg dose of dexamethasone, should be administered at the earliest possible time after injury [23].

Hall demonstrated that high-dose methylprednisolone treatment improves neurological recovery in head-injured mice, and a 60 mg/kg dose of prednisolone was optimal, while lower and higher doses had no effect. Hydrocortisone did not improve recovery over the wide dose-range tested [26]. Steroids prior to the onset of ischemia were effective through inhibiting the release of arachidonic acid from ischemic cells [2]. Megadose steroids may improve the outcome of cerebral damage when they are given as soon as possible after injury, or if possible before the brain insult. In clinical situations, however, pretreatment is impossible, therefore administration immediately after injury or pre- or peri-operative treatment is advisable.

Summary. Barbiturates and steroids are losing their status as brain protectors, however these drugs may still play a role if used immediately after head injury or ischemia or used pre- or perioperatively. A time-lag between the onset of damage and the administration of these drugs can minimize their effect, as shown in the case of delayed neuronal death. An extensive clinical study focusing on the timing of administration may clarify the real power of these drugs.

References

1. Bircher NG (1985) Ischemic brain protection. Ann Emerg Med 14:784–788
2. Hoff JT (1966) Cerebral protection. J Neurosurg 65:579–591
3. Newberg LA (1984) Cerebral Resuscitation; advances and controversies. Ann Emerg Med 13:853–856
4. Braakman R, Shouten HJA, Dishoeck MB, Minderhoud JM (1983) Megadose steroids in severe head injury. J Neurosurg 58:326–330
5. Cooper PR, Moody S, Clark WK, Kirkpatrick J, Maravilla K, Gould AL, Drane W (1979) Dexamethasone and sever head injury, a prospective double-blind study. J Neurosurg 51:307–316
6. Dearden NM, Gibson JS, McDowall DG, Gibson RM, Cameron MM (1986) Effect of high-dose dexamethasone on outcome from severe head injury. J Neurosurg 64:81–88
7. Grafton GT (1988) Steroids after cardiac arrest: A retrospective study with concurrent, nonrandomized controls. Neurology 38:1315–1316
8. Ward JD, Becker DP, Miller JD, Choi SC, Marmarou A, Wood C, Newlon PG, Keenan R (1985) Failure of prophylactic barbiturate coma in the treatment of severe head injury. J Neurosurg 62:283–388
9. Michenfelder JD, Milde JH, Sundt TJ Jr (1976) Cerebral protection by barbiturate anesthesia. Use after middle cerebral artery occlusion in Java monkeys. Arch Neurol 33:345–350
10. Astrup J (1982) Energy-requiring cell functions in the ischemic brain, their critical supply and possible inhibition in protective therapy. J Neurosurg 56:482–497
11. Eisenberg HM, Frankowski RF, Contant CF, Marshall LF, Walker MD, and the Comprehensive Central Nervous System Trauma Centers (1988) High-dose barbiturate control of elevated intracranial pressure in patients with severe head injury. J Neurosurg 69:15–23
12. Kirino T (1982) Delayed neuronal death in the gerbil hippocampus following ischemia. Brain Res 239:57–69
13. Kirino T, Tamura A, Sano K (1986) A reversible type of neuronal injury following ischemia in the gerbil hippocampus. Stroke 17:455–459
14. Akiten PG, Schiff SJ (1986) Barbiturate protection against hypoxic neuronal damage in vitro. J Neurosurg 65:230–232
15. Kirino T, Tamura A, Tomukai N, Sano K (1986) Treatable ischemic neuronal damage in the gerbil hippocampus. Brain and Nerve 38:1157–1163
16. Benveniste HB, Drejer J, Schousboe A, Diemer NH (1984) Elevation of the extracellular concentrations of glutamate and aspartate in rat hippocampus during

transient cerebral ischemia monitored by intracerebral microdialysis. J Neurochem 43:1369–1374
17. Tamura A, Asano T, Sano K, Tsumagari T, Nakajima A (1979) Protection from cerebral ischemia by a new imidazole derivative (Y-9179) and pentobarbital. A comparative study in chronic middle cerebral artery occlusion in cats. Stroke 10:126–134
18. Saito I, Asano T, Ochiai C, Takakura K. Tamura A, Sano K (1983) A double-blind clinical evaluation of the effect of nizofenone (Y-9179) on delayed ischemic neurological deficits following aneurysmal rupture. Neurol Res 5:29–47
19. Ohta T, Kikuchi H, Hashi K, Kudo Y (1986) Nizofenone administration in the acute stage following subarachnoid hemorrhage. J Neurosurg 64:420–426
20. Watanabe H (1985) Brain protection in head injury and cerebrovascular disease by nizofenone. Rinsho-iyaku 1:307–313
21. Ochiai C, Asano T, Tamura A, Sano K, Fukuda T and Nakamura T (1981) An experimental study on the mechanism of the protective action of pentobarbital and Y-9179 against cerebral ischemia. Neurol Med Chir (Tokyo) 21:303–311
22. Sasaki M, Manaka S, Takakura K (1987) Time course of the cerebroprotective effect of dexamethasone in the experimental head injury. Brain and Nerve 39:155–161
23. Braughler JM, Hall ED (1985) Current application of "high-dose" steroid the therapy for INS injury, a pharmacological perspective. J Neurosurg 62:806–810
24. Jane JA, Rimel RW, Pobereskin LH (1982) Outcome and pathology of head injury. In: Grossman RG, Gildenberg PL (eds) head injury: Basic and clinical aspects. Raven, New York, pp. 229–237
25. Giannotta SL Weiss MH Apuzzo MLJ, Martin E (1984) High dose glucocorticoids in the management of severe head injury. Neurosurgery 15:497–501
26. Hall ED (1985) High-dose glucocorticoid treatment improves neurological recovery in head-injured mice. J Neurosurg 62:882–887

15

Anesthetic Protection in Brain Ischemia

Leslie Newberg Milde[1]

Introduction

This review will be confined to the use of anesthetics for protection of the brain
during surgery in which either the disease of the patients or the surgery itself
places the patient at risk of cerebral ischemia. This could include patients with
intracranial pathology or cerebrovascular disease undergoing any surgical procedure, and any patient undergoing cardiovascular or neurosurgery especially
if there may be temporary occlusion of a major cerebral artery. Intraoperative cerebral ischemia may be focal ischemia due to the temporary clamping of
a major cerebral vessel, to emboli produced by blood clot, air, or particulate
matter, or to focal cerebral edema secondary to surgery trauma. Intraoperative
cerebral ischemia may also be global but incomplete as that which may occur
during hemorrhagic or cardiogenic shock, but also could be the result of such
low flow states as induced (deliberate) hypotension or during cardiopulmonary
bypass (Table 1).

Mechanism of Cerebral Protection by Anesthetics

The major mechanism by which the anesthetics might provide brain protection is
to optimize the oxygen/supply demand ratio. This can be done by improving
oxygen delivery or by decreasing oxygen demand. Improved oxygen delivery
might be accomplished by certain anesthetics which maintain cerebral blood flow
(choosing anesthetics that do not decrease cerebral blood flow) by anesthetics
which increase cerebral blood flow or, most importantly, by anesthetics which
favorably redistribute cerebral blood flow. Redistribution of regional cerebral
blood flow in a way that blood is redirected from areas of normal flow to areas in
which there is ischemic flow is known as a "Robin Hood steal" (robbing the rich

---

[1] Department of Anesthesiology, Mayo Clinic, Rochester, MN 55905, USA

Table 1. Intraoperative cerebral ischemia

Focal ischemia	Incomplete global ischemia
Surgical clamping of a vessel	Shock (hemorrhagic, cardiogenic)
Emboli (blood clot, air, particulate matter)	Induced hypotension
	Cardiopulmonary bypass
Edema	

Table 2. Major mechanisms of brain protection by anesthetics

Optimize the oxygen supply/demand ratio	
Supply	Demand
Maintain cerebral blood flow	Prevent increased cerebral metabolism
Increase cerebral blood flow	Decrease cerebral metabolism
Redistribute cerebral blood flow	

to give to the poor). On the demand side of the oxygen supply/demand ratio, certain anesthetics prevent an increase in cerebral metabolism; that is, they can prevent or treat any catecholamine-induced hypermetabolism or hypermetabolism secondary to seizure activity. Most significantly, certain anesthetics are capable of decreasing normal cerebral metabolism (Table 2).

The metabolic demands of the brain might be viewed as subserving two components: one is the energy needed for neuronal function (synaptic transmission of nerve impulses which is indicated by electrical activity on the EEG, Fig. 1) [1]. This component of cerebral metabolism has been termed "activation metabolism" [2]. The other component of cerebral metabolism is the energy needed for the preservation of cellular integrity which includes such processes as membrane stabilization, ion pumping to maintain gradients, the synthesis and handling of neurotransmitters, protein, nucleic acids and lipids, and the initial steps in glucose metabolism. This component of the cerebral energy requirement has been referred to as "residual metabolism," that metabolism which supports the cellular processes which continue to function after synaptic transmission has been abolished. In neurons, approximately 55% of the total energy requirement is used for neuronal function or synaptic transmission and 45% is required for the maintenance of cellular integrity. This division of the energy requirements of neurons is an important concept to remember when we discuss possible therapeutic actions of anesthetics for ischemia. While hypothermia decreases cerebral metabolism by reducing all cellular functions proportionately, the anesthetics have no effect on the neuronal energy requirements for the maintenance of cellular integrity. Most anesthetics decrease cerebral metabolism to the extent that they decrease neuronal function or synaptic transmission of impulses—that electrical activity measured by EEG. Therefore, the anesthetics can only de-

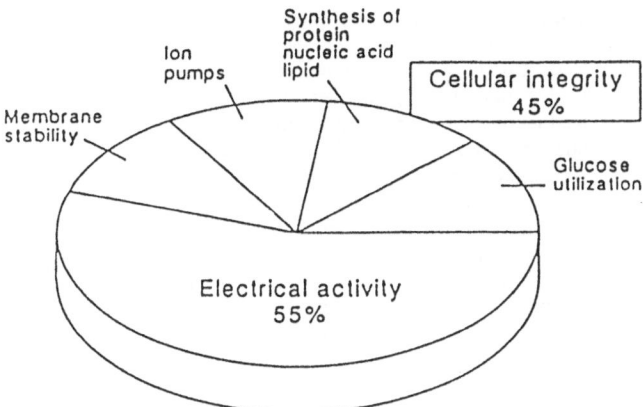

Fig. 1. Components of cerebral oxygen consumption rate ($CMRO_2$). (From [1] with permission)

crease the energy requirements by 55%. In order for anesthetics to be protective by decreasing cerebral metabolism, there must be some neuronal electrical activity present. If neuronal electrical activity has been abolished by ischemia, anesthetics will have no effect on cerebral metabolism. The most effective anesthetics for decreasing cerebral metabolism are the barbiturates and the volatile anesthetic, isoflurane, because they can do so at clinically useful concentrations without overly depressing the heart.

Barbiturates

Mechanism of Protection

Barbiturates have served as the prototype for anesthetic protection against ischemia. Proposed mechanisms of barbiturate protection include a reduction in cerebral oxygen consumption or metabolism so that the reduced oxygen delivery occurring as a result of the ischemia may be better able to supply the oxygen demand, a redistribution of regional cerebral blood flow so that flow is redirected to ischemic areas, and a decrease in intracranial pressure resulting in an increase in cerebral perfusion pressure (Table 3). These three mechanisms have been well documented in animal models of focal ischemia and in humans. Other mechanisms of action of the barbiturates which may have a role in cerebral protection include the suppression of post-ischemic hyperactivity (such as seizures at a time when post-ischemic hypoperfusion exists), and the abolition of thermoregulation in a way that hypothermia produced by the barbiturates may contribute to the protective effect. The subcellular effects of the barbiturates (such as free radical scavenging, membrane stabilization, calcium channel blockade, and alterations of free fatty acid metabolism) have not yet been demonstrated to be of consequence in protecting the brain from ischemia.

Barbiturates produce a dose-related decrease in neuronal function as evi-

Table 3. Mechanisms of barbiturate protection

Reduction in cerebral metabolism
Redistribution of regional cerebral blood flow
Reduction of intracranial pressure
Suppression of convulsive hyperactivity
Abolition of thermoregulation
Scavenging of free radicals
Membrane stabilization
Calcium channel blockade
Alterations in fatty acid metabolism

Table 4. Failure of barbiturates to provide protection during complete cerebral ischemia

Steen et al. [4]	Dogs	Vascular occlusion
Todd et al. [6]	Cats	Ventricular fibrillation
Gisvold et al. [5]	Primates	Neck tourniquet inflation
Abramson et al. [7]	Humans	Cardiac arrest

denced by progressive changes on the EEG from low amplitude fast activity to high amplitude slow waves to burst suppression to an isoelectric EEG. This is accompanied by a dose-related decrease in cerebral metabolism, reaching a plateau when an isoelectric EEG is produced [3]. As stated earlier, the maximal metabolic suppression that can be produced by the barbiturates is 55%. Once an isoelectric EEG is produced, giving additional barbiturates will have no further effect on cerebral metabolism. Because cerebral blood flow and metabolism remain coupled, the decrease in cerebral metabolism is accompanied by a decrease in cerebral blood flow and cerebral blood volume. This can explain the ability of the barbiturates to decrease intracranial pressure (by decreasing intracranial vascular volume) and to redistribute regional cerebral blood flow (by decreasing cerebral blood flow in normal areas of the brain where barbiturates can decrease neuronal function and metabolism) thereby shunting blood to ischemic areas.

The importance of a reduction in cerebral metabolism as a primary mechanism of protection by the barbiturates is evidenced by the many animal studies which demonstrate that barbiturates provide protection from focal cerebral ischemia but fail to provide protection from complete global ischemia [4–7]. In the latter case, the complete ischemia itself will have abolished neuronal function so that the barbiturates will have nothing on which to act (Table 4).

Effects on Focal Ischemia

There is considerable experimental evidence from studies of laboratory animals that the barbiturates provide protection from neurologic injury following focal cerebral ischemia (Table 5). However, the reports are somewhat confusing and contradictory. Demonstration of barbiturate protection in focal ischemia is influenced by the type of barbiturate, dose, timing, duration of administration, the

Table 5. Barbiturate protection following focal ischemia

Author	Species	Barbiturate	Model	Outcome
Smith et al. [8]	Dogs	Pentobarbital (56 mg/kg)	Pretreatment permanent MCA$_O$	↓ Infarction ↓ Neurologic damage
Hoff et al. [9]	Primates	Pentobarbital (60–120 mg/kg)	Pretreatment permanent MCA$_O$	↓ Infarction Neurologic damage
Michenfelder et al. [10]	Primates	Pentobarbital (14 mg/kg + 7 mg/kg q 2 h)	Post-treatment permanent MCA$_O$	↓ Infarction ↓ Neurologic damage
Selman et al. [11]	Primates	Pentobarbital (30 mg/kg + infusion)	Post-treatment permanent MCA$_O$	↓ Neurologic damage
Selman et al. [11]	Primates	Pentobarbital (30 mg/kg ↓ infusion)	Post-treatment temporary MCA$_O$	↓ Neurologic damage ↓ Edema
Hoff et al. [12]	Cats	Pentobarbital (30 mg/kg + 5 mg/kg q 4 h)	Pretreatment permanent MCA$_O$	↓ Infarct size
Gelb et al. [13]	Cats	Thiopental (40 mg/kg)	Pretreatment temporary MCA$_O$	− Infarct size

↓ , decreased; ↑ , increased; − , no effect

experimental model, animal species differences, the permanent or temporary nature of cerebrovascular occlusion, the incidence of post-ischemic seizures, and cardiovascular and respiratory changes occurring as a result of the ischemia and/or the barbiturate administration.

It has been demonstrated in dogs and primates that pretreatment with large doses of pentobarbital [56 mg/kg to dogs [8] and 60–120 mg/kg to primates [9] significantly reduced cerebral infarct size when given prior to permanent occlusion of the middle cerebral artery (MCA) when compared to animals under 1–2 MAC halothane. Neurologic damage was reduced in the dogs but not in the primates. This lack of demonstration of protection by pentobarbital in the primates may have been due to the increased cardiovascular and respiratory complications occurring in the barbiturate animals. Treatment with smaller doses of pentobarbital (14 mg/kg) followed by boluses of 7 mg/kg every 2 h for 48 h to maintain a barbiturate level decreased infarct size and neurologic damage when compared to the halothane control group [10]. However, Selman et al. demonstrated that treatment with barbiturates significantly worsened the neurologic damage when given after permanent occlusion but significantly improved the neurologic outcome following temporary focal ischemia [11].

Timing of the administration of barbiturate after the onset of ischemia is crucial. Protection has been demonstrated when administration of barbiturate is delayed up to 1 h after the onset of ischemia but no protection is demonstrated if the barbiturate is delayed further. These studies have reported that barbiturates decrease cerebral oxygen demand, redistribute cerebral blood flow, and decrease intracranial pressure so that a combination of these effects probably acts to provide protection from focal cerebral ischemia in these animal models.

However, extending the results of these studies to clincial trials of stroke or intraoperative focal ischemia to assess the effects of the barbiturates in humans may be difficult because of the hemodynamic and pulmonary complications accompanying the anesthetic effect of the barbiturates given at the doses needed for protection as suggested by the animal studies.

Nevertheless, prophylactic barbiturate treatment has been recommended for brain protection during periods of focal ischemia when temporary clamping of a major cerebral vessel is necessary during carotid endarterectomy, extracranial to intracranial bypass procedures, and aneurysm clipping. Relatively small single doses of barbiturate have been recommended because of concern that larger doses would cause hypotension, thereby decreasing cerebral perfusion pressure, and result in a drug-induced coma delaying examination of neurologic function postoperatively. It was speculated that with a small single bolus of barbiturate given immediately prior to vessel clamping, sufficient drug would be trapped in the ischemic area to be protective [14]. However, in an animal study by Gelb et al. [13] a single bolus of thiopental in a dose sufficient to produce burst suppression on the EEG was given just prior to temporary MCA occlusion. This bolus of barbiturate failed to attenuate the histologic neuronal injury resulting from the temporary focal ischemia. If cerebral preservation during temporary focal ischemia is due primarily to a decrease in cerebral metabolism, then maximal protection should be achieved by doses of barbiturate sufficient to produce burst suppression on the EEG. Normal anesthetizing doses of thiopental (3–5 mg/kg) will produce burst suppression on the EEG lasting less than 5–10 min. Even larger doses of thiopental (10–25 mg/kg) produce burst suppression lasting only 10 min.

If barbiturates are to be given for protection prior to the temporary clamping of a major cerebral vessel, the following recommendations are made. If possible, identify patients at risk for ischemia by temporarily occluding the vessel while monitoring the EEG for signs of ischemia. If these occur, the vessel should be reopened. When the EEG has returned to baseline, a large enough bolus of thiopental (3–5 mg/kg) should be given to produce burst suppression on the EEG. The vessel can then be temporarily occluded for the surgical procedure. Ideally, the occluded vessel should be briefly opened every 5–10 min for reperfusion to allow subsequent small doses of thiopental (1–2 mg/kg) to reach the ischemic area. If this cannot be done, a continuous infusion of thiopental (approximately 3–5 mg/kg per hour) should be administered throughout the ischemic period. Blood pressure should be maintained within normal limits with a phenylephrine infusion, if necessary, to maintain optimal cerebral blood flow to the penumbral ischemic area. If induced hypotension is required during an aneurysm clipping, a larger bolus of thiopental (6–8 mg/kg) should decrease both arterial pressure and cerebral metabolism.

Despite these recommendations, clinical studies on the use of barbiturates for protection from intraoperative ischemia are only anecdotal. There is little clinical evidence on the efficacy of the barbiturates to prevent or treat focal ischemia.

Table 6. Neuropsychiatric complications after cardiopulmonary bypass

	Complications at day 1		Complication at day 10	
	Thiopental ($n = 89$)	Control ($n = 93$)	Thiopental	Control
Patients with dysfunction	5	8	0*	7
Neurologic	0	4	0	4
Psychiatric	4	2	0	1
Both	1	2	0	2

(From [15] with permission by J.P. Lippincott Company)

Several variables that influence clinical outcome other than barbiturate administration include patient variation, coexisting medical disease (especially cardiovascular disease), variations in the types of vascular occlusion, and the incidence
of seizures. The high doses of barbiturates required to provide a protective
effect also produce severe cardiovascular and respiratory depression that
necessitate intensive care with extensive hemodynamic monitoring. Determination of the efficacy of barbiturates for protection during cerebrovascular surgery
may be difficult because of the already low incidence of morbidity and mortality
associated with these procedures and the large variation in patient population.

There is one clinical study by Nussmeier et al. (Table 6) [15] which has reported that barbiturates provide a significant improvement in neurologic function after focal ischemia in a highly selected group of patients undergoing normothermic open cardiac surgery. Embolism by either air, clot, or particulate
matter occurring while being placed on and taken off cardiopulmonary bypass
was the presumed cause of the sensorimotor neurologic dysfunction occurring
after surgery in these patients. Thiopental in doses sufficient to produce burst
suppression on the EEG throughout the bypass period (40 mg/kg) significantly
improved postoperative neurologic dysfunction. At seven days postsurgery,
none of the patients treated with thiopental showed neurologic dysfunction
while 7 out of 93 untreated patients continued to have some neurologic dysfunction. However, administration of the large doses of thiopental was associated
with a prolonged anesthetic period and an increased requirement for inotropic
support in the postbypass period.

Isoflurane

Effects on Focal Ischemia

Isoflurane is also used when there is a risk of focal ischemia: for example, during
carotid endarterectomy and when there is a risk of incomplete global ischemia
during induced hypotension for aneurysm clipping or resection of arteriovenous
malformation (AVM). During carotid endarterectomy, intraoperative focal
EEG changes occur most commonly in association with clamping on the carotid
artery. This occurs in approximately 18%–25% of the time depending upon the
anesthetic. The changes usually occur within 20–30 sec after clamping and are

associated with a reduced cerebral blood flow. The cerebral blood flow at which cerebral ischemia occurs depends on cerebral metabolism or demand. The higher the metabolism, the higher the cerebral blood flow demand. Cerebral metabolism is altered to a different extent by the different anesthetics. This is an important fact that is often ignored by nonanesthesiologists who report absolute blood flows at which signs of ischemia occur without regard to the background anesthetic or cerebral metabolism.

A comparison of cerebral blood flows at which signs of ischemia occur on EEG under four anesthetic regimens in patients undergoing carotid endarterectomy from three studies done at the Mayo Clinic demonstrates that background anesthetic does influence the incidence of ischemia. Under halothane-N_2O anesthesia, signs of ischemia occurred on EEG at cerebral blood flows below 18–20 ml/min per 100 g [16]. With enflurane or Innovar (a mixture of droperidol and fentanyl) + N_2O slightly lower cerebral blood flows were tolerated, 15–18 ml/min per 100 g, before signs of ischemia occurred [17]. However, under isoflurane-N_2O anesthesia, the blood flow could decrease significantly lower to 8–10 ml/min per 100 g before ischemic changes occurred on the EEG [18]. From this comparison, it was concluded that isoflurane increased the tolerance of the brain to low cerebral blood flows so that ischemia did not occur. It was hypothesized that isoflurane, even at the low concentrations administered— approximately 0.5 MAC—significantly lowered cerebral metabolism more than the other anesthetics, thereby lowering oxygen demand so that oxygen delivery even with these low flows was adequate to meet demand. This hypothesis has been confirmed by animal studies of CBF and $CMRO_2$ with equiMAC concentrations of the volatile anesthetics.

This prevention of ischemia by isoflurane was confirmed by a large retrospective study by Michenfelder et al. from the Mayo Clinic [19] in which more than 2000 patients were studied. They found a significantly lower incidence of EEG changes of ischemia under isoflurane anesthesia (18%) than under either halothane or enflurane anesthesia (25%–26%). This decreased incidence of intraoperative ischemia in the patients under isoflurane occurred despite the fact that those patients in the isoflurane group had a significantly greater preoperative neurologic risk (as demonstrated by major medical or angiographically determined risks) than the patients under halothane or enflurane anesthesia. Because of a decreased incidence of EEG changes of ischemia, the number of shunts placed in the patients under isoflurane anesthesia was significantly less than in those patients under halothane or enflurane. Therefore, because of the lower incidence of ischemia, not only is there less risk for the ischemia per se, but there may be decreased morbidity because of the lesser number of shunts placed.

As a result of this experience with the use of volatile anesthetics for patients undergoing carotid endarterectomy, we routinely use isoflurane anesthesia for patients undergoing surgery on the cerebral vasculature. Despite these clinical studies, however, there is some controversy concerning the use of isoflurane when there is a risk of focal cerebral ischemia. Nehls et al. reported in 1987 [20] that thiopental in doses sufficient to produce burst suppression on the EEG and therefore sufficient to produce maximal metabolic suppression provided a better neurologic outcome and significantly less cerebral infarction than did comparable doses of isoflurane in baboons exposed to temporary focal ischemia produced by 6 h of middle cerebral artery occlusion. However, the mean arterial pressure was significantly different between the two groups, the thiopental animals having a mean arterial pressure of 100–109 mmHg despite the use of nitroprusside and hydrazine to lower it, while the animals receiving isoflurane had a mean arterial pressure of 83–89 mmHg throughout the ischemic period despite the use of phenylephrine and metaraminol to increase it. It is reasonable to conclude that the differences in the neurologic outcome and infarct size reported in this study may have been influenced not only by proposed differences in the cerebral hemodynamic and metabolic effects of isoflurane and thiopental but also by significant differences in blood pressure and differences in the use of vasoactive drugs which may have redistributed regional cerebral blood flow.

When this study was repeated and the blood pressure was equally maintained, there were no differences in the neurologic outcome or in infarct size between the animals receiving thiopental and the animals receiving isoflurane [21]. Infarct size correlated with neurologic deficit. It was concluded that in situations of temporary focal ischemia when the hemodynamic values are equally maintained, cerebral metabolic suppression by thiopental offers no advantage over isoflurane for the preservation of neurologic function or the prevention of cerebral infarction.

Effects on Incomplete Global Ischemia

Isoflurane has also been used for deliberate hypotension during neurosurgical procedures and may provide an advantage over other hypotensive agents when very low levels of blood pressure are required, producing a risk of incomplete global ischemia. Deliberate hypotension is indicated to reduce the risk of vessel rupture in such cases as leaking thoracic or abdominal aortic aneurysm or cerebral aneurysm clipping. Rupture is dependent upon intraluminal pressure, aneurysm size, and wall thickness of the aneurysm. Deliberate hypotension has been shown to decrease intraluminal pressure and aneurysm size and to increase the wall thickness of the aneurysm.

The characteristics of the ideal hypotensive agent include its being efficacious and producing predictable and reliable results. It should produce a rapid onset and then a rapid recovery when no longer needed. There should be no accumulation of the drug and no toxicity from the concentrations used. Isoflurane fulfills many of these characteristics. Isoflurane, being a potent systemic vasodilator in addition to a general anesthetic, is a good agent for induced hypotension. In a study by Lam and Gelb [22] of patients undergoing cerebral aneurysm clipping, isoflurane, in concentrations of 2%–3% was used to produce hypotension to a MAP of 40 mmHg. These concentrations of isoflurane had little effect on the heart with maintenance of cardiac output and adequate peripheral perfusion due to the lowered vascular resistance (Table 7).

It has little effect on cerebral blood flow in man because its intrinsic cerebrova-

Table 7. Isoflurane-induced hypotension

	Prehypotension	Hypotension	Posthypotension
Inspired isoflurane (%)	0.67	2.20	0.62
MAP (mmHg)	73	40	76
C.I. (L/min/m²)	2.8	2.7	3.1
SVR (dyne·cm/s⁵)	1270	651	1162

(Reprinted with permission from the International Anesthesia Research Society [22])

Table 8. Effect of isoflurane on cerebral hemodynamics and metabolism

	Control	2.5% Isoflurane
CBF (ml/min/100g)	49 ± 14	45 ± 12
CMRO₂ (ml/min/100g)	2.0 ± 0.6	1.5 ± 0.5

(From [23] with permission by J.P. Lippincott Company)

sodilatory effects to produce an increase in CBF are offset by its producing a decrease in cerebral metabolism with a parallel decrease in cerebral blood flow, the net result being little change in CBF. In a study by Newman et al. [23] of patients receiving 2.5% isoflurane CBF was maintained at approximately awake levels while cerebral metabolism was decreased significantly (Table 8).

Isoflurane may prevent ischemia better than other hypotensive agents when profound levels of hypotension are required. In a dog study comparing different hypotensive agents (Fig. 2), when isoflurane was used to induce hypotension to a MAP of 40 mmHg (with a cerebral perfusion pressure of 22 mmHg),CBF was decreased about 60%, similar to that produced by trimethaphan, nitroprusside or halothane. However, isoflurane reduced cerebral metabolism significantly more than did the other three agents. It is assumed that the ability to decrease cerebral metabolism might increase the tolerance of the brain to lower CBF without the occurrence of ischemic damage. This is demonstrated by the preservation of cerebral metabolites during hypotension to 40 mmHg. ATP and phosphocreatine concentrations decreased significantly, while lactate increased significantly during induced hypotension with trimethaphan, nitroprusside, or halothane. These changes are indicative of anaerobic metabolism occurring in response to cerebral ischemia. However, during isoflurane-induced hypotension, the metabolites remained normal. This indicates that normal aerobic metabolism was maintained without evidence of ischemia. From these studies it is concluded that if very low blood pressures are required during the clipping of a fragile cerebral aneurysm with difficult access, isoflurane is the hypotensive agent of choice.

Hypotension can easily be achieved within 6 min by increasing the inspired

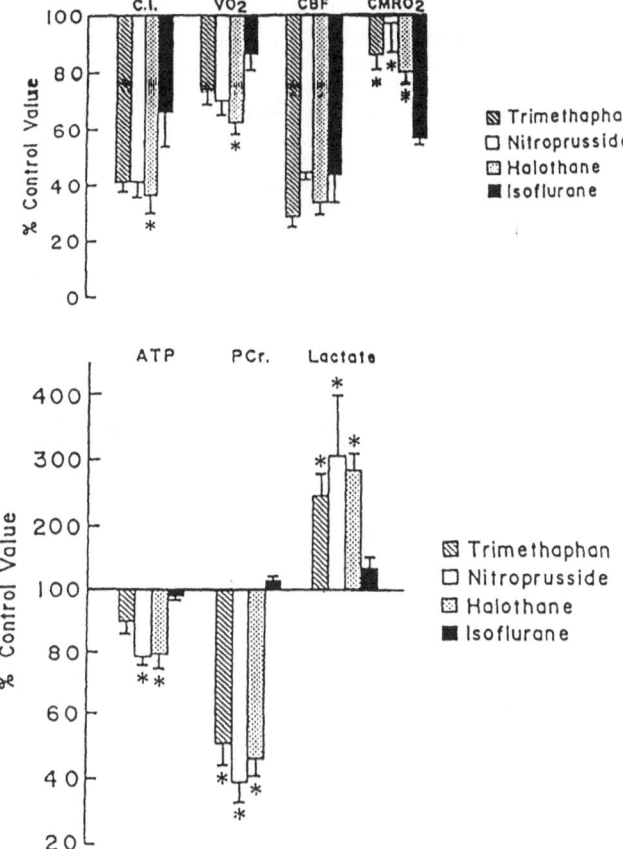

Fig. 2. Systemic and cerebral hemodynamic and metabolic effects of different hpotensive agents. (From [24] with permission by J.P. Lipponcott (Company)

isoflurane concentration to 4% until the desired level of hypotension is reached. This level of hypotension is then maintained with 2%–2.5% inspired isoflurane. In most cases, no adjustment of the isoflurane concentration is required during the course of hypotension. The inspired isoflurane concentration is reduced to 0.25% when hypotension is no longer needed. It takes approximately 6 min for the MAP to return to 90% of its prehypotensive level. The inspired isoflurane concentration is then reset to its prehypotensive value for the remainder of the surgery. Even with long periods of hypotension with high concentrations of isoflurane, patients awaken rapidly at the end of surgery.

Summary. I have tried to present the rationale for our use of anesthetics as protective agents when there is a risk of intraoperative cerebral ischemia. We use barbiturates prior to the temporary clamping of a major cerebral vessel and isoflurane during carotid endarterectomy when we use continuous monitoring

EEG for signs of ischemia and for induced hypotension during cerebral aneurysm clipping or resection of an AVM.

References

1. Messick JM Jr, Milde LN 1987 Brain protection. In: Stoelting RK, Barash PG, Gallagher TJ (eds) Adv Anesthesiol 4:47–88
2. Astrup J (1982) Energy-requiring cell functions in the ischemic brain. Their critical supply and possible inhibition in protective therapy. J Neurosurg 56:482–497
3. Michenfelder JD (1974) The interdependency of cerebral functional and metabolic effects following massive doses of thiopental in the dog. Anesthesiology 41:231–236
4. Steen PA, Milde JH, Michenfelder JD (1979) No barbiturate protection in a dog model of complete cerebral ischemia. Ann Neurol 5:343–349
5. Gisvold SE, Safar P, Hendrickx HHL, Rao G, Moossy J, Alexander H (1984) Thiopental treatment after global brain ischemia in pigtailed monkeys. Anesthesiology 60:88–96
6. Todd MM, Chadwick HS, Shapiro HM Dunlop BJ, Marshall LF, Ducck R (1982) The neurologic effects of thiopental therapy following experimental cardiac arrest in cats. Anesthesiology 57:76–86
7. Abramson NS and the Brain Resuscitation Clinical Trial I Study Group (1986) Randomized clinical study of thiopental loading in comatose survivors of cardiac arrest. N Engl J Med 314:397–403
8. Smith AL, Hoff JT, Nielsen SL, Larson CP (1974) Barbiturate protection in acute focal ischemia. Stroke 5:1–7
9. Hoff JT, Smith AL, Hankinson HL, Nielsen S (1975) Barbiturate protection from cerebral infarction in primates. Stroke 6:28–33
10. Michenfelder JD, Milde JH, Sundt TM Jr (1976) Cerebral protection by barbiturate anesthesia. Use after middle cerebral artery occlusion in Java monkeys. Arch Neurol 33:345–350
11. Selman WR, Spetzler RF, Roessmann VR, Rosenblatt JI, Crumrine RC (1981) Barbiturate-induced coma therapy for focal cerebral ischemia, effect after temporary and permanent MCA occlusion. J Neurosurg 55:220–226
12. Hoff JT, Nishimura M, Newfield P (1982) Pentobarbital protection from cerebral infarction without suppression of edema. Stroke 13:623–628
13. Gelb AW, Floyd R, Lok P, Farrell M (1986) A prophylactic bolus of thiopental does not protect against prolonged focal cerebral ischemia. Can Anaesth Soc J 33:173
14. Michenfelder JD (1985) Cerebral preservation for intraoperative focal ischemia. Clin Neurosur 32:105
15. Nussmeier NA, Arlund C, Slogoff S (1986) Neuropsychiatric complications after cardiopulmonary bypass: Cerebral protection by a barbiturate. Anesthesiology 64:165
16. Sharbrough FW, Messick JM Jr, Sundt TM Jr (1973) Correlation of continuous electroencephalograms with cerebral blood flow measurements during carotid endarterectomy. Stroke 4:674
17. McKay RD, Sundt TM, Michenfelder JD, Gronert GA, Messick JM, Sharbrough FW, Piepgras DG (1976) Internal carotid artery stump pressure and cerebral blood flow during carotid endarterectomy: Modification by halothane, enflurane, and Innovar. Anesthesiology 45:390
18. Messick JM Jr, Casement B, Sharbrough FW, Milclc LN, Michenfelder JD, Sundt TM (1987) Correlation of regional cerebral blood flow (rCBF) with EEG changes

during isoflurane anesthesia for carotid endarterectomy: Critical CBF. Anesthesiology 66:344

19. Michenfelder JD, Sundt TM, Fode N, Sharbrough FW (1987) Isoflurane when compared to enflurane and halothane decreases the frequency of cerebral ischemia during carotid endarterectomy. Anesthesiology 67:336

20. Nehls DG, Todd MM, Spetzler RF, Drummond JC (1987) A comparison of the cerebral protective effects of isoflurane and barbiturates during temporary focal ischemia in primates. Anesthesiology 66:453

21. Milde LN, Milde JH, Lanier WL, Michenfelder JD (1988) Comparison of the effects of isoflurane and thiopental on neurologic outcome and neuropathology following temporary focal cerebral ischemia in primates. Anesthesiology 69:905–913

22. Lam AM, Gelb AW (1983) Cardiovascular effects of isoflurane-induced hypotension for cerebral aneurysm surgery. Anesth Analg 62:742–748

23. Newman B, Gelb AW, Lam AM (1986) The effect of isoflurane-induced hypotension on cerebral blood flow and cerebral metabolic rate for oxygen in humans. Anesthesiology 64:307–310

24. Newberg LA, Milde JH, Michenfelder JD (1984) Systemic and cerebral effects of hypotension induced with isoflurane in dogs. Anesthesiology 60:541–546

II

Clinical Aspects

16

The Role of Calcium Entry Blockers in Brain Ischemia

Petter Andreas Steen[1]

Introduction

Drug treatment of cerebral ischemia is still controversial after decades of research, and no treatment appears to be firmly established. This is probably not so surprising when even the mechanisms causing the damage are not firmly established. Many good working hypotheses have been considered during the years, however, and it is exciting that recently a scientifically based connection between many of them appears to have evolved. Disturbances in the calcium homeostasis appears to be present in most cases [1,2], and it was therefore natural to start looking at possible protection against ischemic damage with calcium entry blockers a decade ago. In this paper some of the mechanisms whereby calcium theoretically can be involved in ischemic/hypoxic pathophysiology will be briefly discussed.

The next step in the development of a possible clinical treatment, namely in vitro or in vivo animal studies exploring mechanisms looking at mainly biochemical or pathological indices of a possible beneficial effect will not be discussed. The emphasis will instead be on outcome studies. There are many well-performed studies full of information that are not outcome studies, but in the area of calcium entry blockers, human, placebo-controlled, randomized, and blinded studies are becoming available, and these do, after all, represent the bottom line.

To know that a drug causes changes in cerebral blood flow that are likely to be beneficial, increases the survival time of hypoxic mice or reduces an increased intracranial pressure (ICP) is not enough. We must also know how it affects the long-term outcome. Let us use a reduction of an increased ICP as an example: the most effective way of reducing ICP is cardiac arrest with zero blood pressure, but that will certainly not improve the patient's outcome.

---

[1] Department of Anesthesiology, Ullevaal University Hospital, 0407 Oslo 4, Norway

Classification of Cerebral Hypoxic Insult

The pathophysiology of the different types of cerebral hypoxia varies greatly, and a drug can therefore not be expected to have the same protective effect in all conditions. One way to subdivide cerebral hypoxic insults is into three main categories: (1) global cerebral hypoxia (incomplete) where the whole brain is involved and where it still receives some oxygen, but not sufficient for normal function, (2) regional ischemia which, because of the nature of the cerebral circulation, always seems to be incomplete in man, and (3) global cerebral anoxia (complete) such as occurs with cardiac arrest.

In the first two categories the degree of hypoxia/ischemia can vary, but there is still some oxygen and other substrates being supplied to the brain. This makes a continuous production of, for instance, lactate and oxygen free radicals possible. In global cerebral anoxia the supply to the brain is completely shut off. Thus the production of these potentially tissue-damaging compounds must stop when the tissue stores of glucose and oxygen have been spent.

Possible Causes of Cerebral Damage Triggered by Calcium

As mentioned previously, increased intracellular calcium concentrations have been incriminated as a major trigger of the tissue damage occuring during hypoxia, resulting in reduced CBF and deleterious intracelluar events. Three factors might be involved in a possible decrease in CBF:

1. Vasoconstiction from an increase calcium concentration in the smooth muscle of the cerebral vasculature.
2. Blood viscosity might increase due to decreased deformability of the red blood cells secondary to an accumulation of calcium in the red cell membrane.
3. Calcium might be involved in platelet aggregation.

It also appears that an increase in free cytosolic calcium in the brain cells can activate many catabolic reactions with breakdown of proteins and probably most important, of phospholipids. The latter causes release of free fatty acids, including arachidonic acid, with production of ecosanoids such as prostaglandins, thromboxane, and possibly leucotriens—factors that might induce further tissue damage by further vasoconstriction, platelet aggregation [3] etc., depending on the balance between the various substances.

Large concentrations of calcium might also uncouple mitochondrial oxidative phosphorylation during hypoxia in that the oxygen supplied is used for pumping calcium directly into the mitochondria. The production of oxygen free radicals also appears to be influenced by the intracellular content of calcium in that increasing intracellular calcium levels might catalyze the production of xanthin oxidase from xanthin dehydrogenase [4]. To end this list, but probably not the in vivo list of possibilities, calcium appears to be involved in the hypoxic damage caused by excitatory amines such as glutamate [5].

The Role of Calcium Entry Blockers in Amelorating Brain Damage

There is a wide field of possible damaging effects of increased free calcium; but even if calcium might be involved in pathologic processes, it is not certain that the use of a calcium entry blocker might have any effects. The following variables must be considered:

1. Which calcium is involved? Is it only calcium from intracellular depots or also calcium which can be blocked from entering the cells by calcium entry blockers. There are numerous calcium channels—do the blockers block those responsible for calcium induced cell damage?
2. If calcium entry/release affected by the drugs is involved, can the drug reach the target in sufficient amounts to affect the pathophysiology? This point includes crossing the blood brain barrier and getting into the part of the brain where it is required. Causing vasodilatation in normal parts of the brain would, for instance, not be very helpful.
3. Finally, if the drug can reach the right place, can it be given early enough to effect the pathophysiology?

Results

There are good studies trying to answer some of these questions in detail on the mechanistic side. This review will concentrate on outcome and human data where available. If a drug improves the eventual clinical result, we know that it is possible to reduce tissue damage caused by increased levels of free calcium. There are numerous calcium entry blockers available. They are often quite different chemically, and the results with one drug can therefore not automatically be extrapolated to another. We are starting to gather some clinically valuable information on three types of hypoxia. Two are mainly regional, that is, subarachnoid hemorrhage (SAH) and stroke, and one is global cerebral anoxia which occurs during cardiac arrest.

Subarachnoid Hemorrhage (SAH)

Prospective, randomized, double-blind, and placebo-controlled data on the use of the calcium entry blocker nimodipine in man, are now available. Allen et al. [6] first reported on 116 neurologically normal patients with SAH due to rupture of an aneurysm. Nimodipine 0.7 mg/kg or placebo was given orally followed by 0.35 mg/kg every 4 h for 3 weeks starting within 96 h of SAH. The groups of patients were similar with regard to demographic data, previous illness, blood pressure, location of the aneurysm, and the amount of blood seen on the CT scan before treatment. The investigators found that the number of patients developing spasm that could be recognized angiographically was not different in the two groups: 13/56 in the nimodipine group vs 16/60 in the placebo group. There were significantly fewer patients with severe neurologic deficits in the nimodipine group that in the placebo group: 1/56 vs 4/60. If those with mild/

moderate deficits were also included, the difference, 5/56 vs 10/60, was no longer significant. There was no side effect attributable to nimodipine.

Allen et al. were later criticized because the study included so few patients with angiographically recognized spasms, but they replied that the study was halted by the controller for ethical reasons due to the significant improvement with nimodipine. Uncontrolled enthusiasm for the use of the drug thereafter held the scene until more studies have appeared in the last couple of years. Most important, the results are consistent with those of Allen et al. Phillippon et al. [7] reported on 70 patients studied within 72 h of SAH due to aneurysm rupture. The patients were grade I–III on the Hunt and Hess classification. Thus, the patient sample was fairly small which would make it difficult to achieve significant differences unless the drug effects were really striking. In this study 31 patients received 6 mg nimodipine orally every 4 h for 3 weeks while 39 patients received placebo. The groups were comparable with regard to demographic data, neurologic status at the time of inclusion, site of the aneurysm, time of surgery, and the number of patients with blood present in the basal subarachnoid space. Again, in this study there was no significant difference in the number of patients with vasospasm verified on angiography: 52% in the nimodipine group vs 72% in the placebo group. Only 6% in the nimodipine group has severe neurologic deficits (Hunt and Hess grade III-IV) related to vasospasm at a 3 weeks follow-up, significantly less than the 26% in the placebo group. There was no difference in side effects such as hypotension.

The third investigation is the Canadian multicenter study published by Petruck et al. [8]. A total of 154 patients were included for analysis, 72 of which received nimodipine 90 mg every 4 h starting within 96 h of SAH and ending 3 weeks after SAH. There were 82 patients who received a placebo. This study was different from the other two in that only poor-grade patients were involved, with Hunt and Hess clasification grade III or worse. All patients had angiography done 8 days post SAH and CT scans at their 3 months' assessment. There were no significant differences in demographic or previous illness data. The only significant baseline difference between the groups concerned intraventricular hemorrhage, in that there were significantly more patients in the nimodipine group who had blood occupying one full lateral ventricle. Thus, if anything, the nimodipine group should be worse off. There was no significant difference in the degree of vasospasm in the nimodipine and placebo groups when evaluated angiographically, 64% and 66% moderate to severe vasospasm, respectively. The degree of vasospasm correlated with the thickness of the subarachnoid clot as seen on the admission CT.

Also in this study the vasospasms were accompanied by significantly less deficits in the nimodipine group than in the placebo group. There were no differences where causes other than spasm could explain the results. This reduction in delayed ischemic deficits corresponded with a significantly better 3 months' outcome in the nimodipine group as evaluated with the Glasgow Outcome Scale. Although this represents a positive trend, there was no significant difference in the number of infarcts or in their size as evaluated on a CT scan at 3 months' follow-up. Again, there were no significant differences in side effects such as hypotension.

Other placebo-controlled studies that have only appeared in abstract form yet seem to confirm these results. All of these studies indicate that nimodipine can reduce neurologic deficits attributable to vasospasm in patients with SAH, but that this effect cannot be correlated with a reduction in angiographically observed vasospasm.

Stroke

Stroke is another regional insult. There are numerous animal studies looking at the effects of calcium entry blockers on cerebral blood flow, edema formation, electrolytes and outcome in stroke, some reporting positive results, some not. To date there is only one double-blind, randomized, placebo-controlled published study in patients, the multicenter study from Gelmers et al. [9]. The data analysis included 93 patients in each group. The patients received nimodipine or placebo every 6 h for 4 weeks starting within 24 h of the onset of stroke. All patients also received low molecular weight dextran and heparin. There were no significant differences between the two groups for demographic data or risk factors such as hypertension or diabetes, and, as in the other studies, no difference in blood pressure or heart rate.

The neurologic outcome at the end of the treatment period was significantly better in the nimodipine group than in the placebo group. When the data were analysed, the improvement was striking in patients with a medium neurologic baseline score, while the subgroups with a very poor starting point or those with only a minor deficit fare the same with and without nimodipine. Parallell to these data, significantly more patients died in the placebo group (20% vs 9%) at the end of the treatment period. The authors noted that the improvement in result with nimodipine was large for males, with no significant difference in results for females.

Cardiac Arrest

The third category of insults is cardiac arrest or complete global cerebral ischemia such as is seen with cardiac arrest. It has been speculated that much of the neurologic damage with this insult occurs in the post-ischemic period where a severe delayed hypoperfusion occurs, a period where all the tissue damaging processes mentioned earlier can act. Initial studies in animals indicated that many of the calcium entry blockers could increase CBF during this period and possibly also improve the neurologic outcome.

In dogs subjected to 10 min of circulatory arrest by occlusion of the aorta and both venae cavae, nimodipine given both before and after the arrest doubled global CBF and improved the outcome [10]. When given only after the arrest, CBF still doubled, but the outcome results were equivocal, not significantly different from either control or pretreated animals [11]. Flunarizine did not appear to have any effect on CBF or the outcome [12] while nicardipine increased the CBF but failed to improve the outcome in the same model [13]. Lidoflazine failed to affect CBF with equivocal outcome results ($P = 0.077$, 9 vs 9 animals) [14]. In dogs with ventricular fibrillation, lidoflazine appeared to improve the outcome, but not after an asphyxial arrest [15]. Lidoflazine failed to improve the outcome of primates when given after 17 min of complete cerebral ischemia obtained by hypotension combined with a neck tourniquet, after which the monkeys were treated in an intensive care unit for 5 days [16]. Nimodipine on the other hand improved the outcome in the same model [17].

Although these are all studies of neurologic outcome, there are numerous problems in evaluating the results. There is no complete consistency in the outcome as illustrated by the Lidoflazine results. Furthermore, animal studies can never exactly mimic the clinical situation. Some are invasive, requiring a thoracotomy for instance, and the use of catecholamines and blood pressure-reducing drugs might not be the same as in the clinical situation. The link between an increased CBF in the post-ischemic hypoperfusion period and an improved neurologic outcome must also be questioned. There is no consistency between the changes in CBF and outcome, and Smith et al. from Siesjö's laboratory found that although nimodipine increased CBF, the blood flow pattern was still very patchy, in some areas had very high and in some very low, although none lower than without nimodipine [18].

We were interested in the latter problem, and decided to investigate the effects of nimodipine on CBF and correlate this with neurologic outcome in patients resuscitated after cardiac arrest [19]. Fifty-one patients were thus studied after successful cardiopulmonary resuscitation by a physician-manned ambulance; 25 were treated with nimodipine and 26 with placebo in a blinded randomized fashion. Nimodipine, $0.25 \text{ mg} \cdot \text{kg}^{-1} \cdot \text{min}^{-1}$ or placebo was started as soon as the patients had an adequate peripheral pulse. After 1 h this infusion was increased to $0.5 \text{ mg} \cdot \text{kg}^{-1} \cdot \text{min}^{-1}$ and continued for an additional 9 h. CBF was evaluated during the first 4 h postresuscitation and 24 h later when feasible, using the the intravenous Xenon method with 10 regional detectors over the brain. Patients were then followed neurologically for 4 months or until death.

The results were as follows: as expected, CBF was very low in the placebo group despite an adequate blood pressure. Nimodipine doubled this flow just as in the dog experiments. Spinal fluid pressure was not different in the two groups, with no single measurement above 20 mmHg during the drug treatment period. Mean arterial blood pressure (MABP) was significantly lower in the nimodipine group during the treatment period, but the MABP did not decreased below 86 mm Hg with nimodipine. There was no significant difference in the use of catecholamines, and it might be worth mentioning that the nimodipine-treated patients required significantly less antiarryhthmic treatment than the placebo-treated group. There was no significant difference in the neurologic outcome. Twelve patients in each group woke up, and one in the nimodipine group was not evaluated since he died after 13 h. Four months later, 8 patients in the nimodipine-treated group and 10 patients in the control group returned to a normal life at home, 6 having died in the meantime of cardiac causes. Of the patients that did not wake up, only one nimodipine-treated patient survived for 6 months with severe brain damage. When neurologic outcome was compared to CBF, another pattern appeared. The highest blood flow values were found in patients receiving nimodipine that did not wake up. Thus it appears unlikely that the CBF effects of nimodipine can be linked to any improvement in the neurologic outcome.

Two other blinded placebo-controlled studies are under way, evaluating the effects of calcium entry blockers on neurologic outcome after cardiac arrest. One is a multicenter study of lidoflazine monitored by Safar's group in Pittsburgh. The other is a study of nimodipine from the University of Helsinki. Both should have sufficient data within a year to make a proper evaluation.

It might be of interest to note that Kaste's group in Finland uses the same doses of nimodipine that was used in monkeys and dogs ($10 \ \mu g \cdot kg^{-1}$ followed by $1 \ \mu g \cdot kg^{-1} \cdot min^{-1}$) appearently without serious side effects. They have reported the preliminary results from 22 patients in a pilot study in the British Medical Journal [20]. The promising 64% survival with nimodipine was twice as good as in a historic control group. This should be taken only for what It is, a study not being randomized or blinded and with a retrospectively studied control group from a previous time period.

Conclusions

It seems established that calcium entry blockers, at least nimodipine, are of value in patients suffering from SAH. Although this is promising, more information is needed to evaluate the effects in the treatment of stroke. The jury is still out in the cardiac arrest case. More conclusive, human data should be available by a year or two. We do know already that the drugs are well tolerated hemodynamically in these sick patients.

Summary. Increased levels of free calcium appears to be involved in most of the pathophysiologic processes occurring in connection with cerebral hypoxia. Possible deleterious effects include reduced cerebral blood flow (CBF), breakdown of proteins and phospholipids, ecosanoid production, free oxygen radical production, and increased levels of excitatory amines. It is possible that calcium entry blockers could reduce some of these effects. This review is concentrated on clinical studies of calcium entry blockers in cerebral ischemia. In subarachnoid hemorrhage it seems established that the calcium entry blocker, nimodipine, has a beneficial effect. In the only clinical study of nimodipine in stroke, the drug also improved the outcome, while animal studies have given variable results. In complete cerebral ischemia such as cardiac arrest, animal studies have again given variable results. Uncontrolled human data are promising, but the increase in CBF seen with nimodipine after cardiac arrest did not correlate with an improved outcome.

References

1. Cheung JY, Bonventre JV, Malis CD, Leaf A (1986) Calcium and ischemic injury. New Engl J Med 314:1670–1676
2. Siesjö BK (1981) Cell damage in the brain: A speculative synthesis. J Cereb Blood Flow Metab 1:155–185
3. Greer IA (1987) Platelet function and calcium channel-blocking agents. J Clin Pharm Ther 12:213–222
4. McCord JM (1985) Oxygen-derived free radicals in postischemic tissue injury. N Engl J Med 312:159–163
5. Rothman SM, Olney JW (1986) Glutamate and the patophysiology of hypoxic-ischemic brain damage. Ann Neurol 19:105–111
6. Allen GS, Ahn HS, Preziosi TJ, Battye R, Boone SC, Chou SN et al. (1983) Cerebral arterial spasm: A controlled trial of nimodipine in patients with subarachnoid hemorrhage. N Engl J Med, 308:619–624
7. Philippon J, Grot R, Dagreou F, Guggiari M, Rivierez M, Viars P (1986) Prevention of vasospasm in subarachnoid haemorrhage. A controlled study with nimodipine. Acta Neurochir (Wien) 82:110–114
8. Petruk KC, West M, Mohr G, Weir BKA, Benoit BG, Gentili F et al. (1988) Nimodipine treatment in poor-grade aneurysm patients. J Neurosurg 68:505–517
9. Gelmers HJ, Gorter K, de Weerdt CJ, Wiezer HJA (1988) A controlled trial of nimodipine in acute ischemic stroke. N Engl J Med 318:203–207
10. Steen PA, Newberg LA, Milde JH, Michenfelder JD (1983) Nimodipine improves cerebral blood flow and neurologic recovery after complete cerebral ischemia in the dog. J Cereb Blood Flow Metab 3:38–43.
11. Steen PA, Newberg LA, Milde JH, Michenfelder JD (1984) Cerebral blood flow and neurologic outcome when nimodipine is given after complete cerebral ischemia in the dog. J Cereb Blood Flow Metab 4:82–87
12. Newberg LA, Steen PA, Milde JH, Michenfelder JD (1984) Failure of flunarizine to improve cerebral blood flow or neurologic recovery in a canine model of complete cerebral ischemia. Stroke 15:666–671
13. Stazabe T, Nagai I, Ischikawa T, Takeshita H, Masuda T, Matsumoto M, Tateishi A (1986) Nicardipine increases cerebral blood flow but does not improve neurologic recovery in a canine model of complete cerebral ischemia. J Cereb Blood Flow Metab 6:684–690
14. Fleischer JE, Lanier WL, Milde JH, Michenfelder JD (1987) Effect of lidoflazine on cerebral blood flow and neurologic outccme when administered after complete cerebral ischemia in dogs. Anesthesiology, 66:304–311
15. Vaagenes P, Cantadore R, Safar P, Mossy J, Rao G, Diven W, Alexander H, Stezoski W (1984) Amelioration of brain damage by lidoflazine after prolonged ventricular fibrillation cardiac arrest in dogs. Crit Care Med 12:846–855
16. Fleischer JE, Lanier WL, Milde JH, Michenfelder JD (1987) Lidoflazine does not improve neurologic outcome when administered after complete cerebral ischemia in primates. J Cereb Blood Flow Metab 7:366–371
17. Steen PA, Gisvold SE, Milde JH,. Newberg LA, Scheithaner BW, Lanier WL, Michenfelder JD (1985) Nimodipine improves outcome when given after complete cerebral ischemia in primates. Anesthesiology 62:406–414
18. Smith ML, Kågström E, Rosen I, Siesjö BK (1983) Effect of the calcium antagonist nimodipine on the delayed hypoperfusion following incomplete ischemia in the rat. J Cereb Blood Flow Metab 3:543–546

19. Forsman M, Aarseth HP, Nordby HK, Skulberg A, Steen PA (1989) Effect of nimo-
    dipine on cerebral blood flow and cerebrospinal fluid pressure after cardiac arrest:
    Correlation with neurologic outcome. Anesth Analg, 68:436–443
20. Roine RO, Kaste M, Kinnunen A, Nikki P (1987) Safety and efficacy of nimodipine
    in resuscitation of patients outside hospital. Br Med J (Clin Res) 294:20

17

Management of Brain Edema in Head Injury

J. Douglas Miller and N. Mark Dearden[1]

Introduction

Strictly speaking, brain edema represents an increase in the bulk of the brain
that has been produced by an increase in its tissue water content. In the context of
the management of severely head-injured patients in the intensive care unit,
however, the term *brain edema* is commonly used in a wider context, to include
all conditions of increased brain volume or brain swelling frequently, but not
always, associated with raised intracranial pressure (ICP). There are many
potential causes of raised ICP after head injury. These include formation of true
brain edema, development of an intracranial mass lesion (such as a subdural
hematoma, which may also be associated with edema formation), acute swelling
of the brain due to an increase in cerebral blood volume, and accumulation of
cerebrospinal fluid in the form of acute hydrocephalus. For optimal management
of the severely head-injured patient and effective treatment of raised ICP and its
adverse sequelae, it is important to be aware of the several possible causes of
intracranial hypertension and different mechanisms by which water content may
increase and ICP rise. Wherever possible, therapy should be chosen that is
appropriate to the underlying pathophysiological mechanism. In the comatose
head-injury patient, the therapeutic "window" is narrow and time does not permit a random sequence of empirically applied therapies for reduction of ICP and
resolution of brain edema.

Types of Brain Edema that May Follow Head Injury

A major advance in understanding the pathophysiology of brain edema was the
differentiation by Klatzo [1] between vasogenic and cytotoxic edema. Stern [2],
Langfitt [3], Fishman [4], and Miller [5] later delineated further types of cerebral

---

[1] Department of Clinical Neurosciences, University of Edinburgh, Western General Hospital, Edinburgh EH4 2XU, Scotland, UK

edema—osmotic, hydrostatic and interstitial—each of which could be distinguished from vasogenic and cytotoxic edema. All of these forms of edema may be found in head-injured patients.

Vasogenic Edema

Vasogenic edema consists of an outpouring of protein-rich fluid into the extracellular space from vessels in which the blood-brain barrier is deficient. The fluid emanates largely from grey matter, where most cerebral vessels are located, but spreads into the white matter of the centrum semiovale, where tissue compliance is greater than in the tightly packed grey matter. The rate of edematous fluid outflow from vessels is increased by hypercapnia, hypoxia, arterial hypertension, and elevated body temperature; it is decreased by hyperventilation, arterial hypotension, and raised ICP, but with a major risk of cerebral ischemia. Resolution of brain edema is impeded by a raised ICP and improved by reducing the ICP. In head injury, vasogenic edema is most often associated with localised brain contusion, visible on CT scan as a lucent zone surrounding the contused area of brain.

Cytotoxic Edema

Cytotoxia edema is caused by impairment of the cell membrane pump mechanisms as a result of energy failure or poisoning. Fluid accumulates within cells, therefore this form of edema is most evident in grey matter. In head injury, cytotoxic edema is most often seen as a complication of cerebral ischemia. This may be due to raised intracranial pressure with reduced perfusion pressure, to cerebral vasospasm or occlusion of a major vessel, or to distortion of cerebral vessels involved in brain shift. As the cerebral blood flow falls below a series of threshold values, astrocytes swell, and neurones stop firing and then become swollen. At this stage there is an efflux of potassium into the extracellular space, followed by a calcium influx that can end in cell destruction under circumstances and by mechanisms that will be described elsewhere. The threshold conditions for edema formation are related not only to flow levels but also to the time over which they apply [6]. Thus, lesser degrees of cerebral ischemia may cause edema if applied for longer time periods.

Hydrostatic Edema

Hydrostatic edema is caused by a rise in cerebral intravascular or transmural pressure sufficient to alter the Starling Equation and to result in an outpouring of water from cerebral capillaries into the extracellular space. Because the blood-brain barrier is intact, the fluid is protein poor, providing the contrast with vasogenic edema. In head injury, hydrostatic edema is most often encountered after surgical decompression of large acute intracranial hematomas. The brain under the hematoma is compressed and autoregulation in this part of the cerebral vascular bed is impaired. When decompression occurs, there is a major fall in ICP, and a corresponding abrupt increase in cerebral perfusion pressure—the difference between arterial pressure and ICP. Because autoregulatory vasoconstriction is impaired, the rise in perfusion pressure is transmitted into the capillary bed, and the transmural pressure gradient results in fluid shift to the extracellular space.

Osmotic Edema

Osmotic edema results when the serum osmolality falls; the water accumulation is intracellular. In head injury, this is most often related to hyponatremia produced by natriuresis or by overinfusion of intravenous dextrose solution as fluid replacement. In this context, renal dialysis of severely head-injured patients carries significant risks.

Interstitial Edema

Interstitial edema occurs in the periventricular white matter as a complication of obstructive high-pressure hydrocephalus. It is very rare in head injury, since even in cases of posterior fossa hematoma the direct effects of the mass are usually manifest before obstructive hydrocephalus becomes marked.

Vascular Brain Swelling Following Head Injury

The bulk of the brain may be increased by a rise in the volume of blood contained in the cerebral vascular bed. In general terms, this may result from active cerebral arterial vasodilatation, from passive distension of vessels associated with an increase in arterial pressure in the presence of impaired autoregulation, or to obstruction of the cerebral venous outflow.

In patients with head injury, episodes of cerebral vasodilatation are common. Because the increase in blood volume often results in a marked increase in ICP, this may act as a flow-limiting factor, so that cerebral blood flow is not increased even though there is dilatation of cerebral vessels [7]. Obstruction of the venous outflow can occur due to injuries of the skull overlying the superior sagittal or transverse sinuses. Venous pressure may be increased by positive end-expiratory pressure (PEEP), elevated central venous pressure (CVP), excessive flexion, or rotation of the neck. As cerebral compliance becomes reduced after severe brain injury, the increases in ICP that result from changes in cerebral blood volume may become very large indeed [8], to the point that cerebral blood flow is reduced below the threshold level of energy delivery required to maintain the membrane pumps [6,9]. At that point, congestive, vascular brain swelling is followed by true cytotoxic brain edema. The ill-effects of brain edema and swelling appear to be mediated via two main mechanisms, reduction in cerebral blood flow and brain shift or distortion, when the causative lesion and the edema are focal. The reduction in cerebral blood flow is largely related to the extent of any increase in ICP.

Diagnosis of Raised ICP and Brain Edema

In the context of head injury, there are no reliable bed-side clinical indicators or the presence of raised ICP. Papilledema, normally a reliable sign of raised ICP, seldom occurs after head injury, even in patients known by direct measurements of ICP to be suffering from intracranial hypertension [10]. Computed tomography is, however, very helpful. Edema is shown as a radiolucent area with shift of surrounding structures [11]. Reliable signs indicating that raised ICP is present, or likely to occur, include the pattern of diffuse brain swelling, with loss of the appearance of the third ventricle and the perimesencephalic CSF cisterns, or of mass effect with compression of one lateral ventricle and enlargement of the opposite lateral ventricle [12].

Clearly, the most accurate information can be obtained from a direct measurement of the ICP from the cranial cavity. Such an extradural, subdural, or intraventricular measurement is invasive but may be necessary to answer important questions about therapy. If ICP is being measured, then it is possible to obtain additional information by charting ICP changes over time to see if a trend develops, or if there is a distinctive pattern of waves of increased ICP [13]. If it possible to make controlled changes in CSF volume, the resultant change in ICP can be measured, and from this data can be derived values that describe lumped intracranial compliance and the resistance to the outflow of CSF [14]. Analysis of the ICP waveform can be performed to estimate the relative weights of low and high frequency components in the compound waveform. If appears likely that the higher frequency components reflect factors related to intracranial compliance. Thus, an increase in the high frequency components of the ICP waveform would suggest that the pressure volume index (PVI) is reduced [15–17].

Measurement of low density areas in CT of the brain has been successfully related to the severity and extent of some types of cerebral edema, but such methods have not worked very well in head injury [11]. Magnetic resonance imaging provides a more effective way of delineating in vivo areas of brain edema and providing a quantitative estimate of the actual value of brain tissue water content [18]. Other techniques of obtaining valuable information on brain function and pathophysiological status in the presence of brain edema include recording of brain electrical activity, measurement of cerebral arteriovenous oxygen content difference (using a catheter placed in the jugular bulb to obtain samples of cerebral venous blood), and transcranial measurements of blood-flow velocity in the middle cerebral and other major intracranial arteries.

Starting Therapy for Raised Intracranial Pressure or Brain Edema

It is perhaps fortunate that therapy applied to the reduction of ICP also tends to reduce the extent or severity of brain edema, and measures aimed at the resolution of brian edema often produce reduction of ICP. In the severely head-injured patient, the decision to treat raised ICP usually has to be made on the basis of the continuous recording of ICP, since the patient at this stage is under the influence of a muscle relaxant and/or sedative drugs and clinical assessment is no longer possible. The threshold level of ICP above which treatment should be started, is a matter of current controversy. It is our practice in Edinburgh to treat ICP levels above 25 mmHg mean, measured at 10° head-up tilt with zero reference at the auditory meatus, in the first 2 days after injury, and levels above 30 mmHg thereafter. However, others have suggested that if the treatment threshold were lowered to 20 or even 15 mmHg then subsequent episodes of severe intracranial hypertension might be averted. This has yet to be proved; lowering the treatment threshold will result in many more therapeutic interventions, and these are not without risk.

General Measures for Treatment of Raised ICP

In the management of severely head-injured patients in the Intensive Care Unit, a number of situations are encountered that may produce increases in ICP. These include extreme flexion or rotation of the neck, airway obstruction, spontaneous respiratory efforts against the ventilator, increase in body temperature, hypoxia or hypercapnia, hyponatremia, pain with arterial hypertension, and epileptic seizures. Meticulous attention must be paid to the detection and correction of these factors before considering specific treatment of intracranial hypertension. When the patient is being artificially ventilated the usual target level of arterial PCO_2 is 30 mmHg. Levels lower than this may cause excessive cerebral vasoconstriction and push the cerebral arteriovenous oxygen content difference into the ischemic range (> 9 ml0_2/dl) [19]. Another possible hazard of extreme hyperventilation is arterial hypotension.

In cases of chronic perifocal brain edema related to brain tumor or abscess, steroid therapy has an established place. The administration of dexamethasone, betamethasone, or methylprednisolone produces clinical improvement in 6–12 h, improvement in brain compliance and loss of ICP waves in 24–36 h, then gradual reduction of the baseline ICP after 48–72 h of treatment [20–22]. In acute head injury the situation is entirely different. Steroid therapy reduces neither the mortality, the incidence, nor the severity of raised ICP [23–25]. Indeed, in the clinical trial reported by Dearden and associates, steroid-treated patients who already had intracranial hypertension had a higher mortality and morbidity than similar patients who received placebo [26]. In Edinburgh, steroid therapy is not used in patients with severe head injury. There is still a question about the value of steroid therapy in patients with focal cerebral contusion associated with perifocal edema. Although there is a stronger theoretical case for the employment of steroids in such cases, there is still no firm data in support of their use, even in this instance.

Specific Therapy for Raised Intracranial Pressure in Head Injury

If intracranial hypertension persists despite general measures, and a space-occupying lesion has been excluded by CT, more specific therapy is required. In most head-injured patients, this comes down to a choice between osmotic agents and hypnotic drugs. Osmotic agents reduce ICP by causing a reduction in brain water [18], but also cause cerebral vasoconstriction under circumstances of pre-served cerebral vasoreactivity and reduced cerebral perfusion pressure [27–29]. Osmotic agents also reduce cranial CSF volume [30]. Hypnotic drugs reduce ICP by causing a fall in cerebral energy metabolism and in brain tissue blood flow. They are most effective in the presence of cerebral hyperemia and with pre-served brain electrical activity and vascular reactivity to CO_2; when a mass lesion is present barbiturates may produce cerebral ischemia [31]. Each approach to treatment, therefore, has advantages and drawbacks and it appears likely that a selective approach to the administration of these therapies is to be preferred. Ward and colleagues reported on a randomised trial in which barbiturate ther-apy was applied to patients with severe head injury soon after admission, and compared it to a standard regimen to determine whether barbiturates would prevent or reduce the severeity of intracranial hypertension [32]. No such effect was observed, but there was a 50% incidence of episodes of arterial hypotension in the barbiturate-treated patients.

Indiscriminate, continued use of mannitol can also cause problems, related to excessive levels of serum osmolality that can result in renal failure [33]. It appears preferable to define specific indications for each form of treatment and proceed directly to use of that agent. This saves time, which may be a critical factor, and probably reduces the risk of ill effects related to treatment.

As a starting paradigm, it can be proposed that in severely head-injured pa-tients, most rises in intracranial pressure result primarily from an increase in cerebral blood volume or in brain tissue water content (edema). When blood volume changes predominate, barbiturate or other sedative therapy should be indicated; when brain edema is the major factor, osmotherapy is the preferred treatment. How can one distinguish rapidly between these two mechanisms? On CT, brain edema is more common in the case of focal lesions such as contusion, or following the evacuation of an extracerebral or intracerebral hematoma. Vascular brain swelling is more common when CT shows only diffuse brain swelling, with loss of the CSF spaces. On the chart record, an ICP trace that has a large pulse pressure compared to the respiratory component measured while employing standardised ventilator settings (pulse: respiratory ratio > 1.4) is sug-gestive of intracranial hypertension that is predominantly due to increased blood volume [34]. Conversely, a pulse: respiratory ratio less than 0.8 suggests brain edema and reduced craniospinal compliance as the main problems. Frequency analysis of the ICP record may also be of help. An increasing preponderance of high frequency components in the ICP waveform compared to the BP waveform as revealed by Fast Fourier Analysis is suggestive of decreasing compliance, and can be correlated with a reducing value of PVI [17].

Preservation of cerebral vascular responsiveness to changes in arterial pressure and PCO_2 appears to be necessary for a response to mannitol and barbiturates, respectively, in terms of ICP reduction [28,35]. Finally, brain electrical activity must be retained if hypnotic therapy is to be successful in reducing ICP. When the EEG is flat, administration of hypnotics is without benefit [36].

When an osmotic agent is indicated, we currently prefer to give intravenous mannitol in a dose of 0.5 G/kg body weight over a period of 20 min. At the end of the infusion we administer the diuretic agent, frusemide 20–40 mg in an adult patient, then 2 ml/kg of plasma protein solution to guard against hypovolemic hypotension as a result of the mannitol and frusemide-induced diuresis.

When hypnotic drugs are indicated, thiopentone 5 mg/kg or gammahydroxy-butyrate 60 mg/kg are given intravenously over 5–10 min. During this time the effect on both arterial and intracranial pressure are measured; therapy is continued by continuous infusion only if CPP is preserved or increased, using the agent that produces the least reduction in arterial pressure [37].

Comparison Study of Hypnotics and Osmotherapy in Head Injury

We are engaged in a continuing study comparing the efficacy of mannitol and hypnotic drugs in reducing ICP and preserving or restoring cerebral perfusion pressure (CCP) in severely head-injured patients. To date, 16 patients have been studied, one on two occasions. All patients were managed in a standardised way, employing artificial ventilation, with constant values for inspiration, pause expiration time, respiratory rate, and inflation pressure, continuous monitoring of ICP, systemic arterial pressure (SAP), CVP and brain electrical activity (cerebral function monitor, CFM). Arterial and jugular venous blood samples have been obtained for hemoglobin and blood gas analysis and calculation of the cerebral arteriovenous oxygen content difference ($AVDO_2$).

Increases in ICP over 25 mmHg that had no obvious correctable primary cause were treated in randomised order either by mannitol (0.5G/kg) or by hypnotic drugs, thiopentone (5mg/kg) in 6 cases, thiopentone and gammahydroxybutyrate in 9 cases, or gammahydroxybutyrate (60 mg/kg) alone in two cases. Based upon the capacity of each therapeutic agent to reduce ICP towards normal while maintaining CPP above 60 mmHg in adults or 50 mmHg in children under 10 years, we have identified four response groups. These are (A) mannitol superior to hypnotic ($\bar{n} = 5$), (B) hypnotic superior to mannitol ($\bar{n} = 3$), (C) both effective ($\bar{n} = 5$) and (D) neither effective ($\bar{n} = 4$).

The data from the study to date are shown in Tables 1, 2 and 3, and can be summarised as follows. Hypnotic agents were more effective in patients with diffuse brain injury, while mannitol was more effective in cases of focal injury (contusion or post hematoma removal). Hypnotics were more effective when brain electrical activity was preserved and the CFM lower border voltage was greater than 5 microvolts. Indeed we have found that the extent of the reduction of ICP produced by hypnotic drugs is related significantly to the CFM voltage [37]. When hypnotics were more effective, the P:R ratio of the ICP waveform was increased over 2, whereas when mannitol was the more effective therapy the P:R ratio was generally below 1.4 (mean 0.99). The pre-treatment value of $AVDO_2$ did not discriminate between preferred therapies, but when either therapy was successful in reducing ICP and restoring CPP there was a decrease in $AVDO_2$. In some individual cases, unsuccessful therapy has been signalled by an increase in $AVDO_2$ into the ischemic range (> 9 ml/dl).

J. Douglas Miller and N. Mark Dearden

Table 1. Characteristics of patients in four treatment groups. Brain injury classified as focal (F) or diffuse (D); Glasgow Coma Score (GCS); Therapy–mannitol (M) thiopentone (T) or gammahydroxybutyrate (G). Outcome classified as good recovery (GR), moderate disability (MD), severe disability (SD), or dead (D)

Treatment Group	Patient	Age (years)	Brain Injury	GCS on Admission	Therapy	Outcome
A Mannitol	GM	24	F	4	M,T	MD
superior to	JM	14	F	4	M,C	D
hypnotic	KL	21	F	9	M,T	MD
	JB	31	F	5	M,T	SD
	GM	44	F	11	M,T	D
B Hypnotic	DG	10	D	5	M,G	MD
superior to	GG	3	D	7	M,T	MD
mannitol	WG	25	D	6	M,T	GR
C Both	JM	14	F	4	M,T	D
effective	CS	8	F	7	M,T	GR
	HW	22	F	7	M,T	GR
	HM	54	F	7	M,T	D
	DM	17	F	4	M,T	D
D Neither	JR	41	F	6	M,T	SD
effective	CG	62	F	5	M,T	D
	FS	43	D	3	M,T	D
	LP	24	F	4	M,T	D

Table 2. Changes in intracranial pressure (ICP) and cerebral perfusion pressure (CPP) following mannitol or hypnotic drugs in four treatment groups (mean $\pm$ SEM)

	Treatment Group	Treatment	ICP pre	ICP post	CPP pre	CPP post
A	Mannitol superior to hypnotic ($n = 5$)	Mannitol	37.8 ± 5.1	20.4 ± 2.2	61.2 ± 4.6	77.6 ± 6.1
		Hypnotic	39.8 ± 6.5	26.4 ± 4.2	54.6 ± 4.8	51.2 ± 3.5
B	Hypnotic superior to mannitol ($n = 3$)	Mannitol	44.7 ± 2.3	33.7 ± 4.6	50.7 ± 6.0	62.3 ± 8.0
		Hypnotic	42.7 ± 5.0	14.3 ± 3.8	46.7 ± 1.4	68.3 ± 2.4
C	Both effective ($n = 5$)	Mannitol	36.8 ± 2.7	23.6 ± 1.4	52.8 ± 4.7	71.4 ± 5.4
		Hypnotic	33.2 ± 4.3	20.2 ± 5.1	57.8 ± 5.6	67.4 ± 7.8
D	Neither effective ($n = 4$)	Mannitol	58.7 ± 10.6	44.2 ± 10.2	50.0 ± 3.5	55.0 ± 2.4
		Hypnotic	56.2 ± 10.7	45.8 ± 13.8	53.7 ± 5.1	52.3 ± 4.6

Table 3. Values for brain electrical activity–cerebral function monitor (CFM) lower border voltage, pulse: respiratory ratio (P:R) of the intracranial pressure waveform, and the cerebral arteriovenous oxygen contect difference before and during therapy

	Treatment Group	Treatment	CFM uV	P:R ratio	A-VO$_2$ Difference (ml O$_2$/dl) pre	post
A	Mannitol superior to hypnotic ($n = 5$)	Mannitol	3.8 ± 0.5	0.98 ± 0.23	6.95 ± 1.66	4.81 ± 1.63
		Hypnotic	4.4 ± 0.4	0.99 ± 0.34	6.66 ± 1.47	8.63 ± 2.43
B	Hypnotic superior to mannitol ($n = 3$)	Mannitol	8.0 ± 0.6	2.9 ± 1.15	6.95 ± 3.72	6.44 ± 3.61
		Hypnotic	8.7 ± 0.3	3.4 ± 0.92	7.11 ± 3.55	5.36 ± 2.48
C	Both effective ($n = 5$)	Mannitol	5.7 ± 0.5	0.9 ± 0.12	5.87 ± 0.29	4.58 ± 0.35
		Hypnotic	5.6 ± 0.5	1.1 ± 0.55	6.53 ± 0.50	6.12 ± 0.20
D	Neither effective ($n = 4$)	Mannitol	2.6 ± 0.5	2.9 ± 1.40	7.14 ± 2.05	6.96 ± 2.32
		Hypnotic	3.6 ± 0.6	2.6 ± 1.18	8.02 ± 3.11	8.34 ± 3.06

While this study is a continuing one, and because of the stringent conditions required to make valid comparisons, only a small proportion of our cases can be so studied, we can already make some conclusions; that a selective approach to therapy is feasible, that analysis of the ICP waveform and monitoring of brain electrical activity provide valuable information, and measurement of the cerebral AVDO$_2$ can indicate when one type of treatment is not only unhelpful, but may even be harmful to the patient. Finally, the preferred therapy for raised ICP in a given patient may change with the passage of time.

In reviewing our overall experience in the management of 228 severely head-injured patients (GCS 8 or less) treated over the past two years, we have used mannitol and/or hypnotic agents in 85 of the 123 severely head-injured patients in whom ICP monitoring was employed (69%). While mannitol alone has been adequate therapy in 52 patients, in only two cases have barbiturates alone been adequate therapy. In the remaining 31 barbiturate-treated patients, mannitol was also required at some time to control raised ICP. If a blind choice of therapy is urgently required then mannitol would, in our opinion, be the drug of choice, but as we have seen, hypnotic drugs can produce superior effects in appropriate, carefully selected cases.

Conclusions

Raised intracranial pressure, brain swelling, and edema continue to present a major challenge in the management of the severely head-injured patient, even when intracranial hematomas have been quickly recognised, promptly removed, and the patient intensively monitored and ventilated. We believe that if raised ICP persists after all remediable causes have been dealt with, the best approach to specific ICP-reductive therapy is a selective one based upon determination whether the main cause of the increase in ICP is vascular or related to brain edema. Analysis of the ICP waveform, monitoring of brain electrical activity and measurement of the cerebral arteriovenous oxygen content difference all provide valuable help in the selection and evaluation of therapy.

Summary. Raised intracranial pressure is common following severe head injury. It is associated with both brain edema and vascular brain swelling. One or more of five types of edema may be present—vasogenic, cytotoxic, hydrostatic, osmotic, and interstitial. Clinical signs are unreliable in the diagnosis of raised ICP and edema. Edema can be detected and raised ICP may be suspected by CT and MR imaging, but the only certain way of detecting intracranial hypertension is to measure ICP. Parallel information from EEG, jugular venous oxygen saturation, and blood flow velocity using transcranial Doppler sonography can indicate those situations in which raised ICP is compromising brain function.

Raised ICP should normally be treated when ICP exceeds 25 mmHg. General measures to correct preventable causes must be meticulously applied before proceeding to specific therapy. Specific treatment in the form of osmotic agents and hypnotic drugs should be directed to the predominant cause of the intracranial hypertension, vascular swelling, or edema. In an ongoing comparison study, we find barbiturate therapy superior to hypnotic drugs in cases of vascular swelling. These are found in diffuse brain injury, in young patients with a high pulse: respiratory amplitude ratio on the ICP trace, with preserved brain electrical activity and a low cerebral arteriovenous oxygen content difference. Mannitol is more effective in cases of focal brain injury with a normal or low P:R ratio, and the status of brain electrical activity is not a critical factor.

References

1. Klatzo I (1967) Neuropathological aspects of brain edema. J Neuropathol Exp Neurol 24:1–14
2. Stern WE, Coxon RV (1967) Osmolality of brain tissue and its relation to brain bulk. Am J Physiol 206:1–7
3. Schutta HS, Kassell NF, Langfitt TW (1968) Brain swelling produced by injury and aggravated by arterial hypertension: A light and electron microscopic study. Brain 91:281–294
4. Fishman RA (1975) Brain edema. N Engl J Med 293:706–711
5. Miller JD (1979) Clinical management of cerebral oedema. Br J Hosp Med 20:152–166
6. Bell BA, Symon L, Branston NM (1985) CBF and time thresholds for the formation of ischaemic cerebral edema and effect of reperfusion in baboons. J Neurosurg 62:31–41.
7. Miller JD, Stanek A, Langfitt TW (1972) Concepts of cerebral perfusion pressure and vascular compression during intracranial hypertension. In: Meyer JS,

Schade JP (eds) Cerebral blood flow: Progress in brain research, vol 35, Elsevier, Amsterdam, pp 411–432

8. Miller JD (1975) Volume and pressure in the craniospinal axis. Clin Neurosurg 22:76–105

9. Astrup J (1982) Energy-requiring cell functions in the ischemic brain. J Neurosurg 56:482–497

10. Selhorst JB, Gudeman SK, Butterworth JF, Harbison JW, Miller JD, Becker DP (1985) Papilledema after acute head injury. Neurosurgery 16:357–363

11. Lanksch W, Oettinger W, Baethmann A, Kazner E (1976) CT findings in brain edema compared with direct chemical analysis of tissue samples. In: Pappius HM, Feindel W (Eds) Dynamics of brain edema, Springer, Berlin Heidelberg New York, pp 283–287

12. Teasdale E, Cardoso E, Galbraith S. Teasdale G (1984) A new CT scan appearance with raised intracranial pressure in severe diffuse head injury. J Neurol Neurosurg Psychiatry 46:600–603

13. Lundberg N (1960) Continuous recording and control of ventricular fluid pressure in neuro-surgical practice. Acta Psychiatr Scand 36 (Suppl 149):1–193

14. Marmarou A, Maset AL, Ward JD, Choi SC, Brooks J, Lutz HA, Moulton RI, Muizelaar JD, Desalles A, Young HF (1987) Contribution of CSF and vascular factors to elevation of ICP in severely head-injured patients. J Neurosurg 66:883–890

15. Takizawa H, Gabra-Sanders T, Miller JD (1986) Spectral analysis of the CSF pulse-wave at different locations in the craniospinal axis. J Neurol Neurosurg Psychiaty 49:1135–1141

16. Piper IR, Dearden NM, Miller JD (1989) Can waveform analysis of ICP separate vascular from non-vascular causes of intracranial hypertension? In: Hoff JT, Betz AL (eds) Intracranial pressure VII. Springer, Berlin Heidelberg New York, pp 157–163

17. Bray RS, Sherwood AM, Halter JA, Robertson C, Grossman RG (1986) Development of a clinical monitoring system by means of ICP waveform analysis. In: Miller JD, Teasdale GM, Rowan JO, Galbraith SL, Mendelow AD (eds) Intracranial pressure VI. Springer, Berlin Heidelberg New York, pp 260–264

18. Bell BA, Smith MA, Kean DM, McGhee CHJ, MacDonald HL, Miller JD, Barnett GH, Tocher JL, Douglas RHB, Best JJK (1987) Brain water measured by magnetic resonance imaging: Correlation with direct estimation and change following Mannitol and dexamethasone. Lancet I:66–69

19. Obrist WD, Langfitt TW, Jaggi JL, Cruz J, Gennarelli TA (1984) Cerebral blood flow and metabolism in comatose patients with acute head injury. J Neurosurg 61:241–253

20. Brock M, Wiegand H, Zillig C, Zywietz C, Mock P, Dietz H (1976) The effect of dexamethasome on intracranial Pressure in patients with supratentorial tumors. In: Pappius HM, Feindel W (eds) Dynamics of brain edema. Springer, Berlin Heidelberg New York, pp 330–336

21. Miller JD, Leech PJ (1975) Assessing the aspects of mannitol and steroid therapy on intracranial volume/pressure relationships. J Neurosurg 42:274–281

22. Miller JD, Sakalas R, Ward JD, Adams WE, Vries JK, Becker DP (1976) Methylprednisolone treatment in patients with brain tumors. Neurosurgery 1:114–117

23. Gudeman S, Miller JD, Becker DP (1979) Failure of high dose steroid therapy to influence intrancranial pressure in patients with severe head injury. J Neurosurg 51:301–306

24. Cooper PR, Moody S, Clark WK, Kirkpatrick J, Maravilla K, Could AL, Drane W

(1979) Dexamethasone and severe head injury: A Prospective double-blind study. J Neurosurg 51:307–316

25. Braakman R, Schouten HJA, Blauuw-van Dishoeck, Minderhoud JM (1983) Megadose steroids in severe head injury: Results of a prospective double-blind clinical trial. J Neurosurg 58:326–330

26. Dearden NM, Gibson JS, McDowall DG, Gibson RM, Cameron MM (1986) Effect of high dose dexamethasone on outcome from severe head injury. J Neurosurg 64:81–88

27. Muizelaar JP, Wei EP, Kontos HA, Becker DP (1983) Mannitol causes compensatory cerebral vasoconstriction and vasodilation to blood viscosity changes. J Neurosurg 59:822–828

28. Muizelaar JP, Lutz HA, Becker DP (1984) Effect of mannitol in ICP and CBF and correlation with pressure autoregulation in severely head injured patients. J Neurosurg 61:700–706

29. Rosner MJ, Coley I (1987) Cerebral perfusion pressure: A hemodynamic mechanism of mannitol and the post-mannitol hemogram. Neurosurgery 21:147–156

30. Takagi, Saito T, Kitahara T, Morii S, Ohwada T, Yada K (1983) The mechanism of the ICP-reducing effect of mannitol. In: Ishii S, Nagai H, Brock M (eds) Intracranial pressure V. Springer, Berlin Heidelberg New York, pp 729–733

31. Bricolo AP, Glick RP (1981) Barbiturate effects on acute experimental intracranial hypertension. J Neurosurg 55:397–406

32. Ward JD, Becker DP, Miller JD, Choi SC, Marmarou A, Wood C, Newlon P, Keenan R (1985) Failure of prophylactic barbiturate coma in the treatment of severe head injury. J Neurosurg 62:383–388

33. Becker DP, Vries JK (1972) The alleviation of increased intracranial pressure by the chronic administration of osmotic agents. In: Brock M Dietz H (eds) Intracranial pressure: Experimental and Clinical Aspects. Springer Berlin Heidelberg New York, pp 309–315

34. Dearden NM (1985) Intracranial pressure monitoring. Care of the Critically Ill 1:8–13

35. Messeter K, Nordstrom C-H (1986) Sundbarg G Cerebral hemodynamics in patients with acute severe head trauma. J Neurosurg 64:231–237

36. Bingham RM, Procaccio F, Prior PF, Hinds CJ (1985) Cerebral electrical activity influences the effects of etomidate on cerebral perfusion pressure in traumatic coma. Br J Anaesth 57:843–848

37. Dearden NM, Miller JD (1989) Paired comparison of hypnotic and osmotic therapy in the reduction of intracranial hypertension after head injury. In: Hoff JT, Betz AL (eds) Intracranial pressure VII. Springer Berlin Heidelberg New York, pp 474–481

18

Head Injury and Brain Ischemia: Monitoring and Prediction of Outcome

Ross Bullock and Graham M. Teasdale[1]

Introduction

Neuropathological data from the early 1970's demonstrated ischemic brain damage in 91% of severely head-injured patients who died at the Institute of Neurological Sciences in Glasgow. About 40% of these patients had spoken at some time after their injury and before death, suggesting that ischemic brain damage occurred *after* the patient came under medical care, providing an opportunity for prevention of ischemic damage in these patients [1].

Epidemiological studies have shown that poor outcome after head injury is frequently due to clinical episodes of hypoxia, hypotension [3], or high ICP [4], and each of these is known to be a powerful cause of ischemic brain damage. [3,4]. Subsequently, animal studies have demonstrated the vulnerability of the injured brain to ischemic damage [5,6].

This knowledge has led to widespread implementation of intensive head injury management programs aimed at optimising oxygenation and perfusion of the injured brain as soon as possible after injury. In spite of more rapid resuscitation, diagnosis by CT-scanning, and intensive care therapy for head-injured patients, the decline in head injury mortality and morbidity rates has been only slight [7,8]. Recent statistics from the Combined North American Traumatic Coma data bank have shown that head injury mortality rates remain around 34% for patients in coma 6 h or more after injury, despite optimal care [7]. A second neuropathological study carried out during the early 1980s in Glasgow has shown that ischemic brain damage was still present in *88%* of the patients who died after severe head injury, despite the increase in the number of patients accepted for CT scanning, and reduction of the interval between injury and neurosurgical care [8].

---

[1] Institute of Neurological Sciences, University of Glasgow, Glasgow G51 4TF, Scotland, UK

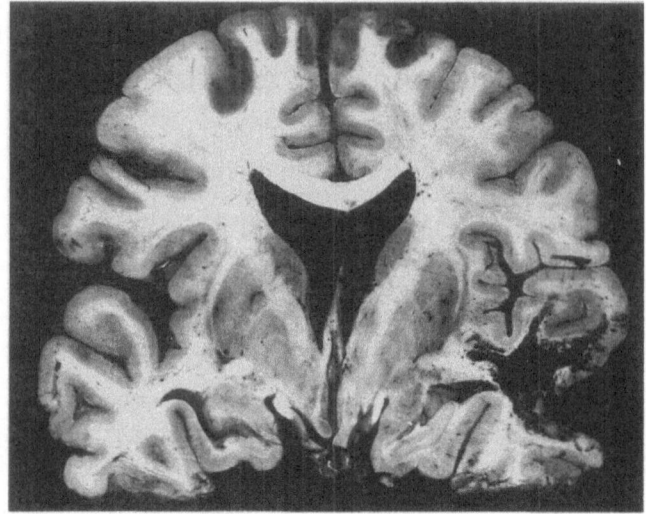

Fig. 1. a. Ischemic neuronal damage, coronal brain section, from a 45-year-old male who survived 4 days after severe head injury. There is bilateral hemorrhagic infarction in the boundary zones between the anterior and middle cerebral arterial territories. There is also a 'burst' right temporal lobe. (From [2]) **b.** Inset: photomicrograph to show cerebral infarction, 15 h old, in cerebral cortex H and E Stain, ×330

Current "state of the art" therapy, therefore, appears unable to prevent ischemia after head injury. In order to accurately target the delivery of newer forms of brain protective therapy, it is essential to understand both the mechanisms and timing of the onset of ischemic damage after head injury. Much knowledge has already been obtained from animal studies, but these cannot fully duplicate the situation in humans. Improved forms of monitoring after head injury may be the key to obtaining this information in the future.

Pathophysiological Mechanisms Causing Ischemic Brain Damage after Head Injury

Events at the Moment of Impact

Animal models have shown that apnea, bradycardia and hypertension occur immediately after impact and may continue for up to several minutes [9]. Ischemic brain damage may be occurring at this time due to apnea. Coma together with transient failure of brain stem reflexes may result in upper airway occlusion, and inhalation of blood, foreign bodies, or gastric contents may cause hypoxemia as well. Pneumothorax and hypotension due to fractures or internal or external hemorrhage may also cause hypoxic ischemic cerebral damage in patients with multiple injuries [10].

The frequency with which ischemic neuronal damage is found at post-mortem, even in patients who have had no demonstrable hypotension or hypoxic episodes during life, suggests that other mechanisms may also be responsible. Recent animal studies have shown that axonal shear injury may worsen the degree of neuronal damage after mild ischemia [11]. Jenkins et al., using a mild fluid percussion model of axonal injury in the rat, have shown that neither mild fluid percussion alone nor a mild ischemic insult alone (bilateral temporary carotid clamping with 6 min of hypotension—MABP 60 mmHg) were sufficient to cause histological ischemic neuronal damage. When both of these insults were administered, 1 or 24 h apart, however, ischemic neuronal damage was marked 7 days after injury, particularly in the hippocampus. This synergistic effect could be blocked by pretreatment with the N. Methyl D aspartate (NMDA) antagonist, phencyclidine, and the anticholinergic drug scopolamine, suggesting that an "excitotoxic" mechanism may be initiated by axonal injury.

Post Impact Events—Delayed Ischemic Brain Damage

1. Airway occlusion and cardiovascular instability may cause secondary ischemic brain damage hours or days after injury, particularly when unconscious patients are not optimally managed [3].

2. Raised intracranial pressure (ICP) is a powerful cause of ischemic neuronal damage because cerebral perfusion pressure is impaired [1]. Raised ICP may manifest itself in the hours or days after injury, due to an intracranial hematoma, or caused by progressively increasing edema within a cerebral contusion or due to "diffuse cerebral engorgement." Little is known about the frequency of

occurrence, and response to therapy of "diffuse cerebral swelling," and "diffuse brain edema" after severe head injury, yet effective management of raised ICP due to diffuse injury is dependent upon an improvement in our knowledge of the pathophysiology of these conditions [12,13].

Surgical removal of a space-occupying lesion is the most effective form of therapy available to reduce ICP and improve perfusion pressure, but surgery is not applicable when diffuse brain swelling or multiple cerebral contusions are the causes for high ICP. Barbiturate therapy to control ICP may *worsen* ischemic brain damage by reducing systemic blood pressure and thus jeopardising cerebral perfusion pressure [14]. This limitation may explain the failure of barbiturates to improve outcome when used after head injury [15]. Raised ICP may initiate a vicious cycle: reduced cerebral perfusion pressure (CPP) induces ischemic neuronal damage which induces cellular swelling which in turn induces a rise in ICP. Injudicious therapy, seizures, or a fall in blood pressure (BP) may worsen ischemic damage under these circumstances. This cycle will eventually lead to death, and is the most common sequence of events preceding brain death certification in modern head injury (HI) intensive care units. In a certain group of patients with uncontrollable ICP, brain damage may be so severe that worthwhile survival cannot occur. Monitoring techniques need to dissect out this group, for whom intensive therapy may be fruitless.

3. Cerebral vasospasm occurs in about 25%–40% of patients after severe head injury, yet is seldom demonstrated because modern head injury management does not often include angiography. When vasospasm occurs, ischemic brain damage is significantly more common in the same hemisphere [16]. Its effects may become apparent several days after injury, despite adequate ICU care.

4. Cerebrovascular autoregulation to changes in systemic blood pressure and arterial hypoxemia is frequently impaired following severe head injury [6], and this may reduce cerebral blood flow and oxygenation to levels below threshold for ischemic neuronal damage when blood pressure and oxygen saturation are marginal [17]. This effect may be particularly marked when raised ICP is also present. Mean arterial pressure levels around 60–80 mmHg—considered acceptable for many ICU patients—may be insufficient to prevent ischemic brain damage in patients with high ICP and impaired autoregulation. Similarly, neuronal oxygenation may be insufficient in the injured brain when arterial PaO_2 levels fall below ±70–80 mm Hg (±10 kpa).

5. Delayed neuronal death. Necrosis of neurons, particularly in areas with high "selective vulnerability," such as the hippocampus, has been shown to occur 2–4 days after an ischemic insult in both gerbil and rat models of global brain ischemia, where the animals have been both functionally and histologically normal during the period prior to developing ischemia [18]. The factors leading to this delayed hippocampal neuronal death are poorly understood, but probably include the release of high concentrations of the excitatory amino acid neurotransmitter, glutamate, in hippocampal extracellular fluid in response to global ischemia [19,20]. Although this form of delayed neuronal ischemia has not been demonstrated in larger animals, and may be related to seizure activity

or undocumented hypoxic or hypotensive insults in the gerbil and rat models, the high frequency of hippocampal ischemic damage in human head injury post-mortem material (81%, [1]) and the frequency of clinically observed delayed deterioration after human head injury, accords with the hypothesis of delayed neuronal death and invites further studies in this area.

Why is Clinical Monitoring Seldom Able to Detect Events Leading to Ischemic Neuronal Damage after Head Injury?

Failure to prevent ischemic brain damage in spite of monitoring and manage-ment may be occurring because of several possible factors:

1. The ischemic insults may be occurring before the patient is under medical care or while the patient is in transit between institutions. Arterial hypoxemia may therefore be occurring undetected. A recent study in Glasgow has shown that up to a third of head-injured patients in coma arriving at the neurosurgical centre from other trauma units are hypoxemic, despite intubation and oxygen [21] (Table 1).

2. Ischemic brain damage may be occurring during monitoring and therapy but is undetected by conventional monitoring methods because monitoring is inter-mittent. For example, arterial blood gas analysis and intermittent blood pressure measurement or cerebral blood flow measurement may fail to detect troughs of low cerebral perfusion associated with patient suction, movement, CT-scanning, etc, particularly in those with multiple injuries or high ICP.

3. Monitoring techniques which are insensitive may fail to detect ischemic brain damage while it is occurring. For example, cerebral blood flow studies using the non-tomographic Xenon 133 washout method have only seldom shown CBF levels likely to cause ischemic neuronal damage (below ± 15 mm/100 gm per min) [17,22]. Recently, tomographic regional cerebral blood flow (CBF) studies using SPECT and 99 mT$_c$ HMPAO (CERETEC) as a cerebral blood flow (CBF) marker have revealed marked reductions in CBF around focal injuries such as

Table 1. Frequency of hypoxaemia (PaO$_2$ < 8.5 kpa) in severely head-injured patients on arrival at the neurosurgical unit. (From [21])

	Hypoxemia present (%)
Intubated and ventilated	7%
Intubated, spontaneous ventilation, on oxygen therapy	26%
Oral Guedel airway and oxygen therapy, spontaneous ventilation	13%
Oxygenation face mask only	20%
Total sample ($n = 164$)	15%

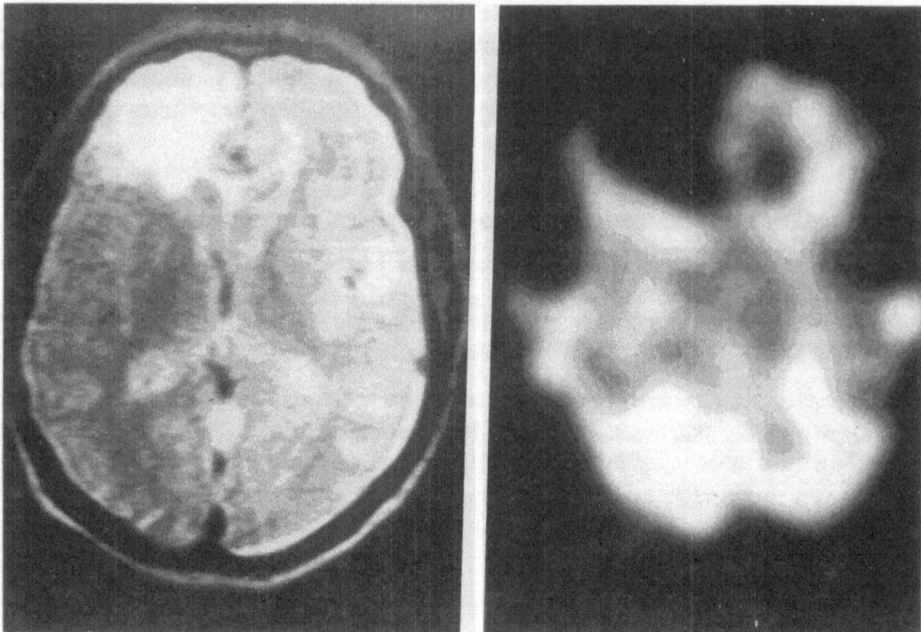

Fig. 2. Reduced rCBF following frontal contusion. (*Left*) T$_2$-weighted MRI image 6 days after head injury, showing left frontal contusion and surrounding edema (*right*) SPECT 99 mTc HMPAO image, 6 days after head injury in the same patient, showing extensive zone of reduced CBF, corresponding to the contusion (note also bilateral temporal lobe reduction in CBF)

intracerebral hematoma and contusion. Simultaneously however, other brain regions may demonstrate hyperemic CBF [23].

The close proximity of hyperperfused and hypoperfused regions of brain (Fig. 2) may summate to produce hemispheric values close to normal, as seen by a single detector Xe^{133} study. We have studied rCBF in 22 patients with focal cerebral contusion at various times after injury, and in each case a zone of reduced CBF has been shown around the lesion. Acute subdural hematoma appears to reduce CBF in the hemisphere ipsilateral to the lesion, but not contralaterally. This supports the neuropathological finding that ischemic neuronal damage is particularly marked adjacent to a focal injury [1,16]. It thus appears that ongoing ischemic neuronal damage may occur in some brain regions while perfusion in other areas is normal or even increased.

Animal studies using a rat intracerebral hematoma model have shown a similar zone of perifocal reduced CBF, and ischemic neuronal damage, which could be reduced by the use of the calcium antagonist nimodipine [24].

4. Cerebral vasospasm may be initiated due to subarachnoid hemorrhage or vessel trauma at the time of injury, but the effects of vasospasm may only become apparent five to ten days *after* injury [16]. A "sub-threshold dose" of ischemic damage sufficient to disturb neurotransmitter function—but not sufficient to

disturb cellular integrity and membrane structure—may initiate *delayed* neuronal destruction which occurs several days later. The mechanisms for this remain poorly understood, but may include the release of excito-toxic amino acids such as glutamate from damaged tissue which then act to worsen neuronal damage in surrounding cells [11]. Likewise, vasoconstrictor substances may be released from blood which escapes from small vessels which were damaged at the time of injury. This damage may be amenable to therapeutic measures.

Monitoring of Brain Ischemia after Head Injury

For a monitoring technique to be useful it must be performed frequently and improve patient management.

Monitoring CNS Damage

Clinical

Objective and reproducible assessment of conscious level and neurological deficit are the mainstay of management after head injury (the Glasgow Coma Scale) but monitoring should aim to detect potential causes of secondary ischemic brain damage *before* they cause a change in conscious level. When patients are paralysed and ventilated, non-clinical monitoring becomes especially vital.

Electrophysiological

Continuous monitoring of cerebral electrical activity and evoked responses have been advocated to assess function in the injured brain [25]. Multi-modality evoked potential measurement has been used in prognosis after head injury, but such data have largely *failed* to influence day-to-day patient management, except in the determination of brain stem death.

Monitoring Cerebral Oxygenation

Continuous Transcutaneous Oximetry

Simple, portable, and cheap monitors can detect a fall in oxygen saturation, and can be used together with arterial pressure monitoring from the earliest possible time after injury. This will demonstrate the effects of patient transfer, movement, tracheal suctioning, anesthesia, etc, upon oxygen delivery. Unfortunately, oximetry becomes inaccurate when peripheral vasoconstriction is present, and this may be a severe limitation in the multiple-injured patient. [26,27]. Nevertheless, transcutaneous oximetry should be used as frequently and as early as possible in head injured patients.

Cerebral Arteriovenous Oxygen Differences (A-VDO$_2$)

Placement of a jugular bulb catheter allows A-VDO$_2$ measurements, which may help diagnose "whole brain" ischemia if AVDO$_2$ exceeds 6–7 mmPO$_2$. When venous PO_2 falls below ± 32 mm Hg or when jugular vein lactate production rapidly increases, then ischemia is present [12]. When brain engorgement or hyperemia is present after head injury, $AVDO_2$ falls below ± 3 ml/100 ml, and such patients may be those who will benefit the most from hyperventilation therapy [12] however, $AVDO_2$ measurements only reveal averaged "whole brain" events, and do not reflect *regional* changes.

Continuous Polarographic Oxygen Electrode Measurement of Cerebral Oxygenation

Animal studies have shown that oxygen-sensitive polarographic electrodes may be implanted in brain tissue to measure oxygen delivery at tissue level [28]. In the future, implantation of an ICP-monitoring transducer may be augmented by simultaneous implantation of oxygen electrodes for this purpose.

Continuous Biochemical Monitoring

Pilot studies have shown that microdialysis electrodes may be implanted coaxially with intracranial pressure (ICP) monitors, to give information about events in the extracellular fluid compartment of the brain in humans after head injury [29]. This technique can provide information about release of excitatory amino acids and vasoactive substances, parallel with changes in ICP and oxygen delivery.

Monitoring the Cerebral Circulation

ICP Monitoring

The purpose of monitoring is to prevent secondary ischemic brain damage (IBD) by *optimising cerebral perfusion pressure*, and this may be achieved by removing a mass, by osmotic or hyperventilatory therapy, CSF drainage, or even by *induced hypertension*, if cerebral autoregulation is intact [30] *ICP waveform analysis* has been used to predict patients with poor cerebral compliance before ICP rises.

Transcranial Doppler Velocity

Transcranial Doppler velocity measurements in the middle cerebral artery may be a useful indicator of cerebral perfusion and the Doppler waveform may non-invasively indicate brain compliance [31]. Transcranial Doppler may also help to diagnose post-traumatic vasospasm, although the interdependence of flow volume and vessel diameter in producing the final waveform makes interpretation difficult.

CBF Measurements

Although continuous CBF monitoring in humans is not currently possible, and the value of Xe^{133} CBF measurements has been limited (vide supra) transcranial

Doppler may become useful in the future, although Doppler velocity is not determined by flow alone. SPECT and PET tomographic techniques may resolve fundamental pathophysiological problems, but are unlikely to be useful for routine use after head injury.

MRI Imaging and Spectroscopy

Once structural neuronal damage is established after an ischemic insult, T_2–weighted MRI imaging shows an abnormality in the majority of cases. At the Institute of Neurological Sciences in Glasgow, 50 patients have undergone acute MRI imaging after head injury [32]. In the majority of patients with diffuse injury, multiple small "bright lesions" were seen in the T_2–weighted images, but larger lesions correlating with zones of ischemic neuronal necrosis demonstrated at post–mortem were not seen. MRI imaging thus appears to be of limited value in detecting brain ischemia early, although serial MRI may demonstrate the development of delayed, secondary ischemic damage.

MRI spectroscopy has demonstrated marked changes in the 31P spectrogram in animal models of ischemia [33] but currently, acutely ill patients on multiple forms of monitoring or ventilators, cannot be safely nursed in a high field strength MRI system. Transducer-tipped ICP monitors induce massive MRI artefacts, and cannot be used in such MRI systems.

Prediction of Outcome after Head Injury (HI)

Many of the parameters useful in monitoring after HI are also powerful prognosticators. Evaluating the effects of therapy and comparing series of patients from different centers, requires a knowledge of factors which determine outcome, and such knowledge also allows us to advise the patients' relatives of prognosis. Clinically the best determinants of outcome are coma depth, pupil reaction, eye movements, motor response to pain, and the patient's age. Using these determinants and data from over 2000 patients who have been entered into the Glasgow Traumatic Coma Data Bank, a computer prediction model of outcome has been devised which, when compared prospectively with real life events in a separate prospective set of patients, allows 80% accuracy in prediction of prognosis (Table 2) [34]. Non-clinical predictors of outcome include ICP, CT features of high ICP, absence of brain electrical activity on evoked response, and the demonstration of an ischemic or hypoxic episode [3,25].

Brain Protection and Resuscitation after Head Injury—the Future

Several aspects of head injury make it a useful clinical entity for clinical trials and pilot studies using brain protective pharmacological agents. These include the following:

1. Much secondary ischemia occurs in head injured patients after the patients have come under medical care [1].

Table 2. Outcome at six months related to age, best coma score and best pupil reaction at 24 h

	Dead/vegetative %	Moderate disability/ good recovery %
Age		
0–29	39	50
30–59	49	54
Over 60	81	11
Coma score		
3–5	84	11
6–7	56	29
8–10	28	58
11–15	16	72
Pupils		
Both fixed	86	6
One or both reacting	16	72

2. The emergence of rapid response trauma teams and rapid transfer facilities in many urban centers, means that head-injured patient are seen by medical attendants very soon after injury. This is not usually the case with stroke.
3. Head injured patients are usually cared for with multiple monitoring in an intensive care unit. The vascular and systemic effects of cerebral protective drugs would thus be easy to evaluate.

Disadvantages Associated with the Use of Head-injured Patients in Evaluating Brain Protective Agents

1. Although 80%–90% of severely head-injured patients sustain ischemic cerebral damage, about 25% also sustain severe diffuse axonal injury, and this may not be amenable to modification by pharmacological agents. The proportion of such diffuse axonal injury patients in any trial population would therefore need to be closely monitored, and improved techniques must be developed for its detection. NMDA antagonist drugs may retard the recovery of neurons after mechanical axotomy by inhibition of synapse formation and long-term potentiation [35]. Further studies of the safety of these agents are therefore needed in chronic animal models of axotomy.

2. Because of the wide variety of mechanisms of cerebral damage after head injury, the therapeutic effect of any protective agent may be "diluted". Therefore large numbers of patients (around 200–300) need to be enrolled, necessitating multiple centers. In such trials, the endpoints need to be rigorously

determined, such as the outcome (i.e., the Glasgow Outcome Scale) and neuropsychology.

Synergistic Therapy

The variety of mechanisms of cerebral damage which operate after head injury may mean that therapy needs to be directed from a variety of different approaches. For example, NMDA antagonists may need to be given as early as possible (e.g., by the rapid response team) at the site of the accident. Some of these agents do not detrimentally affect mean arterial blood pressure, and would probably be safe for administration at this time. Subsequently, once blood pressure and cerebral perfusion have been stabilised, calcium antagonists may be given, and their effect may synergise with that of the NMDA antagonists. Similarly, lazeroids (amino steroids) could be given at the time when reperfusion is thought to be at its maximum—usually when the patient has been stabilised in the intensive care unit. Lazeroids would be directed at preventing free radical generation and membrane damage concomitant with reperfusion.

Such synergistic therapy is, however, dependent upon an improvement in our knowledge of the factors which determine the onset and timing of cerebral ischemia after head injury, and this knowledge can only come with improved monitoring of the head-injured patient.

Summary. Ischemic brain damage occurs in over 90% of patients who die of head injuries, and its incidence has not been appreciably reduced, even with modern therapy and optimal intensive care. Much of this damage is secondary, occurring hours or days after the injury and is therefore amenable to prevention, yet its occurrence is largely undetected by current monitoring techniques. Mechanisms for secondary damage include airway occlusion, hypotension, raised ICP "vasospasm" "delayed neuronal death", and synergy between diffuse axonal injury and mild hypoxia. Improvements in monitoring techniques and reliable predictions of outcome are necessary prerequisites for future studies on neuroloprotective drugs in head injury. New techniques, such as transcranial Doppler ultrasound to monitor ICP and CPP, continuous oxygen saturation and $AVDO_2$ measurement, and sensors to detect changes in the cellular microenvironment, may be available in the future. The Glasgow head injury outcome prediction program has been shown to achieve an accuracy of 80% in comparison to actual outcome, and may be useful for comparison of outcome in trials of future forms of therapy.

References

1. Graham DI, Adams JH, Doyle D (1978) Ischaemic brain damage in fatal and non-missile head injuries. J Neurol Sci 39:213–234
2. Adams JH, Graham DI (1975) In: Vinken PJ, Bruyn GW (eds) Handbook, part 1 vol 23. North Holland, Amsterdam
3. Gentleman D, Jennett B (1981) Hazards of interhospital transfer of comatose head

injured patients. Lancet II:3853–3855

4. Miller JD, Butterworth JF, Gudeman SK (1981) Further experience in the management of severe head injury. J Neurosurg 54:289–200

5. Lewelt W, Jenkins LW, Miller JD (1980) Autoregulation of cerebral blood flow after experimental fluid percussion injury of the brain. J Neurosurg 53:500–511

6. Lewelt W, Jenkins LW, Miller JD (1982) Effects of experimental fluid percussion injury of the brain on cerebrovascular reactivity to hypoxia and hypercarbia. J Neurosurg 56:332–337

7. Marshall LF (1988) The Traumatic Coma Data Bank: Monitoring of ICP In: Hoff JT, Betz Al (eds) Intracranial pressure VII. Springer, Berlin, pp 549–551

8. Graham DI, Ford I, Adams JH, Doyle D, Teasdale GM, Lawrence A, Mc Lellan D (1988) Ischaemic brain damage is still common in fatal non-missile head injury. J Neurol Neurosurg Psychiatry

9. Gennarelli TA, Segawa H, Wald U, Czernicki Z, Marsh K, Thompson C (1982) Physiological response to angular acceleration of the head. In: Grossman RG, Gildenberg PL (eds) Head Injury: basic and clinical aspects. Raven, New York, pp 129–140

10. Miller JD, Sweet RC, Narayan R, Becker DP (1978) Early insults to the injured brain. JAMA 240:439–442

11. Jenkins LW, Lyeth BG, Hayes RL (1988) The role of agonist receptor interactions in the pathophysiology of mild and moderate head injury. In: Hoff JT (ed) Mild head injury. Contemporary issues in neurological surgery, vol 3. Blackwell Scientific, Boston

12. Obrist WB, Langfitt TW, Jaggi JL, Cruz J, Gennarelli T (1984) CBF and metabolism in comatose patients with acute head injury. J Neurosurg 61:241–253

13. Teasdale GM, Graham DI, Lawrence A (1988) Brain swelling in fatal head injuries. In: Hoff JT, Betz Al (eds) Intracranial pressure VII. Springer, Berlin, pp 560–563

14. Miller JD (1979) Barbiturates and raised intracranial pressure. Ann Neurol 6:189–193

15. Ward JD, Miller JD, Choi SC (1986) Failure of prophylactic barbiturate coma in the prevention of death due to uncontrollable intracranial hypertension in patients with severe head injury. In: Miller JD, Teasdale GM, Rowan JO, Galbraith SL, Mendelow AD (eds) Intracranial pressure VI. Springer, Berlin, pp 766–768

16. Macpherson P, Graham DI (1978) Correlation between angiographic findings and the ischaemia of head injury. J Neurol Neurosurg Psychiatry 41:122–127

17. Jones TH, Morawetz RB, Crowell RM, Marcoux FW, Fitzgibbon SJ, Degirolami U, Ojemann RG (1981) Thresholds of focal cerebral ischaemia in awake monkeys. J Neurosurg 54:773–782

18. Kirino T (1982) Delayed neuronal death in the gerbil hippocampus following ischaemia. Brain Res 239:57–69

19. Benveniste H, Drejer J, Schousboe A, Diemer NH (1984) Elevation of the extracellular concentrations of glutamate and aspartate in rat hippocampus during transient cerebral ischaemia monitored by intracranial microdialysis. J Neurochem 43:1369–1374

20. Olney JW, Ho OC, Rhee V (1971) Cytotoxic effects of acidic and sulphur containing amino acids on the infant mouse central nervous system. Exp Brain Res 14:61–76

21. Gentleman D, Jennett B (1989) Avoidable secondary brain damage to head injured patients transferred to neurosurgical units. In: Proceedings of the society of British neurological surgeons. J Neurol Neurosurg Psychiatry 52:927

22. Overgaard J, Mosdal C, Tweed WA (1981) Cerebral circulation after head injury. J Neurosurg 55:63–74
23. Bullock R, Stratham P, Patterson J, Teasdale GM, Teasdale E, Wyper D (1988) Tomographic mapping of CBF, CBV and blood brain barrier changes in humans after focal head injury using SPECT. In: Hoff JT, Betz Al (eds) Intracranial pressure VII. Springer, Berlin, pp 637–639
24. Sinar EJ, Mendelow AD, Graham DI, Teasdale GM (1988) Experimental intracerebral haemorrhage: The effect of Nimodipine pretreatment. J Neurol Neurosurg Psychiatry 51:651–662
25. Greenberg RP, Becker DP, Miller JD, Ward J (1977) Evaluation of brain function in severe head trauma with multimodality evoked potentials. J Neurosurg 47:163–177
26. Zorah JSM (1988) Who needs pulse oximetry? Br Med J 296:685–659
27. Bryan-Brown CW (1988) Blood flow to organs: Parameters of function and survival in critical illness. Crit Care Med 16:170–176
28. Farrar JK, Orange LE (1987) Treatment of cerebral ischaemia with Fluosol-DA. Influence of severity and duration of ischaemia. J Cereb Blood Flow (suppl 1) vol 7:S174
29. Persson L, Hillered L, Ponten U (1989) Intracerebral microdialysis for continuous monitoring of neurosurgical patients: Preliminary methodological considerations. J Cereb Blood Flow Metab 9 (suppl 1):S584
30. Muizelaar JP (1988) Induced hypertension in treatment of high ICP. In: Hoff JT, Betz AL (eds) Intracranial pressure VII, Springer, Berlin, pp 508–510
31. Saunders FW, Cledgett RVT (1988) Intracranial blood velocity in head jury. Surg Neurol 29:401–09
32. Jenkins A, Teasdale GM, Hadley MDM, McPherson P, Rowan JO (1986) Brain lesions detected by magnetic resonance imaging in mild and severe head injuries. Lancet II:445–446
33. Komatsumoto S, Nioka S, Greenberg JH, Yoshizaki K, Subramanian VH, Chance B, Reivitch M (1985) In vivo 31P NMR study in MCA occlusion in the cat. Cereb Blood Flow 5: (suppl 1):S413
34. Barlow P, Murray L, Teasdale GM (1987) Outcome after severe head injury: The Glasgow model. In: Corbett WA (ed) Medical applications of microcomputers. Wiley, New York, pp 105–125
35. Harris EW, Ganong AH, Cotman CW (1984) Long-term potentiation in the hippocampus involves activation of NMDA receptors. Brain Res 323:132–137

19
Multimodal Neuromonitoring of Patients with Severe Brain Damage

Toshiyuki Shiogai and Kazuo Takeuchi[1]

Introduction

Neurophysiological brain monitoring (neuromonitoring), is being increasingly applied in the intensive care setting [1]. Neuromonitoring includes conventional EEG, compressed spectral arrays (CSA), measurement of brainstem auditory evoked potentials (BAEP) and short-latency somatosensory evoked potentials (SSEP), also measurement of intracranial pressure (ICP), cerebral perfusion pressure (CPP) [2], and cerebral blood flow, as well as transcranial Doppler sonography (TCD).

The aim of this study was to identify the neuromonitoring findings that indicate irreversibility of severe brain damage. To this end, we attempted to delineate the evolution of neuronal dysfunction in terms of the relationships among neuromonitoring and neurological findings. From the neuromonitoring data, we selected BAEP, SSEP, and CSA as the parameters indicative of brain and upper spinal cord function and TCD as the parameter reflecting intracranial circulation. We evaluated their reliability in predicting outcome and diagnosing brain death and assessed the overall usefulness of neuromonitoring in the management of patients with severe brain damage.

Subjects and Methods

Multimodal neuromonitoring was applied to 276 severely brain-damaged patients whose Glasgow Coma Scale scores were 8 or less. This study was carried out from August, 1983 to July, 1988. The causes of brain damage were cerebrovascular accident in 177 patients, head injury in 85, cerebral anoxia in seven,

---
[1] Department of Neurosurgery, Kyorin University School of Medicine, 60-20-2 Shinkawa, Mitaka, Tokyo, 181 Japan

meningitis in three, brain tumor in two, encephalitis in one)and drug intoxication in one. Their mean age was 50 years (range, 5–91 years). No children under 5 years of age were included. Also, there were no hypothermic patients with a body temperature below 32°C. The data of patients given barbiturates were analyzed separately. Apnea testing with $PaCO_2$ confirmation was not performed in every case.

BAEP Monitoring

To clarify the relationships between BAEP and neurological and conventional EEG findings, a total of 154 severely brain-damaged patients were divided into three groups. Forty-eight patients in group A were diagnosed as brain dead on the basis of neurological and conventional EEG findings prior to intermittent BAEP recordings. Sixty-one patients in group B underwent continuous BAEP monitoring every 10–30 min before and/or after the diagnosis of brain death. Group C, consisting of 45 non-brain dead patients, served for comparison and also underwent continuous BAEP monitoring every 10–30 min.

A total of 1000–2048 alternating polarity clicks, duration 100 μs, were presented to the patient binaurally at 10 Hz and a stimulus intensity of 90–100 dBHL. Recordings were obtained from vertex (Cz)and earlobes (Al and/or A2) at a 100–3000-Hz bandpass.

SSEP Monitoring

Fifty-three patients were divided into two groups. The 32 patients in group I fulfilled the neurological criteria for brain death. The remaining, surviving 21 patients constituted group II. In addition, 22 normal control subjects were analyzed for reproducibility of SSEP components. Taking into account the sites of the reference electrodes, which were frontal scalp (Fz) in 29 patients and 12 normal subjects and earlobes (A) in 24 patients and 10 normal subjects, we compared the intergroup SSEP findings. We also compared the SSEP findings and conventional EEG and BAEP data. In 26 patients, SSEP were monitored automatically every 10–30 min. A total of 1000 electrical square pulses (100 μs, 2 or 5 Hz) were presented to the unilateral median nerve at the wrist. Stimulus intensity was controlled by thumb twitching. The bandpass was 5–3000 Hz. The recording electrodes were placed on the scalp (C3' or C4' or Cz) contralateral to the median nerve stimulation site and above the spinous process (Cv2).

CSA Monitoring

Fifty-nine patients who had undergone CSA monitoring were selected and divided into two groups according to outcome. Group α comprised 37 brain dead patients, while group β consisted of 22 non-brain dead patients. The relationships between CSA and neurological findings, conventional EEG, BAEP, and SSEP were evaluated. Power spectral analysis (0–16 Hz) of two channels (Cz-Al and Cz-A2) of EEG activity (gains, 5 or 7 μV/mm) was carried out at 10–120-sec epochs.

TCD Monitoring

In order to define the relationship between TCD findings and loss of brain neuronal function, we divided 105 patients who had been monitored by TCD into group F, which comprised 58 fatal cases, and group N, composed of 47 non-fatal cases. TCD findings were assessed both before and after loss of brain function had been determined by the following neurological and neurophysiological criteria: (1) neurological determinants of brain death (deep coma, absent brainstem reflexes, and apnea), (2) loss of biological activity as demonstrated by conventional EEG (grade Va or Vb) [3], (3) loss of the BA pattern on CSA in 38 patients, (4) loss of BAEP waves after wave **II** in 44 patients, and (5) loss of SSEP P13-14 or P15 and following components in 28 patients.

Velocity waveform (VW) analyses were mainly performed transtemporally or transorbitally for the middle cerebral artery (MCA), internal carotid artery (ICA), and/or cervical common carotid artery (CCA). If different patterns were obtained in the two MCAs they were recorded as different results. Additionally, if VWs could not be obtained from the MCA or ICA, the CCA VW was recorded.

Results

Brainstem Auditory Evoked Potentials (BAEP)

After brain death had been confirmed by neurological criteria and conventional EEG, BAEP waves beyond wave **II** were not observed in groups A and B (Table 1). All but two of the 61 group B patients were followed until the final loss of BAEP (Fig. 1). The mean follow-up time after clinical brain death was 14.6 h. After the neurological criteria for brain death had been met, BAEP wave **III** was present in three of 48 group B patients not given barbiturates. Temporary bilateral reappearance of wave **II** was observed after the loss of wave **I** in one patient. One 53-year-old anoxic patient demonstrated a unilateral wave **I** 175 h after the diagnosis of clinical brain death (Fig. 1). There were no correlations between absence or presence of BAEP waves **I** and/or **II**, and cause of brain death, age, body temperature, or systolic blood pressure in group A or B.

In all but five of the group C patients, BAEP waves **I–V** were preserved (Table 1). The two patients who retained waves **I–IV** and **I–III** had brainstem hemorrhages, and the latter patient died. Only three patients lost all BAEP waves within 1 month of onset. In one case, all BAEP waves disappeared within 17 days of head injury, and waves **I–V** recovered 3 months later (Fig. 2a). The remaining two patients eventually died.

Analysis of conventional EEG and BAEP findings in group B revealed electrocerebral silence with preservation of BAEP waves **I–V** in two patients not given barbiturates, **I–III** in one patient, and **I** and **II** in four patients. However, two patients with no BAEP waves continued to exhibit EEG activity and brainstem reflexes. The causes of brain damage in these two cases were brainstem hemorrhage and meningitis.

Table 1. Last BAEP recordings in 154 patients with severe brain damage

Waves recorded	Group A No.	(%)	B No.	(%)	C* No.	(%)
None	40	(83)	59	(97)	3	(7)**
I	7	(15)	2	(3)	0	
I–II	1	(2)	0		0	
I–III	0		0		1	(2)
I–IV	0		0		1	(2)
I–V	0		0		40	(89)
Total	48	(100)	61	(100)	45	(100)

The differences among the three groups were significant at $P < 0.0001$ (H = 119, Kruskal-Wallis test). The difference between groups A and B was significant at $P < 0.05$ (U = 2.40, Mann-Whitney test).
* The last BAEPs were recorded within 1 month after onset.
** One patient lost all BAEP waves 17 days after onset but recovered waves I–V 3 months later

Short-Latency Somatosensory Evoked Potentials (SSEP)

In normal control subjects, cervical components (Cv2) Nll and N13 and scalp components (C3′ or C4′) Pll with earlobe reference (A), P13-14 (A), P15 with frontal reference (Fz), and N18-20 were consistently identified on both sides. Scalp P13, having the same latency as cervical N13, and P14 components were clearly identified in all but one subject. The final SSEP recordings indicated the absence of scalp P13-14 or P15 and later components in all 32 group I (brain death) patients. Cervical N13 and N14 were preserved in 18 (56%) and six (19%) patients, respectively, in group I. On the other hand, P13-14 (A) was identified in all 12 group II patients. However, P15 (Fz) was absent in three out of nine group II patients, and one of the three survived (Fig. 2b, Table 2). After loss of biological activity according to conventional EEG, scalp P13-14 or P15 and subsequent components were not observed.

Compressed Spectral Array (CSA)

We discovered three CSA patterns in patients with severe brain damage (Fig. 3). These were: (1) electrocerebral silence (ECS) which is bilateral absence of the CSA power spectra for periods of longer than 40 seconds, and (2) biological activity (BA) or continuous peaks of activity, and (3) equivocal (EQ) or intermittent peaks of activity alternating with unilateral loss of power spectra or bilateral loss lasting not longer than 30 sec. If ECS was identified once during CSA monitoring, we recorded the result as ECS.

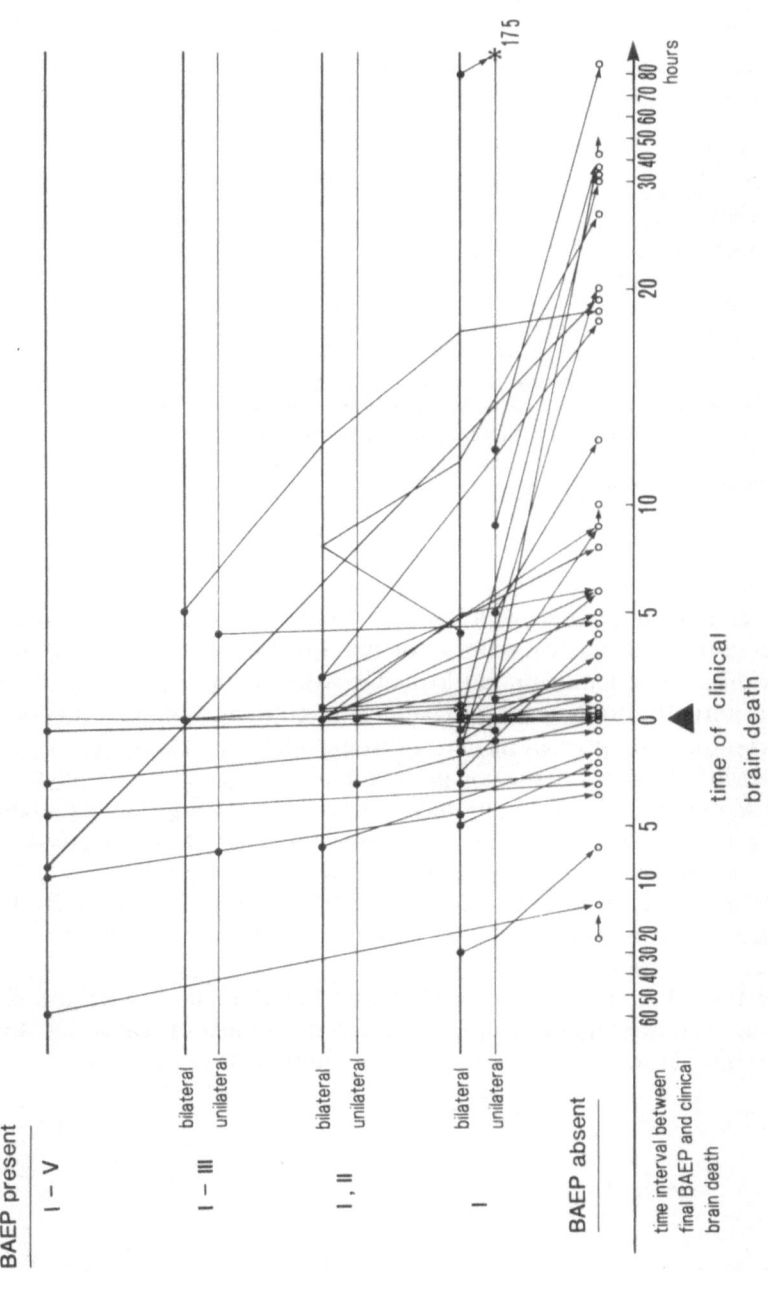

Fig. 1. Temporal relationship between course of BAEP findings and time of neurologically diagnosed brain death (*solid arrowhead*) in 48 group B patients. Open circles denote points at which BAEPs disappeared. The *asterisk* indicates the last BAEP recording and the time interval from the neurological diagnosis of brain death. Thirteen patients given barbiturates are excluded. All but two patients were followed until final loss of BAEP waves. (From [8] with permission)

Table 2. Last SSEP recordings in 53 patients with severe brain damage

SSEP	Group and site of reference electrode*					
Components	I			II		
Cervical (scalp)	A	Fz	Total	A	Fz	Total
None	1	4	5	0	0	0
N9 only	0	2	2	0	0	0
N9-N11 (P9-P11)	3	4	7	0	0	0
N9-N13 (P9-P11)	6	6	12	0	0	0
N9-N14 (P9-P11)	2	4	6	0	3**	3
(-P13-14 or -P15)	0	0	0	2**	2	4
(-N18 and -N20)	0	0	0	10**	4	14
Total	12	20	32	12	9	21

The difference between groups **I** and **II** was significant at $P < 0.0001$ (U = 6.06, Mann-Whitney test). *Reference electrodes were placed on the contralateral earlobe with median nerve stimulation or the linked earlobes (A), or on the frontal scalp (Fz). **One patient fulfilled the criteria for brain death after the last SSEP recording

In group α, all but two patients with loss of BA (ECS in 17 patients and EQ in 19) had satisfied the neurological criteria for brain death during CSA monitoring. These two patients demonstrated loss of BA preceding fulfillment of the neurological criteria for brain death. No patients in group α showed recovery of BA. BA was consistently present in the 22 group β patients (including two patients given barbiturates) and in the one group α patient who had not fulfilled the criteria for brain death. The conventional EEG results of all group α patients demonstrating ECS were classified as Hockaday's Va or Vb. Comparison of CSA and BAEP data in 33 group α patients (Table 3) indicated that loss of BA coincided with loss of all BAEP waves in nine patients and with loss of BAEP waves **II–V** in 12 patients. However, in 12 patients BAEP waves beyond wave **II** were preserved after loss of BA.

Analysis of the relationship between CSA and SSEP findings in 20 group α patients (Table 4) showed that loss of BA coincided with loss of scalp P13-14 or P15 in 17 patients. However, the scalp P13-14 component persisted in the re-

Fig. 2a–c. Neuromonitoring of a 47-year-old patient with a traumatic intracerebral hematoma in the left parietal lobe **a** BAEP monitoring. Bilateral waves **I–V** completely disappeared 3 weeks after the accident. However, 3 months later waves **I–V** reappeared on the left side and waves **I**, **IV**, and **V** on the right. **b** SSEP monitoring with a mid-frontal

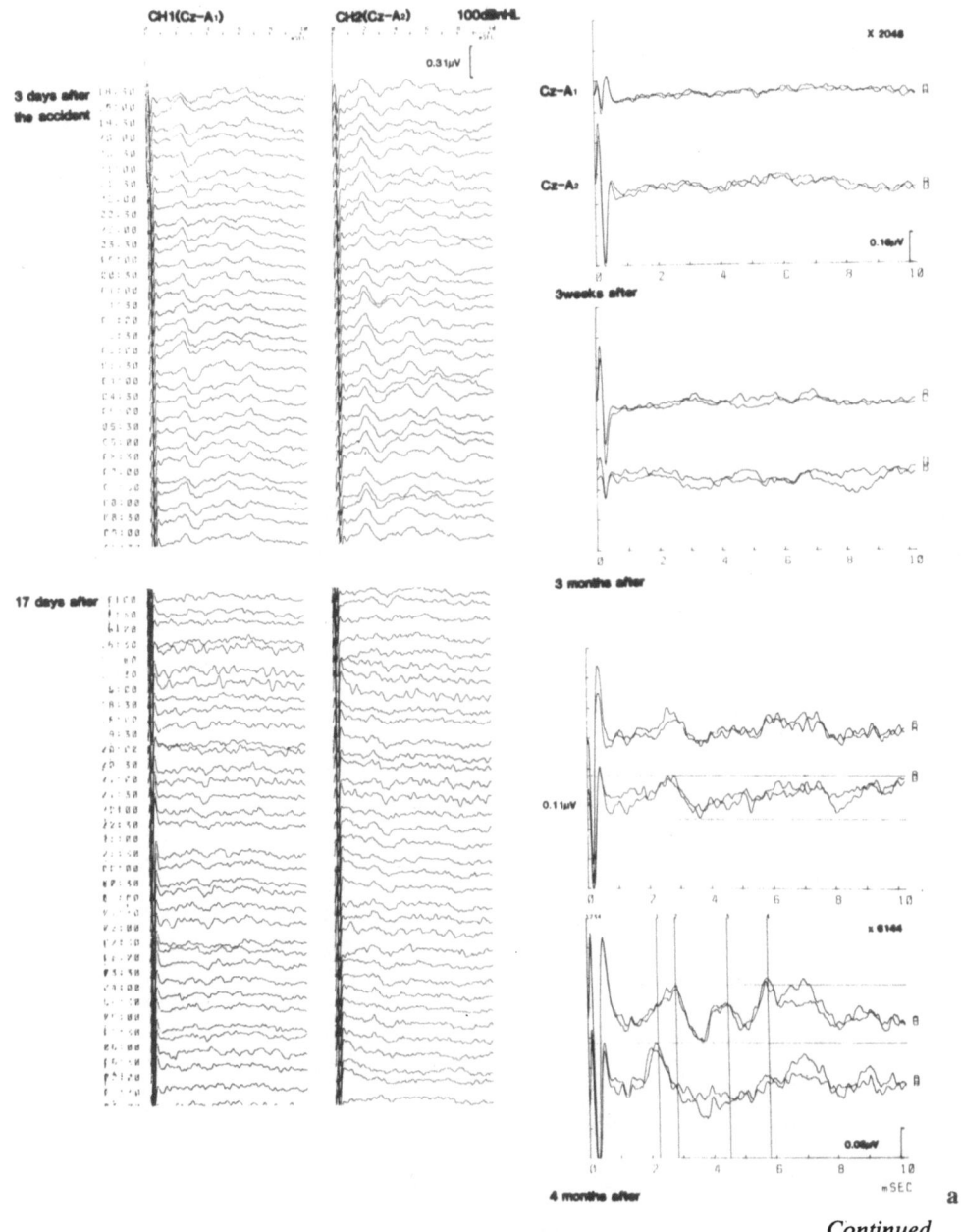

Continued

(Fz) reference. Scalp P15 and N20 were not observed at all during follow-up. However, in cervical recordings, the only initial negative components (N11?) were observed 4 days after the accident. Three months later, cervical N13 was detected and, 4 months later, N13 and N14. (*Continued*)

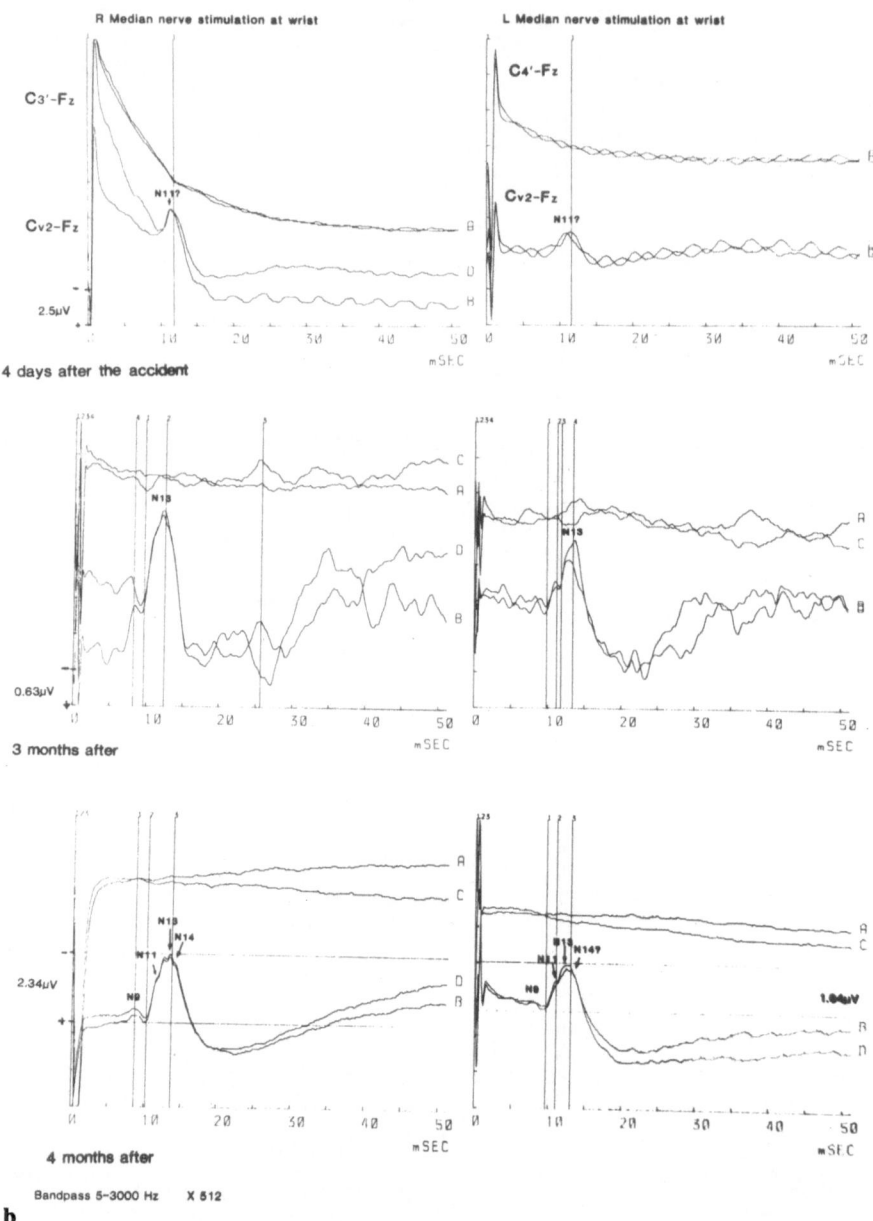

Fig. 2 (*continued*)

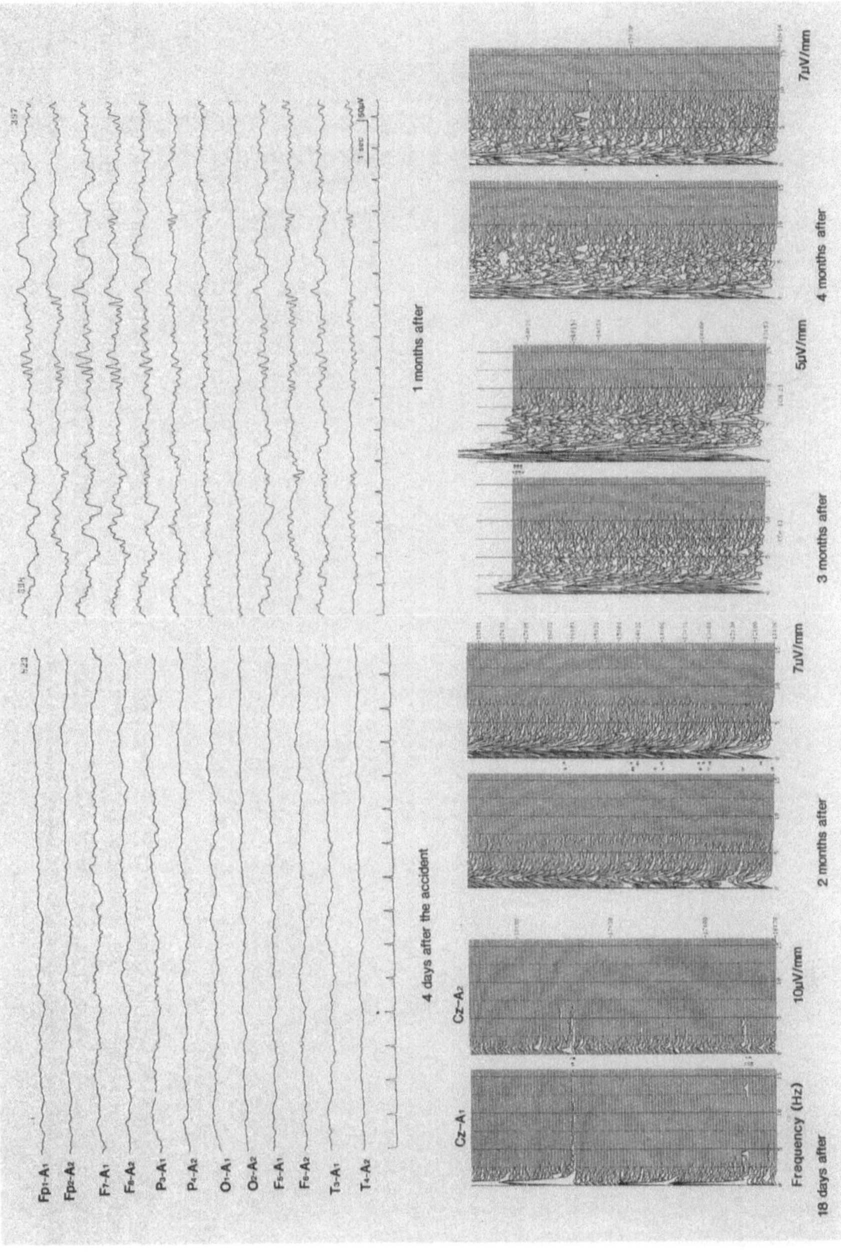

Fig. 2c. CSA and conventional EEG monitoring. Typical biological activity was observed on CSA and conventional EEG during follow-up. (From [22] with permission)

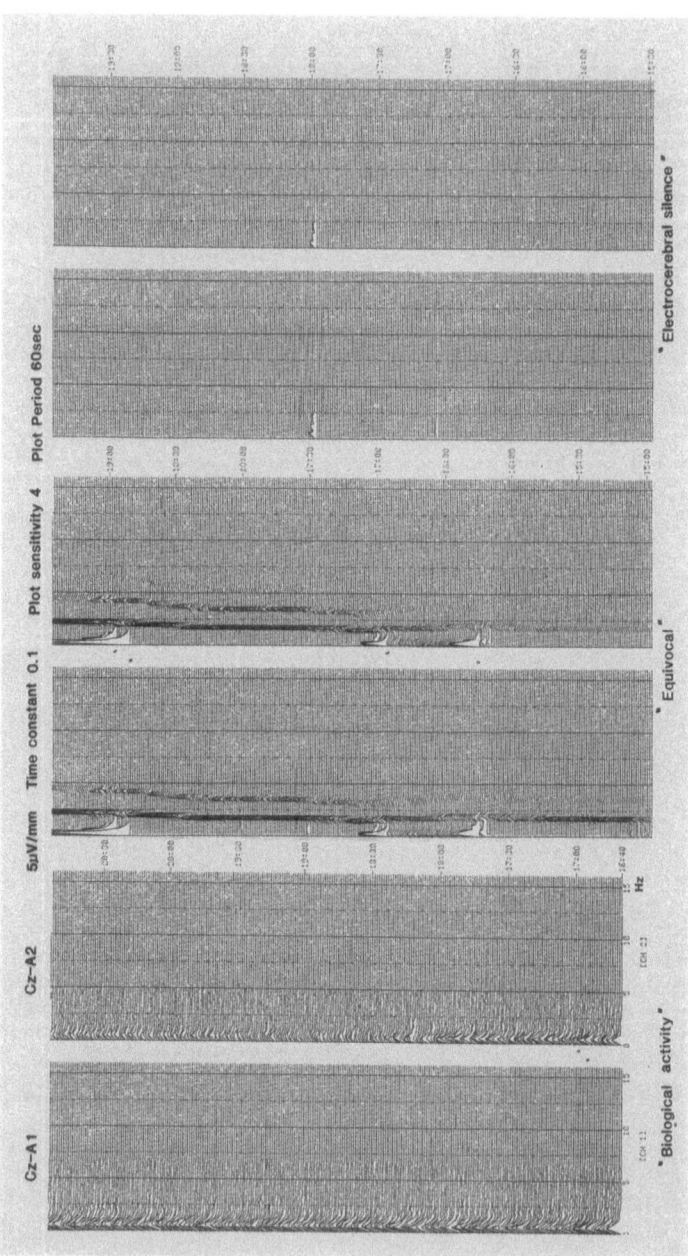

Fig. 3. Three CSA patterns were obtained from a 55-year-old patient with intracerebral hemorrhage: (1) electrocerebral silence (ECS)—bilateral absence of the CSA power spectra for periods of longer than 40 s, (2) biological activity (BA)—continuous peaks of activity, and (3) equivocal (EQ)—intermittent peaks of activity alternating with unilateral loss of power spectra or bilateral loss lasting not longer than 30 s. (From [22] with permission)

Table 3. BAEP recordings after loss of BA on CSA in group α patients

BAEP waves recorded	ECS	CSA patterns EQ	BA	Total
None	2	7	0	9
I	4	8 (1)	0	12 (1)
I–II	0	2	0	2
I–III	0	4 (1)	0	4 (1)
I–V	2 (2)	4 (2)	0	6 (4)
Total	8 (2)	25 (4)	0	33 (6)

$r = 0.02$ (ns), Spearman's test. ECS, electrocerebral silence; EQ, equivocal; BA, biological activity; (n), patients given barbiturates

Table 4. SSEP recordings after loss of BA on CSA in group α patients

SSEP components* Cervical (scalp)	ECS	CSA patterns EQ	BA	Total
None	1	1	0	2
N9 only	0	1	0	1
N9-N11 (P9-P11)	2	2	0	4
N9-N13 (P9-P11)	3 (2)	2	0	5 (2)
N9-N14 (P9-P11)	2 (1)	3	0	5 (1)
N9-N14 (-P13-14)	0	1 (1)	0	1 (1)
N9-N14 (-N18)	0	1	0	1
N9-N14 (-N20)	0	1 (1)	0	1 (1)
Total	8 (3)	12 (2)	0	20 (5)

$r = 0.21$ (ns), Spearman's test. *(n), patients given barbiturates. Reference electrodes were placed on the earlobes in 9 patients and the frontal scalp (Fz) in 11

maining three patients. One patient in group β lost all BAEP waves, cervical N13 and N14, and scalp P15 (Fz) but exhibited biological activity on CSA and conventional EEG (Fig. 2c).

Transcranial Doppler Sonography (TCD)

VWs exhibited the following six patterns in patients with severe brain damage (Fig. 4): continuous forward flow (FF), no flow between systolic and diastolic forward flow (NF), reverse flow between systolic and diastolic forward flow (RF), diastolic reverse flow without diastolic forward flow (DRF), brief systolic forward flow (SFF), and no detectable (U) VWs. The VW patterns before and after loss of neuronal function in groups F and N are summarized in Table 5.

In the MCA, disappearance of the FF pattern and development of the U pattern were obvious after loss of brain function in group F ($\chi^2 = 16.1$, $P < 0.01$).

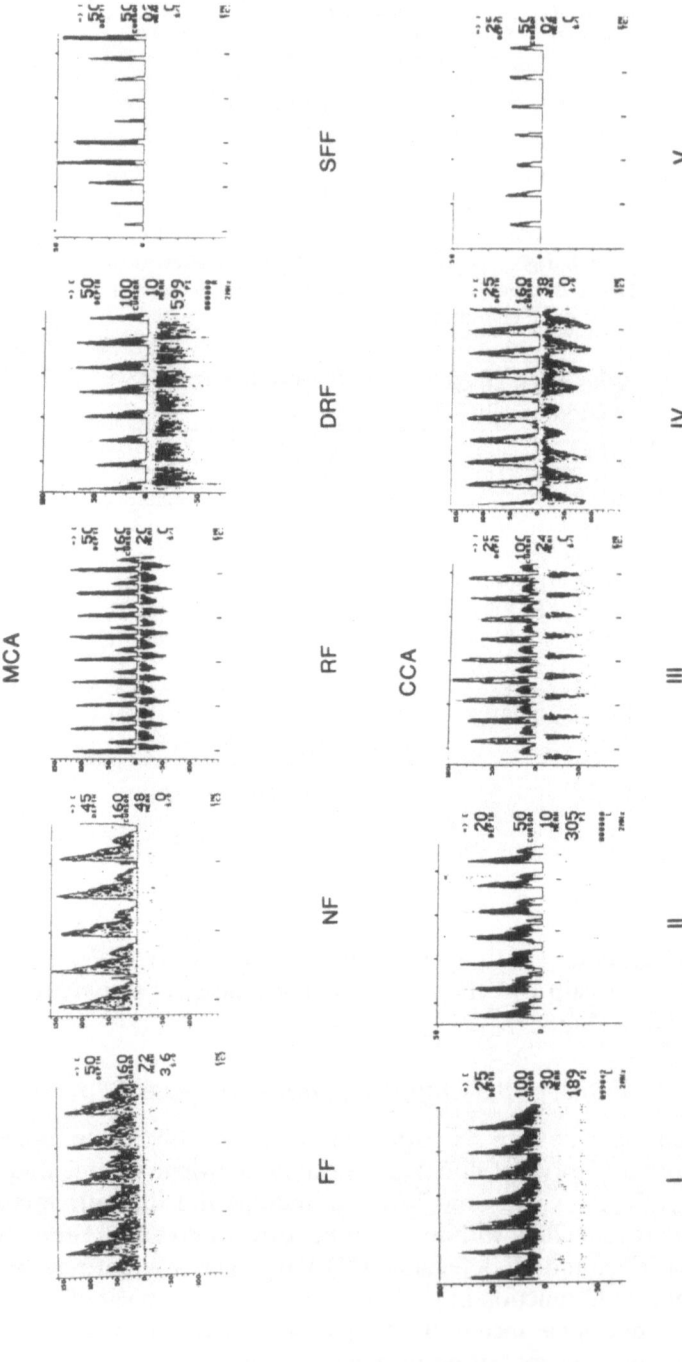

Fig. 4. Velocity waveforms (VWS) exhibited six patterns in patients with severe brain damage: (I) continuous forward flow (FF), (II) no flow between systolic and diastolic forward flow (NF), (III) reverse flow between systolic and diastolic forward flow (RF), (IV) diastolic reverse flow without diastolic forward flow (DRF), (V) brief systolic forward flow (SFF) and undetectability (U) of VWs

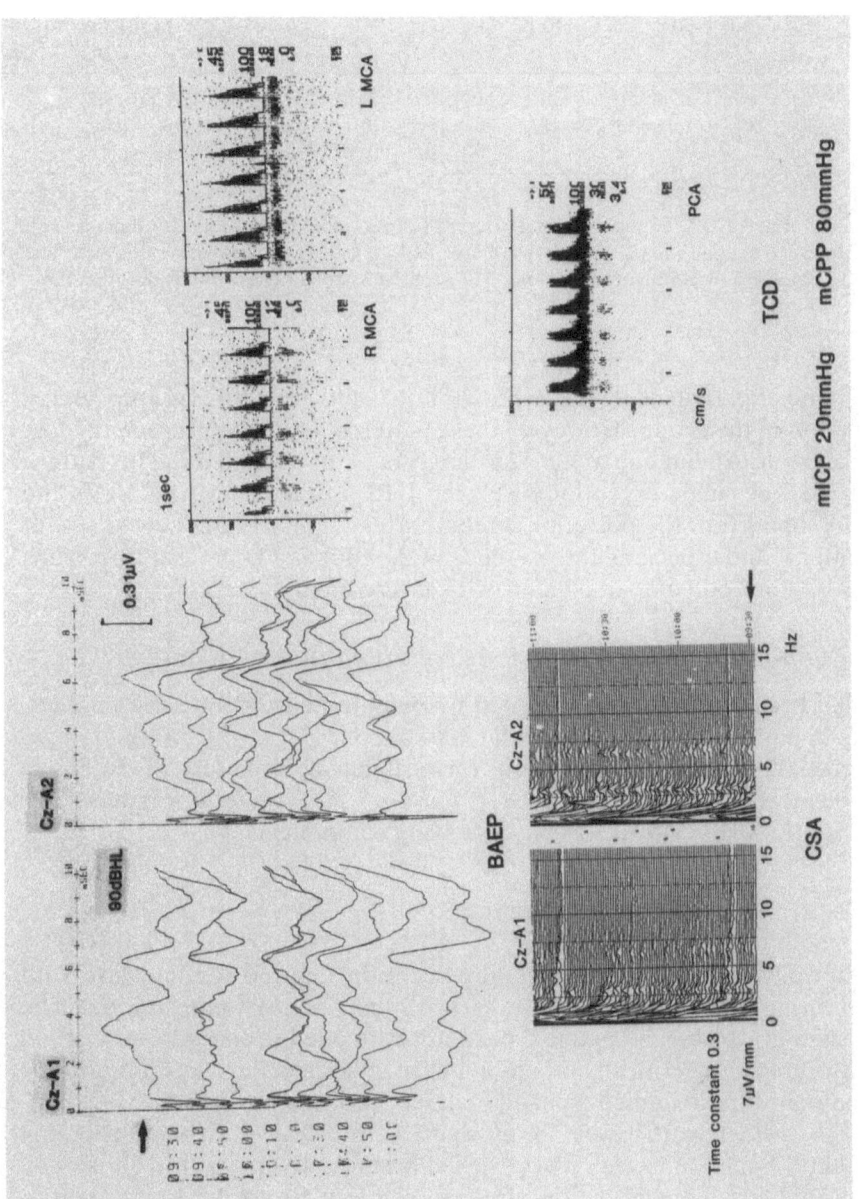

Fig. 5. Simultaneous BAEP, CSA, TCD, ICP, and CPP monitoring of a 79-year-old patient in group N with a left acute subdural hematoma. The DRF pattern in the left MCA and NF pattern in the right MCA were associated with the FF pattern in the left PCA. During the TCD recording (*arrow*), BAEP waves I to V and CSA BA were preserved, and mean ICP and PP were 20 and 80 mmHg, respectively

Table 5. Relationship between velocity waveform patterns and loss of brain function in 105 patients with severe brain damage

		Velocity waveform patterns											
	Brain			MCA						CCA			Patients
Group	function	FF	NF	RF	DRF	SFF	U	(FF	NF	RF	DRF	SFF)	(n)
F	Absent	3	4	2	10	10	11	(0	2	3	2	4)	33
	Present	22	6	1	13	8	5	(1	0	2	2	0)	39
N	Present	46	1	0	1	0	0	(0	0	0	0	0)	47

$\chi^2 = 77$, $P < 0.001$
MCA, middle cerebral artery; CCA, common carotid artery; FF, continuous forward flow throughout cardiac cycle; NF, no flow between systolic and diastolic forward flow; RF, reverse flow between systolic and diastolic forward flow; DRF, reverse flow throughout diastole; SFF, a brief systolic peak without diastolic flow; U, undetectable

There were no apparent changes in NF, RF, DRF, and SFF patterns upon loss of brain function in the group. However, these patterns tended to precede the loss of brain function. All but one of the 47 group N patients exhibited the FF pattern in the MCA. The single exception displayed DRF in the left MCA, NF in the right MCA, and FF in the posterior cerebral artery (Fig. 5). Obviously, loss of the FF pattern, and especially the presence of the SFF and U patterns, portends an extremely poor outcome in patients with severe brain damage.

Temporal Relationship Among Neuromonitoring Findings

Irreversible brain damage was preceded by definite neuromonitoring findings, especially in patients with supratentorial mass lesions (Figs. 6, 7), in the following sequence: (1) loss of the continuous forward flow (FF) pattern on TCD, (2) loss of biological activity (BA) on CSA, (3) loss of BAEP waves beyond wave **II**, and (4) loss of SSEP scalp P13-14 and following components.

Discussion

Many neurophysiological techniques have been introduced for intensive care unit monitoring [1]. In the intensive care context, neuromonitoring must be accomplished at bedside, so patients need not be moved to special examination rooms. Moreover, in critically brain-damaged patients, the neuromonitoring should not only offer immediate and reliable information, but also be non-invasive. In patients with severe brain damage, the final decision concerning irreversibility should be based on established brain death criteria [4]. Loss of all brainstem function, demonstration of electrocerebral silence by electrophysiological methods, and absence of intracranial circulation are very important elements in the diagnosis of brain death [5]. If the diagnosis is to be based on neuromonitoring data, it will be necessary to prove that complete loss of detectable function in the cerebrum, brain stem, and upper spinal cord and/or complete cessation of intracranial circulation in the brain and upper spinal cord [6].

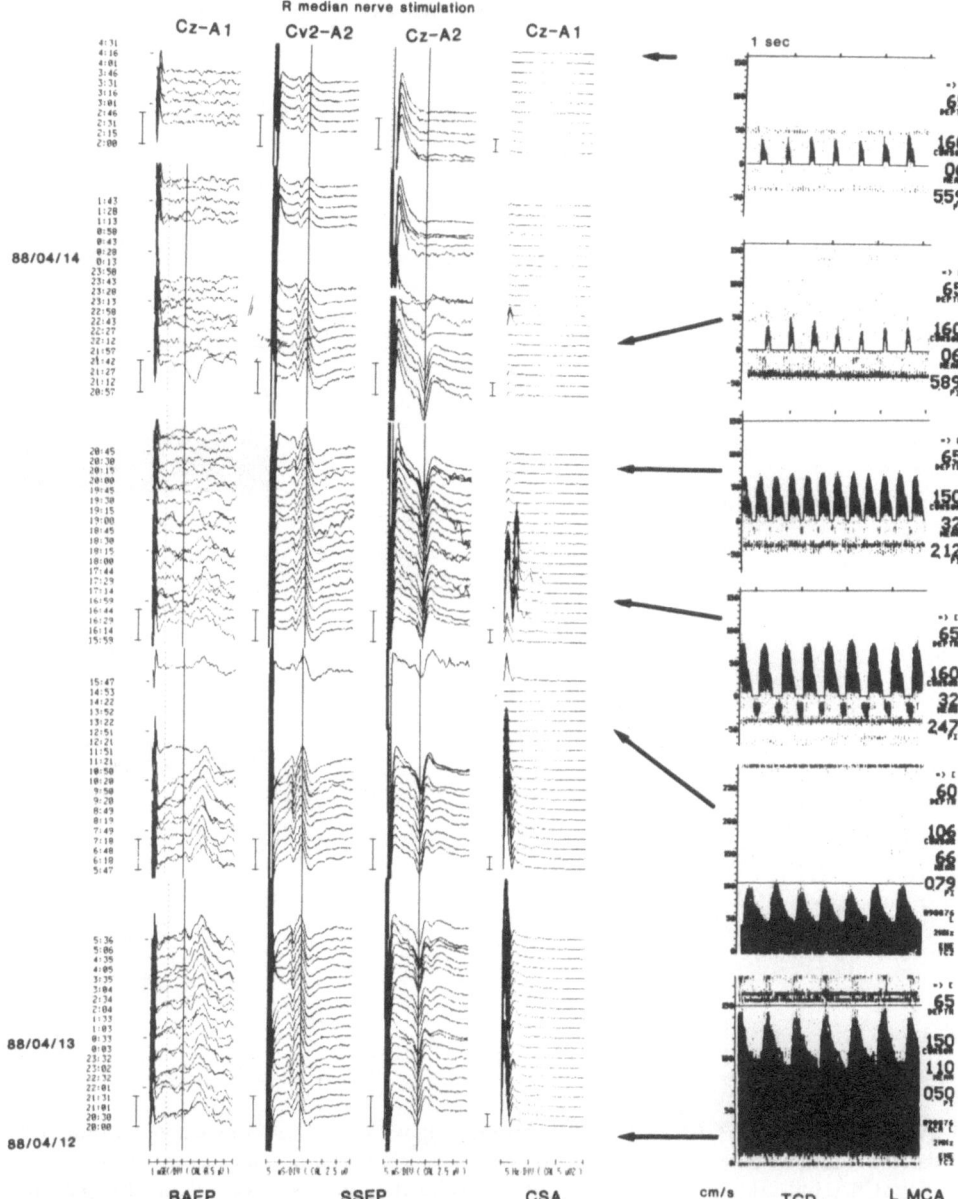

Fig. 6. A 67-year-old patient with a cerebral aneurysm underwent simultaneous monitoring of BAEP, SSEP, CSA, and TCD. Initially, loss of the FF pattern and decreased velocity in the left MCA on TCD indicate supratentorial ischemia. The ensuing loss of BA on CSA preceded the loss of BAEPs beyond wave **II**. Finally, after loss of all BAEP waves, the SSEP P13-14 component (*vertical solid lines in Cz-A2 recordings*) disappeared. At that time, TCD indicated the SFF pattern

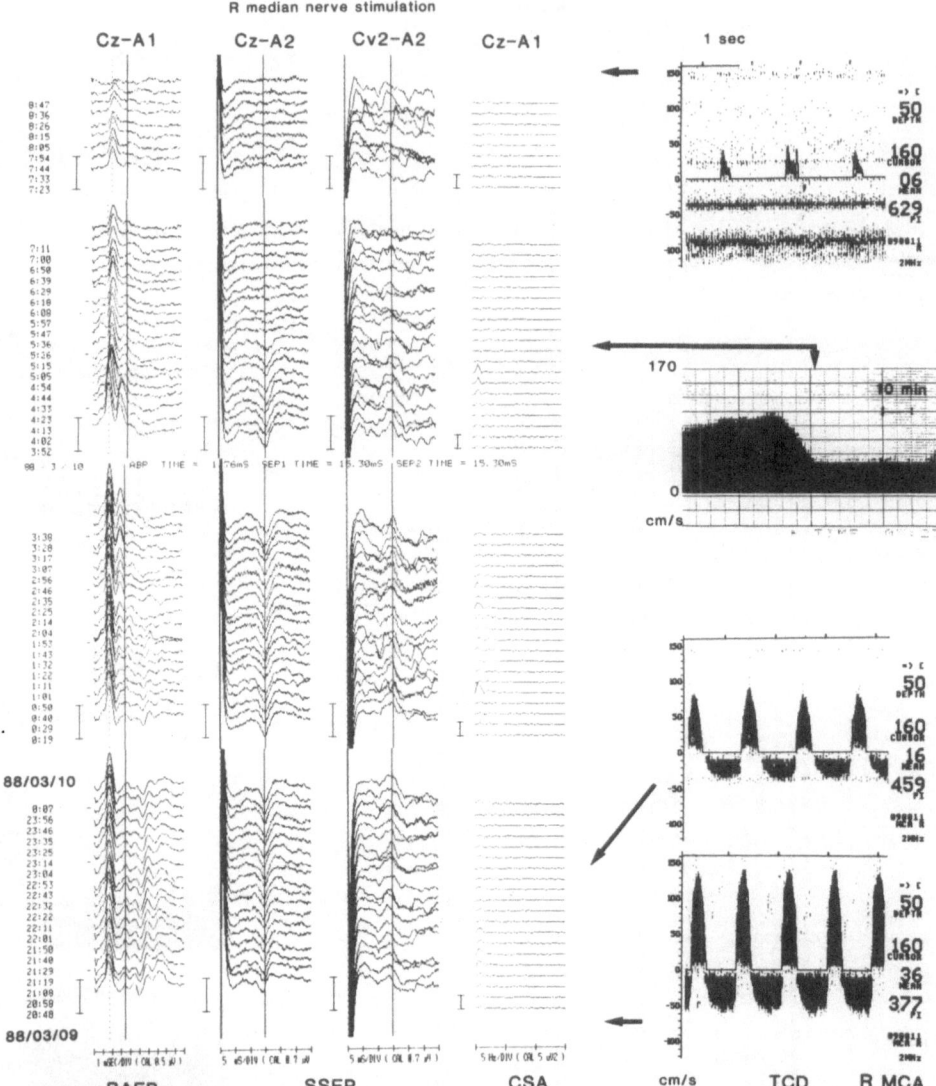

Fig. 7. BAEP, SSEP, CSA, and TCD monitoring were simultaneously and continuously carried out in a 60-year-old patient with an intracerebral hematoma. In the bottom tracings, CSA indicates the loss of BA, while TCD demonstrates loss of the FF pattern. However, BAEP waves **I–V** and SSEP scalp P13-14 components (*vertical solid lines in Cz-A2 recordings*) are preserved. At the time when the SSEP P13-14 scalp component disappeared, which coincided with an abrupt fall in MCA flow velocity on TCD, BAEP waves **I** (*vertical dotted lines*) and **II** persisted. After the loss of BAEPs beyond wave **I**, TCD revealed the SFF pattern

Evaluation of Brain Stem and Upper Spinal Cord Function by BAEP and SSEP

It is known that in predicting the outcome of coma by BAEP, absence of all BAEP components or of those beyond wave **I** or **II** and following waves [7]. Our findings support this [8]. It is not surprising that BAEP wave **II** may originate in the intracranial acoustic nerve [9]. Moreover, in the early period after the clinical criteria for brain death have been met, waves **I–III** may be identified in a very few cases. Preservation of wave **III** may indicate residual brain stem function when such is not detectable by neurological examination. However, waves **II** and **III** always gradually disappear as posterior fossa ischemia progresses following clinical brain death. Furthermore, in these circumstances wave **I** nearly always disappears as well. Persistence of wave **I** may indicate remaining posterior fossa circulation [10]. Therefore, the diagnosis of brain death by BAEP requires the disappearance of BAEP wave **I**, since the feeding arteries of the cochlea derive from the intracranial internal auditory artery.

Loss of all SSEP components or of those beyond P15 portend a very poor outcome [7]. However, there is controversy concerning the origins of cervical components N13 and N14 and scalp components P13, P14, and P15. Also, the positions of the reference electrodes must be taken into account. The SSEP P15 component, with frontal reference, generated from the subthalamus to the thalamus [11,12], and SSEP P14, with earlobe or noncephalic reference, originate in the medial lemniscus [13]. Therefore, with loss of all components beyond P15, and even P14, the patients may survive. In fact, in our series there were two survivors with loss of components beyond P15 and one with loss beyond P14 [14].

The scalp P13 component, with earlobe reference, was recorded in all but one normal subject and showed the same latency as the cervical N13 component. The origin of the scalp P13 remains uncertain, and has been suggested that it differs from that of the cervical P13 and N13 having the same latency [15]. Moreover, scalp P13-P14 dissociation in a patient with a pontine lesion has been reported [16]. Additionally, it appears that cervical N13 and N14, with scalp reference, may be affected by the brain stem far-field potential P14 [17,18].

Therefore, if SSEP findings are applied to the diagnosis of brain death, both the scalp P13-14 component, with earlobe reference, and the cervical N13-14 component, with scalp reference, should disappear since these components may originate in the spinomedullary junction.

Evaluation of Cerebral Cortical Function by CSA

Electrocerebral silence per se does not necessarily denote irreversible brain dysfunction or brain death. Moreover, the limitations of conventional EEG in predicting outcome and determining brain death have been pointed out [19]. It has been reported that CSA is useful in the prognosis of comatose patients [20]. However, its value in the diagnosis of brain death remains uncertain. Technical difficulties in assessing power spectra, such as noise from an electrode or the amplifier, are major problems. Theoretically, all CSA power spectra should be absent in the event of brain death, although this is not always the case. Additionally, the persistence of EEG activity following brain death has been noted [21]. In our study, loss of the BA pattern was relatively well correlated with neurological brain death [22]. However, in determining brain death with CSA, temporal discrepancies between cortical, brainstem, and upper spinal cord function must take into consideration.

Evaluation of Intracranial Circulation by TCD

It is well known that complete interruption of the intracranial circulation, even for a short period, irreparably damages the brain [23]. Total brain infarction as confirmed by aortocranial four vessel angiography implies brain death [24]. However, neither the ischemic threshold for total brain infarction nor the maximum ischemic interval that will allow partial recovery is known. Theoretically, intracranial circulatory arrest implies no cerebral perfusion and, therefore, no CPP and no cerebral blood flow. However, the relationship between CPP and cerebral blood flow may differ in different pathophysiological conditions.

CPP directly affects TCD-measured velocity and velocity waveform patterns. Among the six TCD VW patterns we identified, no flow or reverse flow during the diastole reflected decreased CPP. Reverse flow, in particular, has been observed both extra- and intracranially in patients with angiographically proven intracranial circulatory arrest [25,26]. However, it has been suggested that reverse flow is not necessarily the same as no flow [26]. One would expect complete cessation of intracranial circulation to be accompanied by the U pattern in the intracranial arteries. In this study, it was clear that, in the MCA, VW patterns other than FF—in particular DRF, SFF, and U patterns—herald an extremely grave outcome in comatose patients and, in fact, precede the loss of all brain functions. Therefore, if TCD is applied to the diagnosis of brain death, it is necessary to consider not only the difference between supratentorial and infratentorial perfusion but also temporal factors affecting intracranial perfusion.

Conclusion

The use of only one monitoring modality is not always sufficient to determine irreversible loss of neuronal function because false-positive and false-negative results may be obtained. Our results suggest that multimodal neuromonitoring that takes into consideration the temporal relationships between functional loss in the cerebrum, brain stem, and upper-spinal cord is required for the determination of brain death.

Summary. To evaluate the reliability of neuromonitoring findings in determining irreversibility of brain damage, we applied multimodal neuromonitoring to 276 severely brain-damaged patients. Brainstem auditory (BAEP) and somatosensory (SSEP) evoked potentials, compressed spectral arrays (CSA), and transcranial Doppler ultrasonographic (TCD) data were compared with evidence of clinical brain death (BD). BAEP waves beyond wave **II** were preserved or recovered in all survivors, but not in BD. SSEP scalp P13-14 or P15 and later components disappeared in BD and impending BD, and P15 was lost in one survivor. Biological activity (BA) on CSA disappeared coincident with or prior to BD, but was consistently observed in survivors. Loss of diastolic forward flow and/or appearance of diastolic reverse flow in the middle cerebral artery tended to precede or accompany BD which also occurred in one survivor. Irreversible brain damage was preceded by loss of the following, in order, especially in cases of supratentorial lesions: continuous forward flow on TCD > BA on CSA > BAEP beyond II > SSEP scalp P13-14.

Multimodal neuromonitoring, with consideration of the clinico-pathological features of brain damage and the temporal relationship between loss of brain function and intracranial circulatory arest, is essential for accurate assessment of irreversible brain damage.

References

1. Hacke W (1985) Neuromonitoring. J Neurol 232:125–133
2. Shiogai T, Sakuma T, Nakamura M, Maemura E, Kadowaki C, Hara M, Ogashiwa M, Takeuchi K (1989) Neuronal dysfunction and intracranial pressure: Multimodal monitoring of severely brain-damaged patients. In: Hoff JT, Betz AL (eds) Intracranial pressure VII. Springer Berlin Heidelberg, New York, pp 665–667
3. Hockaday JM, Potts F, Epstein E, Bonazzi A, Schwab RS (1965) Electroencephalographic changes in acute cerebral anoxia from cardiac or respiratory arrest. Electroencephalogr Clin Neurophysiol 18:575–586
4. Takeuchi K, Takeshita H, Takakura K, Shimazono Y, Handa H, Gotoh F, Manaka S, Shiogai T (1987) Evolution of criteria for determination of brain death in Japan. Acta Neurochir (Wien) 87:93–98
5. Walker AE (1985) Cerebral Death, 3rd edn. Urban and Schwarzenberg, Baltimore-Munich, pp 157–195
6. Ingvar DH (1974) Report of the committee on cessation of cerebral function. Electroencephalogr Clin Neurophysiol 37:530–531
7. Stone JL, Ghaly RF, Hughes JR (1988) Evoked potentials in head injury and states of increased intracranial pressure. J Clin Neurophysiol 5:135–160
8. Shiogai T (1989) Evaluation of the diagnosis of brain death: Part I. Brainstem auditory evoked potentials (in Japanese). Brain Nerve 41:73–83
9. Møller AR, Jannetta PJ, Bennett M, Møller MB (1981) Intracranially recorded responses from the human auditory nerve: New insights into the origin of brainstem evoked potentials (BSEPs). Electroencephalogr Clin Neurophysiol 52:18–27
10. Sohmer H, Gafni M, Havatselet G (1984) Persistence of auditory nerve response and absence of brain stem response in severe cerebral ischemia. Electroencephalogr Clin Neurophysiol 58:65–72
11. Allison T, Hume AL (1981) A comparative analysis of short-latency somatosensory evoked potentials in man, monkey, cat and rat. Exp Neurol 72:592–611
12. Momma F, Sabin HI, Symon L, Branston NM (1987) Clinical evidence supporting a subthalamic origin of the P15 wave of the somatosensory evoked potentials to median nerve stimulation. Electroencephalogr Clin Neurophysiol 67:134–139.
13. Desmedt JE (1985) Critical neuromonitoring at spinal and brainstem levels by somatosensory evoked potentials. Cent Nerv Syst Trauma 2:169–186

14. Shiogai T (1989) Evaluation of the diagnosis of brain death: Part II. Short-latency somatosensory evoked potentials (in Japanese). Shinkei Kenkyu no Shinpo 33:507–519
15. Belsh JM, Chokroverty S (1987) Short-latency somatosensory evoked potentials in brain-dead patients. Electroencephalogr Clin Neurophysiol 68:75–78
16. Delestre F, Lonchampt P, Dubas F (1986) Neural generator of P14 farfield somatosensory evoked potential studied in a patient with a pontine lesion. Electroencephalogr Clin Neurophysiol 65:227–230
17. Buchner H, Ferbert A, Hacke W (1988) Serial recording of median nerve stimulated subcortical somatosensory evoked potentials (SEPs) in developing brain death. Electroencephalogr Clin Neurophysiol 69:14–23
18. Stör M, Riffel B, Trost E, Ullrich A (1987) Short-latency somatosensory evoked potentials in brain death. J Neurol 234:211–214
19. Hughes JR (1978) Limitation of the EEG in coma and brain death. Ann NY Acad Sci 315:121–136
20. Bricolo A, Turazzi S, Faccioli F, Odorizzi F, Sciarretta G, Erculiani P (1978) Clinical application of compressed spectral array in long-term EEG monitoring of comatose patients. Electroencephalogr Clin Neurophysiol 45:211–225
21. Grigg MM, Kelly MA, Celesia GG, Ghobrial MW, Ross ER (1987) Electroencephalographic activity after brain death. Arch Neurol 44:948–954
22. Shiogai T (1989) Evaluation of the diagnosis of brain death: Part III. Compressed spectral arrays (in Japanese). Brain Nerve 41:309–318
23. Siesjö(1984) Cerebral circulation and metabolism. J Neurosurg 60:883–908
24. The Swedish Ministry of Health and Social Affairs (1984) The concept of death. Report of the Swedish committee on defining death, Stockholm
25. Lindegaard K-F, Grip A, Nornes H (1980) Precerebral haemodynamics in brain tamponade: Part 1. Clinical studies on blood-flow velocity. Neurochirurgia (Stuttg) 23:133–142
26. Hassler W, Steinmetz H, Gawlowski J (1988) Transcranial Doppler ultrasonography in raised intracranial pressure and in intracranial circulatory arrest. J Neurosurg 68:745–751

20

Short-Latency Somatosensory Evoked Potentials (SSEPs) in Comatose Patients

Akifumi Suzuki, Nobuyuki Yasui[1], and Kenji Nakajima[2]

Introduction

Recently, short-latency somatosensory evoked potentials (SSEPs) have been used diagnostically in a variety of neurologic conditions. Their reliability has been studied by many investigators. The present study was undertaken to clarify the clinical usefulness of SSEPs in patients who are comatose due to severe intracranial organic lesions.

Subjects and Method

Subjects consisted of 26 comatose patients. The 20 males and 6 females studied had a mean age of 63 ± 13 years. The intracranial pathology was brain stem infarction in 7 patients, pontine hemorrhage in 5, cerebral infarction with severe edema in 4, massive cerebral hemorrhage in 2, severe subarachnoid hemorrhage in 5, and severe head trauma in 3.

Thirty-nine SSEP recordings were done in 26 patients and continuous recordings of SSEPs were undertaken in 10 patients. Recording electrodes were placed on the scalp according to the international 10/20 system. Reference electrodes were placed on Erb's point and 4 cm below the inion (the portion over the spinous process of cervical spine II). Electrical stimuli were applied percutaneously to the unilateral median nerve at the wrist. Using a frequency band-pass filter of 0.3-3000 Hz, 1000 responses were averaged. Averaging summation and SSEP analysis were carried out by a Signal Processor 7T17F (NEC Sanei Co., Ltd.) and a Neuropack 8 (Nihon-koden Co., Ltd.). Of the several components of SSEPs, only the P_{13} and N_{16} components were analysed in the present study, because those two components have been considered to be generated in and around the brain stem.

[1]Departments of Surgical Neurology, and [2]Neurology, Research Institute for Brain and Blood Vessels-Akita, 6-10 Senshu-Kubota-machi, Akita, 010 Japan

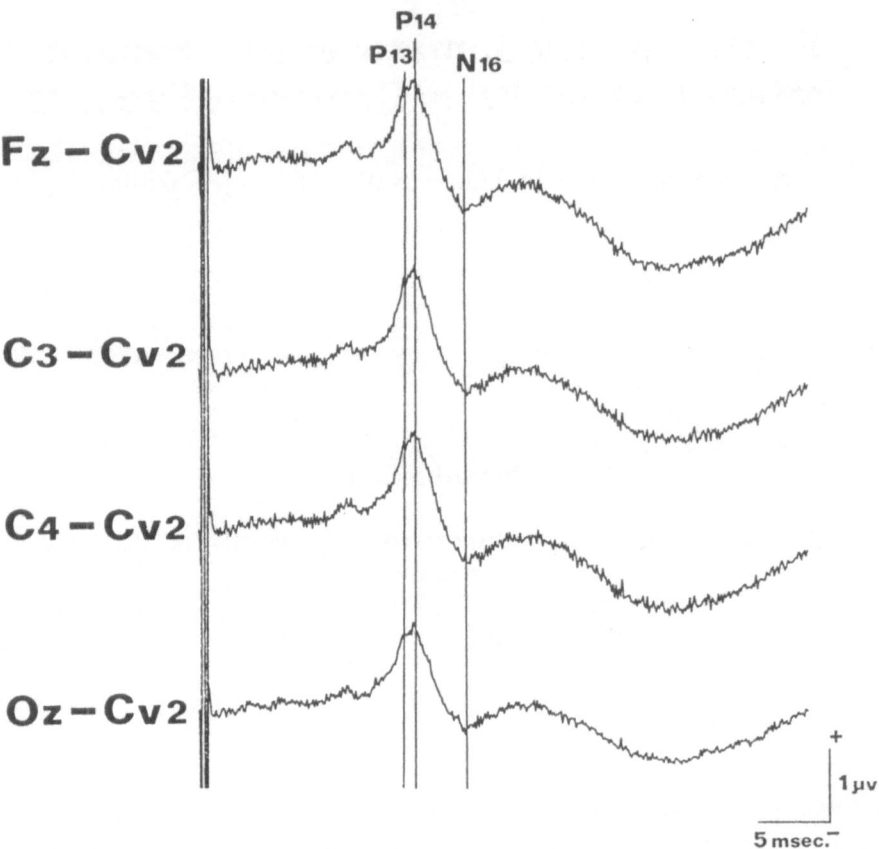

Fig. 1. SSEPs in case 14. Cv_2 is 4 cm below the inion (the level of cervical spine II)

The present study was based on an analysis of the relationship between SSEP findings and clinical outcomes. SSEPs were analysed through the evaluation of the presence of each component and the peak latency. Compared with normal variation, interpeak latency was analysed, and the component of which inter-peak latency was over 2SD of normal variation was evaluated as peak latency prolonged. Interpeak latency between Erb's potential and P_{13} was 3.7 ± 1.2 (2SD) msec, and between P_{13} and N_{16} was 3.0 ± 0.6 (2SD) msec in normal subjects.

Case Presentation

Case 14. A 76-year-old male who suffered from right hemiplegia and total aphasia. After admission, his consciousness level deteriorated gradually, and 3 days after onset, he became comatose. Anisocoria (left mydriasis) with absence of left light reflex was also revealed. Respiration was regular. A CT scan showed severe brain swelling with a low density area in the left hemisphere due to cerebral infarction. After the patient deteriorated to the comatose state, SSEP recordings were made. Figure 1 shows the SSEP recording on right median nerve stimulation. P_{13} and N_{16} were defined as normal, but later components could not be differentiated. The patient was treated conservatively and survived. His level of consciousness improved gradually but the right hemiplegia and total aphasia have continued.

Case 25. A 52-year-old male suffered a severe head trauma. One hour after the accident, he presented in a comatose state with right mydriasis, bilateral absence of light reflex and decerebrated posture. Tracheal intubation was performed because of irregular respiration. Blood pressure was 100/70 mmHg. A CT scan showed bilateral cerebral contusion and right epidural and subdural hematoma. Brain stem injury was also suspected. Continuous recordings of SSEPs are shown in Fig. 2. At the start of recording, Erb's potential, and the P_{13} and N_{16}

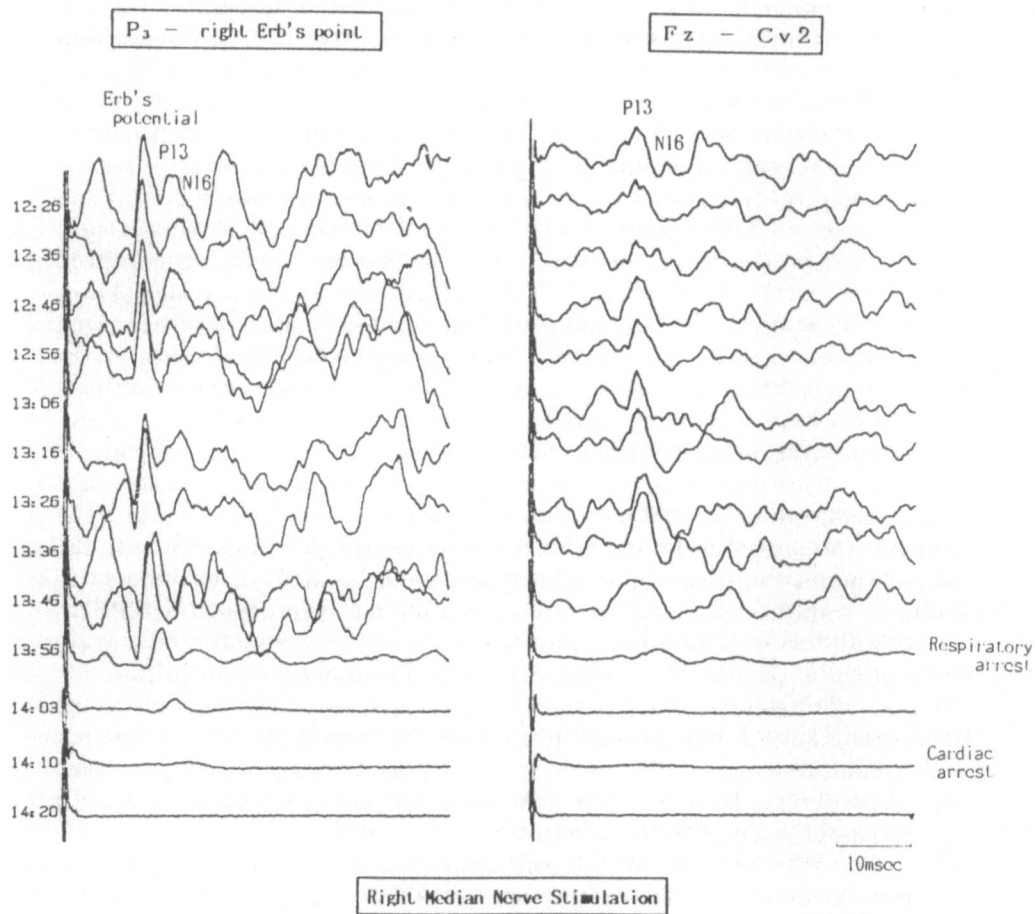

Fig. 2. Continuous monitoring of SSEPs in case 25

components were seen, but the N_{16} component was irregular and its peak latency prolonged. During continuous recordings, the reproducibility of the N_{16} component was poor. The respiratory pattern was Cheyne-Stokes and pupils showed bilateral mydriasis with light fixed. Blood pressure was 50.30 mmHg. At 13:46, the P_{13} component in the SSEPs became irregular in accordance with the reduction of spontaneous respiration. At 13:56, respiratory arrest occurred, and the peak latency of the P_{13} component was prolonged as was the reduction of its amplitude. Artificial respiration was not performed because of the family's insistence. Blood pressure could not be measured. Subsequently, the P_{13} and N_{16} components could not be detected. At 14:10, cardiac arrest occurred, and the patient died about 4 h after the accident.

Results

In Table 1, SSEP and clinical findings as well as outcomes are shown. In patients with brain stem infarction or pontine hemorrhage, 9 out of 12 patients (patient nos. 1-12) died and 3 patients survived. One of the 3 surviving patients improved with the recovery of consciousness. SSEPs on dead patients revealed some abnormalities of the P_{13} and/or N_{16} components such as prolonged latency and absence of the component. Surviving patients with a prolonged impairment of consciousness revealed abnormality of the N_{16} component. However, surviving patients who recovered consciousness revealed normal responses in the P_{13} and N_{16} components. Of the 6 patients with cerebral infarction or hemorrhage (nos. 13–18), 3 patients died and 3 survived. Two of the 3 surviving patients improved with the recovery of consciousness. SSEPs on patients who died revealed abnormalities of the P_{13} and/or N_{16} components. One surviving patient with prolonged impairment of consciousness revealed a prolonged peak latency of the N_{16} component on the affected side, but surviving patients who recovered consciousness had normal responses in the P_{13} and N_{16} components. In patients with subarachnoid hemorrhages due to rupture of aneurysms (nos. 19–23), 4 out of 5 patients died and only one patient survived with the recovery of consciousness. The surviving patient showed normal responses in P_{13} and N_{16} components, but 2 of the 4 patients who died also showed normal responses in those components. In the patients who died in spite of normal responses in P_{13} and N_{16} components, rerupture of intracranial aneurysms occurred after the recording of SSEPs. In patients with severe head trauma (nos. 24–26) all of the patients died and responses in the P_{13} and/or N_{16} components revealed various abnormalities.

Thus, abnormalities in the P_{13} and/or N_{16} components were observed in all fatal cases except for patients with subarachnoid hemorrhages. In surviving patients without recovery of consciousness, only abnormalities in the N_{16} component were observed. However, in patients who recovered consciousness, normal responses in the P_{13} and N_{16} components were recorded.

On the other hand, 7 out of 8 recordings under artificial respiration due to respiratory arrest revealed no responses of the P_{13} and N_{16} components. In almost all of those patients, recordings of SSEPs were not performed con-

Table 1. Results of clinical and SSEP findings in case materials

Intracranial Lesion	No.	Side	Resp.	Pupils Rt	Pupils Lt	L.R. Rt	L.R. Lt	SSEP Rt.stim P13	Rt.stim N16	Lt.stim P13	Lt.stim N16	Rec. Day	Outcome
BRAIN STEM INFARCTION	1	Bil	R / I / R	N / N / N	N / N / N	+ / + / +	+ / + / +	N / N / N	N / P / P	N / N / N	P / N / P	2 / 12 / 78	ALIVE (C)
	2	Bil	R / R	Mi / Mi	N / N	+ / +	+ / +	N / N	N / N	N / N	N / N	1 / 3	ALIVE (A)
	3	Bil	R	N	N	+	+	N	-	N	N	62	ALIVE (C)
	4	Bil	R / R	D / D	D / D	- / -	- / -	N / N	P / P	N / N	P / P	0 / 2	DIED (5)
	5	Bil	-	Mi	Mi	-	-	-	-	-	-	2	DIED (6)
	6	Bil	R	Mi	Mi	+	+	N	P	N	P	1	DIED (34)
	7	Bil	I	D	D	-	-	N	P	N	P	1	DIED (3)
PONTINE HEMORRHAGE	8	Bil	I	Mi	Mi	+	+	N	P	N	P	0	DIED (1)
	9	Bil	I / I	Mi / Mi	Mi / Mi	+ / +	+ / +	N / N	- / -	N / N	- / -	0 / 0	DIED (1)
	10	Bil	I	Mi	Mi	+	+	-	-	N	-	0	DIED (4)
	11	Bil	I / -	Mi / D	Mi / D	- / -	- / -	N / -	P / -	N / -	P / -	2 / 6	DIED (7)
	12	Bil	R	Mi	Mi	+	+	N	N	N	P	1	DIED (2)
CEREBRAL INFARCTION (SEVERE EDEMA)	13	Lt	R	N	N	+	+	N	N	N	N	6	ALIVE (A)
	14	Lt	R	N	D	+	-	N	N	N	N	6	ALIVE (A)
	15	Rt	R / R / R	D / D / D	N / D / D	- / - / -	- / - / -			N / N / N	N / P / P	3 / 4 / 5	DIED (8)
	16	Rt	R	D	N	-	+	N	N	N	P	0	DIED (2)
CEREBRAL HEMORRHAGE (MASSIVE)	17	Lt	R	N	D	+	-	N	P	N	N	1	ALIVE (C)
	18	Rt	-	D	D	-	-	-	-	-	-	0	DIED (0)
SUB-ARACHNOID HEMORRHAGE (RUPTURED ANEURYSM)	19	BA	R	D	D	-	-	N	N	N	N	0	ALIVE (A)
	20	?	R	N	N	+	+	N	P	N	P	35	DIED(113)
	21	VA	R	N	N	+	+	N	N	N	N	2	DIED (10)
	22	ACO	I / -	D / D	N / D	- / -	- / -	N / -	N / -	N / -	N / -	0 / 0	DIED (0)
	23	ACO	R / R / I	N / D / D	N / N / N	- / - / -	- / - / -	N / N / N	N / N / N	N / N / N	N / N / N	2 / 4 / 9	DIED (11)
SEVERE HEAD TRAUMA	24	Bil	- / -	D / D	D / D	- / -	- / -	N / -	- / -	N / -	- / -	0 / 0	DIED (3)
	25	Bil	I / -	D / D	N / D	- / -	- / -	N / -	P / -			0 / 0	DIED (0)
	26	Bil	-	D	D	-	-	-	-	-	-	0	DIED (1)

Resp.:Respiration, R:Regular, I:Irregular, N:Normal, P:Peak latency prolonged, -:No response
(numbers):Death day, (A):Alert consciousness, (C):Prolonged consciousness disturbance.

tinuously. Thus, the relationship between respiratory arrest and SSEP changes could not be determined in detail.

Discussion

Some investigators have reported SSEP findings in brain-dead patients [1–3]. Belsh and Chokroverty [2] reported that the P_{13} component could be recorded in three patients with brain death. In apneic patients receiving artificial respiration in the present study, the P_{13} component was recorded only in one patient. However, absolute apnea in this patient was not confirmed by an apneic oxygen test. In previous reports, the P_{13} component was considered to be generated in the cervical spinal cord or brain stem [4–8]. Recent reports supported the belief that the P_{13} component was generated in the cervical spinal cord. Thus, it is suggested that the P_{13} component may be recorded even in a brain-dead patient.

On the other hand, the N_{16} component has been considered to be generated in the medial lemniscus of the brain stem, or in the thalamus or thalamocortical radiations [5,7,9–12]. One of the author's results based on the current study suggest that the N_{16} components is generated in the brain stem [11], and observation of the N_{16} component of SSEPs was considered to be a reliable tool for evaluating brain stem function. SSEPs could provide information about the function of the somatosensory tract. However, in patients with brain stem compression due to raised intracranial pressure or severe brain stem destruction due to organic lesion, the functional disturbance of the brain stem did not occur solely in a localized region such as the medial lemniscus. In those patients, the brain stem was damaged diffusely, and the degree of functional disturbance of somatosensory tract would be correlated with other functional disturbances of the brain stem. Thus, the investigation of the N_{16} component of SSEPs is considered to be conducive for evaluating the degree of functional disturbance of the brain stem in such severely involved patients. From the results of the present study, SSEP monitoring in comatose patients is a reliable tool for evaluating the prognosis of brain stem dysfunction.

Summary. The findings of short-latency somatosensory evoked potentials (SSEPs) were studied in 26 patients who were comatose due to cerebrovascular diseases or head injury. Patients with recovery of consciousness disturbance showed normal response of the N_{16} component, and the brain stem damage was considered to be mild and reversible. However, in patients with normal responses of the N_{16} component, there is the possibility of a poor outcome after recurrence of the disease or progression of intracranial pathology. In patients with an abnormal N_{16} component, brain damage is considered to be severe and irreversible. The clinical outcome of those patients was poor, such as remaining in a vegetative state or dying. Thus, the present study indicated the clinical usefulness of SSEPs to diagnose the reversibility of coma in patients with severe intracranial organic lesions.

References

1. Anziska B, Cracco RQ (1980) Short latency somatosensory evoked potentials in brain dead patients. Arch Neurol 37:222–225
2. Belsh JM, Chokroverty S (1987) Short-latency somatosensory evoked potentials in brain-dead patients. Electroencephalogr Clin Neurophysiol 68:75–78
3. Goldie WD, Chiappa KH, Young RR, Brooks EB (1981) Brain-stem auditory and short-latency somatosensory evoked responses in brain death. Neurology 31:248–256
4. Desmedt JE, Cheron G (1981) Noncephalic reference recording of early somatosensory potentials to finger stimulation in adult or aging normal man: Differentiation of wide spread N_{18} and contralateral N_{20} from the prerolandic P_{22} and N_{30} components. Electroencephalogr Clin Neurophysiol 52:553–570
5. Katayama Y, Tsubokawa T (1987) Somatosensory evoked potentials from the thalamic sensory relay nucleus (VPL) in humans: Correlations with short latency somatosensory evoked potentials recorded at the scalp. Electroencephalogr Clin Neurophysiol 68:187–201
6. Leuders H, Lesser R, Hahn J, Klem G (1983) Subcortical somatosensory evoked potentials to median nerve stimulation. Brain 106:341–372
7. Urasaki E, Matsukado Y, Wada S, Kaku M, Nagahiro S (1984) A clinicophysiological study on the generators of short latency somatosensory evoked potential (in Japanese). No To Shinkei 36:363–374
8. Yamada T, Kimura J, Nitz DM (1980) Short latency somatosensory evoked potentials following median nerve stimulation in man. Electroencephalogr Clin Neurophysiol 48:367–376
9. Desmedt JE, Cheron G (1980) Central somatosensory conduction in man: Neural generators and interpeak latencies of the far-field components recorded from neck and right or left scalp and earlobes. Electroencephalogr Clin Neurophysiol 50:382–403
10. Mauguiere F, Desmedt JE, Courjon J (1983) Neural generators of N_{18} and P_{14} far-field somatosensory evoked potentials studied in patients with lesion of thalamus or thalamocortical radiations. Electroencephalogr Clin Neurophysiol 56:283–292
11. Suzuki A, Yoshioka K, Nagashima M, Yasui N (1988) Clinical study of N_{16} component of short latency somatosensory evoked potentials (in Japanese). Rinsho Noha 30:600–604
12. Yamada T, Graff-Radford NR, Kimura J, Dickins QS, Adams HP (1985) Topographic analysis of somatosensory evoked potentials with well-localized thalamic infarctions. J Neurol Sci 68:31–46

21

Monitoring Evoked Potentials and Cerebral Circulatory Index in Acute Cerebrovascular Disorders and Posthypoxic Brain Damage: Localized Brain Function vs Whole Brain Circulation and Metabolism

Hiroyuki Shimizu, Shingo Ono, and Ippei Watanabe[1]

Introduction

Evoked potentials (EPs) such as auditory brainstem responses (ABRs) and somatosensory evoked potentials (SEPs) reflect localized brain function and have been widely used in the diagnosis of primary lesions in the central nervous system [1] On the other hand, the cerebral circulatory index (CCI), calculated as the reciprocal of the difference in oxygen content between arterial and cerebral venous blood (Fick's principle), can be used as a clinical indicator to evaluate whole brain circulation and metabolism [2].

To date, we monitored EPs and/or CCI in 40 acute comatose patients with cerebrovascular disorders or posthypoxic brain damage. The purpose of this paper is to review our observations of EPs and CCI in these patients, and discuss the significance of brain monitoring using a combination of these methods.

Methods

Forty acute comatose patients served as subjects in this study which included 33 cases with cerebrovascular disorders, and 7 with posthypoxic brain damage (Table 1). On admission to the ICU, all patients had a Glasgow Coma Score of 8 or less. Thirty-five cases were treated with barbiturate therapy to control elevated intracranial pressure. A PC-type computer was used to automatically record ABRs and SEPs from electrodes located in the scalp C_z and a single ear lobe at 5-15 min intervals for 3-20 days until the patient either recovered or reached a stable state. Stimuli for ABR were delivered at a rate of 11Hz through a transducer inserted in the ear. Stimuli consisted of alternating clicks at 90dB. SEPs were obtained by stimulating the ulnar nerve contralateral to the

[1] Department of Neuroanesthesia and Neurointensive Care, Tokyo Metropolitan Neurological Hospital, 2-6-1 Musashidai, Fuchu, Tokyo, 183 Japan

Table 1. Cases subjected to monitoring of evoked potentials and/or CCI

Diagnosis	Monitored			
	Evoked potentials only	CCI only	Both	Total
I. Cerebrovascular disorders				
1) Subarachnoidal hemorrhage	10	3	1	14
2) Intracerebral hemorrhage	5	2	1	8
3) Subdural hematoma	3	0	0	3
4) Pontine hemorrhage	3	0	0	3
5) Others	3	2	0	5
II. Posthypoxic brain damage				
1) 15-20 min (cardiac arrest)	1	0	2	3
2) More than 30 min	3	1	0	4
Total	28	8	4	40

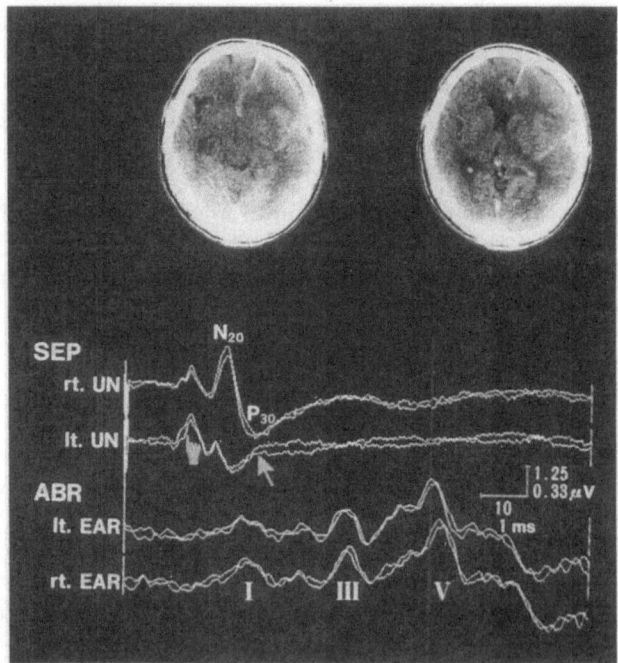

Fig. 1. ABR and SEP record for a patient (52-year-old male) with a subdural abscess. Note that CT scans revealed an abscess in the frontotemporal region, associated with global brain edema of the right hemisphere, and a midline shift. Cortical responses (N_{20} and P_{30}) were completely absent in SEPs elicited ipsilateral to the primary brain lesion in response to left ulnar nerve stimulation (lt. UN), as indicated by the *arrow*. In addition, the fifth component (V) in ABRs evoked ipsilateral to the primary lesions had a longer latency than in ABRs elicited contralateral

recording site. A bandpass filter of 30-3000Hz was chosen for ABR while 10-1000Hz was used for SEP. Interpretation of EP data was made on the following components: short latency components (indicated by the *pointer* in Fig. 1) and cortical responses (Fig. 1, N_{20} and P_{30}) in the SEPs, as well as five components from the first to fifth in the ABRs.

A 30 cm-long urethane catheter (6F) was inserted into the jugular bulb or sigmoid sinus via the internal jugular vein, thus allowing continuous monitoring of dural sinus pressure as well as checking of the CCI three times per day during the acute phase. Since a strong positive correlation between $PaCO_2$ and CCI was seen in patients who had CO_2 reactivity, CCI values at a $PaCO_2$ of 35-40 mmHg were adopted as significant data in the present study; CCI values above and below a $PaCO_2$ of 35-40 mmHg were discarded. EEG, ECG and arterial pressure were also monitored routinely.

Results

Cerebrovascular Disorders

In 4 out of 5 patients with cerebrovascular disorders in whom CCI was investigated, CCI (normal value: 15-25 [2]) ranged from 30-50 in the acute period, during which CT scans showed a midline shift and global brain edema marked by a high amplitude, slow-wave EEG (Figs. 2 and 3). CCI returned to the normal range when EEG showed fast activity (Figs. 2 and 3). Sequential changes in CCI in a patient with subarachnoidal hemorrhage are shown in Fig.2 together with EEG patterns, sigmoid sinus pressure, dosage of pentobarbital and CT scans. The patient was a 46-year-old male who underwent clipping of an aneurysm in the right middle cerebral artery. On the first day following surgery (3/15), the patient showed an increased CCI of more than 40, and CT scans revealed brain edema in the right hemisphere (a midline shift and narrowing of the ventricles). Sigmoid sinus pressure gradually elevated over the next day, suggesting increased intracranial pressure. On the fourth day (3/18), EEG showed fast activity in spite of barbiturate therapy, and CCI returned to the normal range as indicated by the cross hatched zone in Fig. 2. Since CT scans revealed a reduction in brain edema and sigmoid sinus pressure decreased, administration of pentobarbital was discontinued. Eight hours later, the patient regained consciousness and exhibited only minimal neurological sequalae (Fig. 2).

In all 15 patients with cerebrovascular disorders who developed minimal neurological sequelae, both ABRs (I–V) and short latency SEPs evoked bilaterally, as well as elicited cortical SEPs, were recordable throughout the entire course (Table 2). In contrast, all ABR and SEP components, except for the initial spikes in SEPs, were completely lost in all patients whose course ended in brain death (Table 2). Figure 4 shows serial ABR and SEP records in a patient with hemorrhage from the brainstem to the thalamus, together with CT scans made just before the monitoring of evoked potentials was started. All ABR components except the fourth components are present in the top traces, while cortical responses are absent in the SEPs. The next two traces show that the fifth

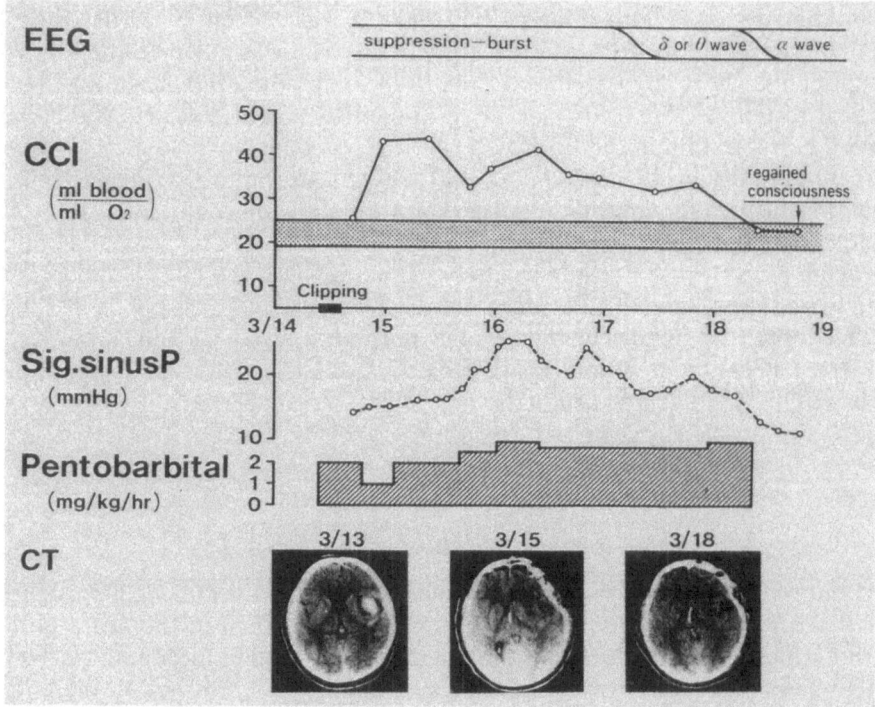

Fig. 2. Sequential changes in CCI in a patient (46-year-old male) with subarachnoidal hemorrhage. EEG patterns, sigmoid sinus pressure, dosage of pentobarbital and CT scans are also shown

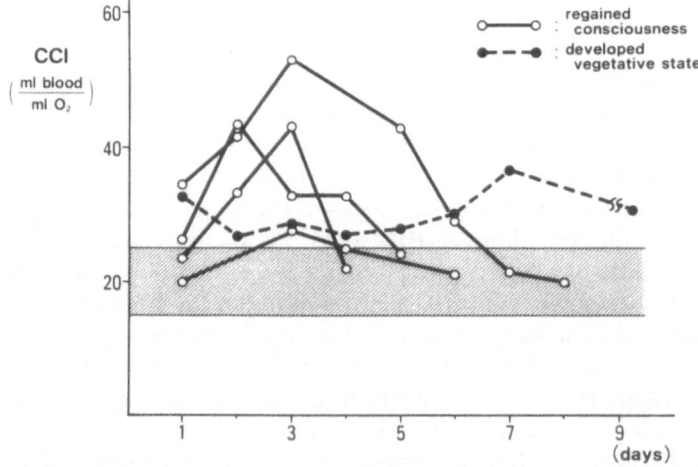

Fig. 3. Sequential changes in CCI in five patients with cerebrovascular disorders

Table 2. Outcome, abnormality in evoked potentials, and changes in CCI in 40 acute comatose patients with various etiologies

Outcome (number of patients)	ABR (I-V)	SEP		CCI $\left(\frac{CBF}{CMRO_2}\right)$	Diagnosis (number of patients)
		Short latency components	Cortical responses		
Recover (15)	+	+	+	temporary increase (30-50)	subarachnoidal hemorrhage (4) cerebral hemorrhage (4) subdural hematoma (3) pontine hemorrhage (1) others (3)
Vegetative (5)	+	+	+	stable at about 40	cerebral hemorrhage (1)
	+	+	−		subarachnoidal hemorrhage (1) posthypoxic brain damage (3)
Brain death (20)	−	−	−	more than 60	subarachnoidal hemorrhage (8) cerebral hemorrhage (4) pontine hemorrhage (4) posthypoxic brain damage (4)

+, each component was recordable;
−, each component was absent or disapperared entirely

ABR components decreased in amplitude and increased in latency while the first ABR components increased in amplitude. Short latency SEPs also decreased in amplitude in the third traces. In the fourth traces from the top, the second to the fifth ABR components as well as short latency SEPs were completely lost. After that, the high-amplitude, long-duration first components in ABR gradually disappeared while the initial spikes in SEP still remained. Thus, the diagnosis in this patient was brain death due to transtentorial herniation.

In a patient whose course terminated in the vegetative state, ABRs and short latency SEPs remained constant while cortical responses eventually disappeared in the SEPs evoked both ipsilateral and contralateral to the primary brain lesions (Table 2).

Posthypoxic Brain Damage

Seven patients with posthypoxic brain damage were divided into two types according to outcome; three patients who experienced a 15-20 min cardiac arrest exhibited the vegetative state while the other 4 exposed to cerebral anoxia for 30-50 min exhibited brain death (Tables 1 and 2). All seven patients showed an absence of cortical SEPs at the beginning of monitoring. Among those patients whose course terminated in the vegetative state, we observed two cases of α-coma, accompanied by temporary inhibition of ABRs in wave components III–V, as well as short latency SEPs (Fig. 5). Repeated CT scans did not show any significant changes and CCI remained approximately at 40 throughout the entire

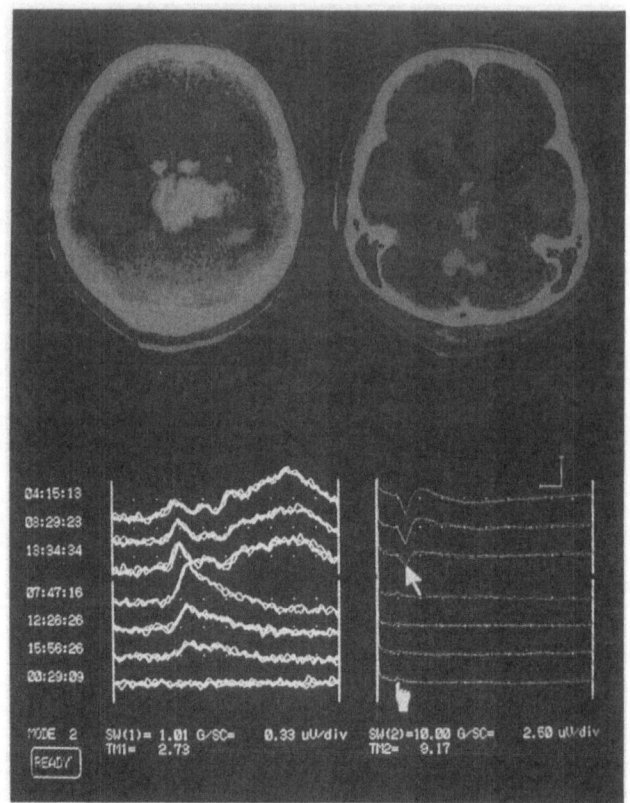

Fig. 4. Serial ABR (left) and SEP (right) records in a patient (72-year-old male) with hemorrhage from brain-stem to thalamus, terminating in brain death. Calibration marks indicate 0.33 μV and 1 msec for ABRs, and 2.5 μV and 10 msec for SEPs. CT scans were made just before EP monitoring was started.

course in these patients. On the other hand, all four brain-dead patients eventually showed a total loss of ABR and SEP components during the post-resuscitative period, marked by high CCI values of 60 or more (Table 2). CT scans showed progressive global brain edema followed by brain tamponade in these patients.

Discussion

CCI, which is calculated as the reciprocal of the difference in oxygen content between arterial and cerebral venous blood, indicates the ratio of cerebral blood flow (CBF) to the cerebral metabolic rate of oxygen ($CMRO_2$). Under non-diseased conditions, CCI ranges from 15-25 at a $PaCO_2$ of about 40 mmHg [2]. A CCI of less than 10 is always associated with an undesirable reduction in CBF,

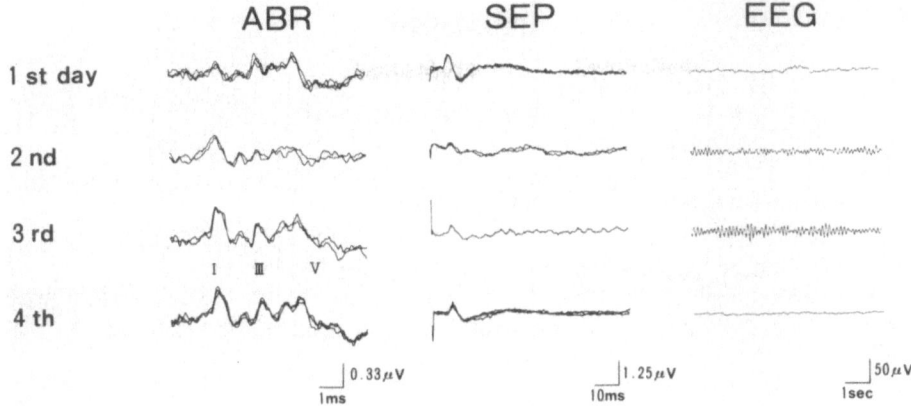

Fig. 5. ABR, SEP and EEG records in a posthypoxic patient (45-year-old female) who developed α-coma, terminating in the vegetative state. The number of days after resuscitation is shown to the left of each record. Note that **III–V** components in ABRs and short latency SEPs were temporarily inhibited with concomitant α-coma

resulting in cerebral ischemia [3]. A CCI of more than 40 under normal ventilation implies a reduction in $CMRO_2$ due to inhibition of cerebral metabolism [4].

In the present study, we observed a temporary increase in CCI, together with a midline shift and global brain edema on CT scans in patients with cerebrovascular disorders who had a good prognosis. ABRs, short latency SEPs, and cortical SEPs elicited contralateral to primary brain lesions were recordable throughout the entire course in these patients. On the other hand, posthypoxic patients following cardiac arrest developed a high degree of brain damage characterized by the vegetative state or brain death and showed a CCI of 40 or more. It is thought that these changes in CCI might directly reflect a reduction in $CMRO_2$, suggesting permanent inhibition of whole brain metabolism with irreversible brain damage, as well as functional changes in whole brain metabolism caused by localized brain lesions.

Summary. Cerebral circulatory index and/or evoked potentials were monitored in 40 acute comatose patients with cerebrovascular disorders or posthypoxic brain damage. Patients with cerebrovascular disorders who had a good prognosis showed a temporary increase in CCI. On the other hand, posthypoxic patients with a high degree of brain damage characterized by the vegetative state or brain death showed a CCI of 40 or more. Changes in EPs may be related to the severity of localized brain lesions which affect whole brain functioning. Whenever all ABR and SEP components were lost, the course always ended in brain death, while in most of the patients terminating in the vegetative state, only cortical SEPs ceased to be recorded. Thus, monitoring both EPs, which reflect localized brain function, and CCI, which is an indicator of whole brain circulation and metabolism, is thought to be useful in the management of acute comatose patients with various etiologies.

References

1. Greenberg RP, Mayer DJ, Becker DP, Miller JD (1977) Evaluation of brain function in severe human head trauma with multimodality evoked potentials: Part 1. Evoked brain-injury potentials, methods, and analysis. J Neurosurg 47:150–162
2. Takeshita H, Okuda Y, Sari A (1972) The effects of ketamine on cerebral circulation and metabolism in man. Anesthesiology 36:69–75
3. Finnerty FA Jr, Witkin L, Fazekas J (1954) Cerebral hemodynamics during cerebral ischemia induced by acute hypotension. J Clin Invest 33:1227–1232
4. Obrist WD, Langfitt TW, Jaggi, JL, Cruz F, Gennarelli TA (1984) Cerebral blood flow and metabolism in comatose patients with acute head injury. J Neurosurg 61:241–253

22

Clinical Evaluation of Brain Damage by Proton T1 Dynamics

Masahiro Furuse, Katsuyoshi Saso, Kaoru Ichihara, Suguru Inao,
Yoshimasa Motegi, Yoshiki Kaneoke, and Minoru Hoshiyama[1]

Introduction

Proton nuclear magnetic resonance (NMR) technique provides a valuable tool to investigate the status of edematous lesions, as well as degenerative processes in the brain owing to its high sensitivity to changes in tissue water. Recent developments of NMR in medicine are directed extensively to studying pathophysiology in the central nervous system. The dynamics of relaxation times are of special importance from the viewpoints of quantitative evaluation of intracranial lesions. In the present study, serial measurements of spin-lattice relaxation time (T1) were directed to assess the status of cerebral infarction, cerebral hemorrhage, and brain injury. Attention was focused upon the possibility of predicting irreversibility of tissue damage by relaxation time measurements.

Subjects and Methods

Seventy patients with cerebral infarction (14 cases), intracerebral hematoma (21 cases) and brain injury (35 cases) were studied to examine the time course of tissue relaxation time. A whole body NMR scanner (Fonar QED 80-α) was used for the studies in vitro and in vivo. The equipment provides dual function modes of image display and focal T1 measurement with a focussing magnetic field of 468 gauss or 2.01 MHz. The T1 values attained in this study were calculated from the direct spin-lattice relaxation curve based upon a τ sequence of 8–13 measuring points.

For in vitro calibration of T1 values, phantoms containing $NiCl_2$ solutions of known concentrations were used. The values of T1 measured in $NiCl_2$ solutions of 5, 7, 10, and 16 mmol/l were 298 ± 10, 216 ± 4.8, 155 ± 7.1, and 106 ± 4.6 ms

[1] Nakatsugawa Municipal General Hospital, 1522-1 Komanba, Nakatsugawa, 508 Japan

Table 1. Factors contributing to T1 in blood and hematoma content examined by multiple regression analysis

Factors	Partial correlation coefficients
1. Free water/solid component	0.8828
2. Hemoglobulin	0.8686
3. Iron	0.6301
4. Water content	0.5092
5. Methemoglobulin	0.4466
6. Albumin	0.3364

(mean ± SD), respectively. It was ascertained before and after each clinical measurement that the variations in T1 values in test phantoms remained within 10% of the standardized T1 values. Standard T1 values for healthy brains in this series were 288 ± 42 and 277 ± 27 ms, respectively, in cerebral gray matter and white matter.

In order to evaluate the degrees and orders of participation of biological factors to the T1 value, the relationship between in vitro T1 and laboratory findings was studied in samples of human blood and fluid taken from chronic subdural hematomas by means of multivariate analysis [1]. Table 1 represents the values of partial correlation coefficient according to multiple regression analysis, which indicates the degrees of contribution to T1 among various biological factors. The factors relating to water amounts and its constitution were revealed as the most significant factors. Among them, the value of free water content [2] was regarded as the most important.

Results and Comments

Cerebral Infarction

The patients of this category were subdivided into two groups: infarction of perforating branches or watershed areas (group A) and occlusion of major cerebral arteries (group B). Changes in in vivo T1 measured in the core of infarcted areas are shown in Table 2. More apparent prolongation of T1 in ischemic lesions was observed in group B as compared to group A, even in the acute stage. In group A, serial measurements of T1 showed little alteration in its time course, with values of 326 ± 29, 317 ± 36, and 331 ± 43 ms, respectively at 1–7 days, 8–30 days, and over 31 days. Contrarily, focal T1 in group B noticeably altered according to the phase after stroke : the T1 values in the acute phase were 474 ± 95 ms, being once reduced to 363 ± 64 ms in the subacute phase, and they thereafter revealed marked elevation in the chronic phase up to 714 ± 319 ms on the average, often exceeding 1,000 ms [3].

The transient decrease of T1 observed in the subacute stage probably reflects the processes of brain edema regression in the ischemic lesions. Marked T1 pro-

Table 2. Changes in T1 values in infarcted area in association with terms after stroke

ms (mean ± SD)

	1–7 days	8–30 days	31 days
Infarction of perforating branches/watershed areas	326 ± 29 (n = 6)	317 ± 36* (n = 17)	331 ± 43 (n = 7)
Occlusion of major cerebral arteries	474 ± 95 (n = 7)	363 ± 64** (n = 12)	714 ± 319** (n = 7)

*$P<0.05$; **$P<0.01$ (to the previous stage)

longation during the chronic stage was considered to originate from tissue degeneration and necrosis. Regarding the prospect of prognostic values, it is our assumption that irreversible damage in ischemic tissue may be manifest by T1 prolongation exceeding 450–500 ms in the acute stage.

Cerebral Hemorrhage

Time courses of T1 in cases with intracerebral hematoma are shown in Table 3. The T1 value of the intracerebral hematoma itself indicated a trend of prolongation with the passage of time. T1 values in the hematoma center were 285 ms on average in the acute stage, whereas they were prolonged to 313 ms by 1 month after the stroke. This seemed to reflect alterations in the properties of the hematoma, such as clot formation in the earlier phase and resolution in the later phase [4]. When intracerebral hematomas separate from their cavities in the chronic stage, the site of hemorrhage revealed much prolonged T1 values, usually exceeding 1,000 ms in the core portion.

In contrast, T1 in the brain tissue surrounding the hematoma was rather prolonged in the earlier phase, with a peak of 312 ± 22 ms around 2–4 weeks following the onset; thereafter they tended to normalize to 296 ± 31 ms in the period over 31 days. This time course probably reflects the natural processes of formation and regression of perifocal brain edema.

Table 3. Changes in T1 values of intracerebral hematomas and their perifocal brain tissue

ms (mean ± SD)

	1–3 days	4–15 days	16–30 days	31 days
Hematoma	285 ± 29 (n = 10)	264 ± 24* (n = 17)	309 ± 32** (n = 14)	313 ± 34 (n = 7)
Perifocal brain	285 ± 28 (n = 10)	307 ± 22** (n = 17)	312 ± 22 (n = 13)	296 ± 31 (n = 7)

*$P<0.05$; **$P<0.01$ (to the previous stage)

Table 4. T1 values of brain parenchyma associated with head injury in acute and chronic phases

ms (mean ± SD)

	Acute phase (1–3 days)	Chronic phase (14 days +)
Contusion	359 ± 121 (*n* = 6)	403 ± 87** (*n* = 11)
Perifocal edema	334 ± 52 (*n* = 9)	283 ± 32** (*n* = 4)
Acute EDH	289 ± 17 (*n* = 4)	281 ± 17 (*n* = 8)
Acute DCS		
White matter	248	223
Cortex	292	280

** *P* < 0.01 (to the previous phase);
EDH, epidural hematoma; DCS, diffuse cerebral swelling

Brain Injury

Changes in tissue T1 in relation to cerebral contusion and its perifocal edema are summarized in Table 4. After head injury, T1 values in the areas surrounding the contusional site were prolonged consistently. Such prolonged T1 values in the perifocal region gradually normalized with time [5]. T1 values corresponding to contusional edema reached 334 ± 52 ms in acute phase, followed by recovery to 283 ± 32 ms 2 weeks after the head injury.

In contrast, parenchymal injury resulted in a progressive T1 elevation, from an average value of 359 ± 121 ms in the acute phase to 403 ± 87 ms 2 weeks later. In cases with markedly prolonged T1 in the initial phase, the value of T1 in the contusional site tended to elevate later and rather steeply beyond the level of 500 ms, often exceeding 1,000 ms in the chronic stage. Such remarkable prolongation of T1 appears to be attributable to post-traumatic degeneration of the brain tissue. It was noteworthy that initial tissue T1 values at the higher levels of 400–500 ms were closely related to such extreme T1 elevations in the later phase.

On the other hand, there were no noticeable changes in tissue T1 over time in patients with acute diffuse cerebral swelling or those who underwent evacuation of acute epidural or chronic subdural hematomas. The underlying pathophysiology in such situations seems to be not brain edema but cerebral hyperemia.

Summary. The numerical values of T1 provide the means to quantitatively assess the alteration of biological water status in intracranial lesions, particularly that relating to progression or regression of brain edema. Progressive and significant elevations of T1 indicate processes leading toward tissue degeneration. A markedly elevated T1 with the ranges of 450–500 ms (2 MHz) during the acute

stage following brain insults is considered to be indicative of the progress towards irreversibility of the damaged tissue.

Such higher ranges of T1 around 500 ms most likely correspond to the tissue
status of abnormal hydration, in which increases of tissue water content may
exceed the limit of binding capacity to normal structures and processes of degeneration may already be occurring in the tissue itself.

References

1. Hoshiyama M, Motegi Y, Yoshida K, Furuse M, Nitta S, Izawa A, Mizuno M (1988)
   Factors influencing T1 values of biological tissue assessed by multivariated analysis.
   JMRM 7 (Suppl 2): 133
2. Furuse M, Gonda T, Inao S, Kuchiwaki H, Hirai N, Kageyama N (1987) Thermal
   analysis of water components in brain tissue: Quantitative determination of free and
   bound water fractions (in Japanese). No To Shinkei 39:761–767
3. Motegi Y, Furuse M, Kanaoke Y, Saso K, Inao S, Yoshida K (1986) Time courses of
   T1 values associated with cerebral infarction. J. NMR Medicine 3:185–192
4. Inao S, Furuse M, Saso K, Yoshida K, Motegi Y, Kaneoke Y, Kamata N, Izawa A
   (1986) Time course of NMR images and T1 values associated with hypertensive intracerebral hematoma (in Japanese). No To Shinkei 38:661–667
5. Inao S, Furuse M, Saso K, Yoshida K, Motegi Y, Kaneoke Y, Izawa A (1987) Significance of focal relaxation times in head injury. Neurol Med Chir (Tokyo) 27:1039–
   1045

23

Advanced Treatment of Prolonged Coma: Selection of Candidates for Deep Brain Stimulation and Clinical Results

Takashi Tsubokawa and Takamitsu Yamamoto[1]

Introduction

In an attempt to facilitate recovery from prolonged coma, deep brain stimulation has been applied to the pallidum, the ventroanterior thalamus, the midbrain reticular formation, and the thalamic intralaminar nucleus [1–4]. This approach is based on the anatomical and electrophysiological finding that brain arousal activity might be produced by stimulation of the ascending activating system [5,6]. Deep brain stimulation might also inhibit an active neural process within the brain stem that is involved in the production of coma [7]. However clinical results of deep brain stimulation have been quite variable and long-term follow-up results have not yet been reported [1–4, 8].

Difficulty in demonstrating the effect of stimulation might be caused by (1) variability of the brain damage which induced prolonged coma and lack of evaluation of the damaged brain before application of the deep brain stimulation, (2) lack of knowledge about physiological mechanisms underlying the loss of consciousness, and (3) lack of knowledge about the effects of the deep brain stimulation on cerebral blood flow, metabolism, and intracranial pressure which might be factors important in the induction of neural plasticity.

Recently, the effects of either thalamic intralaminar nucleus or the midbrain reticular formation on these factors during stimulation were studied on the vegetative state caused by diffuse brain injury [9,10]. EEG showed the arousal pattern, with behavioral arousal response, regional CBF increased markedly and diffusely in both hemispheres (Fig. 1) and in the thalamus associated with increase of glucose uptake [9], and waves of raised intracranial pressure wave immediately inhibited after the onset of MRF stimulation without any change of the base line pressure [10] (Fig. 2). Therefore, it appears likely that chronic deep

[1]Department of Neurological Surgery, School of Medicine, Nihon University, 30-1 Ohyaguchi Kamimachi, Itabashi-ku, Tokyo, 173 Japan

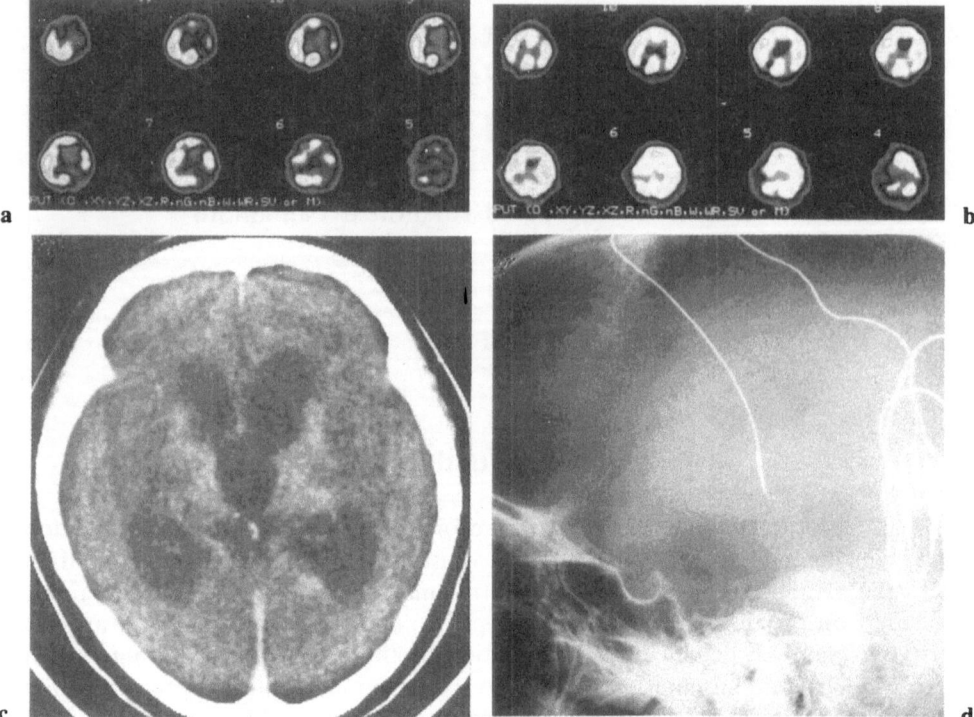

Fig. 1 a–d. The regional cerebral blood flow measured by SPECT increases remarkably and diffusely at the bilateral cortex and the diencephalon during deep brain stimulation at the r-thalamic intralaminal nucleus. **a** Before stimulation; **b** during stimulation; **c** CT findings; **d** chronic implanted electrode located at the thalamic intralaminar nucleus (CEM)

brain stimulation can be an advanced treatment for the relief of prolonged coma, depending upon how the severity of the damaged brain which caused the prolonged coma was evaluated.

In this presentation, a classification of the severity of the cases suffering from prolonged coma was developed using CT findings and electrophysiological studies. The long-term clinical results of deep brain stimulation on such classified groups were studied in order to define the most suitable candidates for deep brain stimulation as an advanced treatment.

Method

Twenty-four patients suffering from prolonged coma without any localized gross brain damage revealed on CT and lasting more than 2 months were studied. EEG was recorded from C_3 and C_4 by using compressed spectral array for 48 h, and classified as a slow monotonous and changeable spectrum. Brain stem audi-

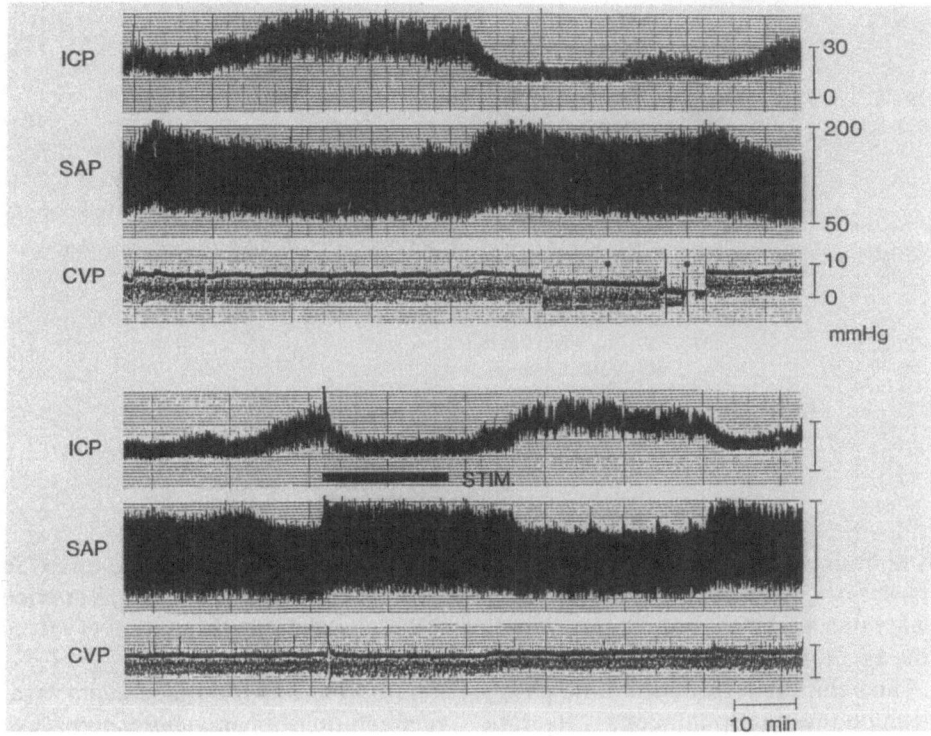

Fig. 2. Baseline intracranial pressure did not change during or after deep brain stimulation, in spite of behavioral sympathetic excitation appearing during stimulation. However, spontaneous pressure waves ceased immediately after the onset of MRF stimulation. The pressure waves did not occur as long as the stimulation was continued

tory evoked responses (BSR) and somatosensory evoked responses (SER) were recorded by conventional methods. To evaluate alterations of the BSR, four grading were used; elongation of the latency of the V wave, disappearance the V wave, disappearance of the IV and V waves, and no response. Alteration of the SER was evaluated by three grades; elongation of central conduction time, absence of the N_{20} wave, and no response.

Regional cerebral blood flow and glucose uptake were studied by positron emission computed tomography (PET) scanning in some cases and in other cases, regional cerebral blood flow was detected by using single photon emission CT using N-isopropyl-$_{123}$I iodoamphetamine. For chronic deep brain stimulation [10], a flexible platinum-iridium electrode which consisted of four twisted wires connected with four uninsulated contact points was used. Under local anesthesia, the electrode was implanted into the thalamic intralaminar nucleus or the midbrain reticular formation (nuclei cuneiformis). The stimulation was applied by using a high frequency transmitter-receiver system or an implantable pulse generator (Medtronic; Model 3380, Model 3360). This stimulator can deliver an

Table 1. Neurological and behavioral scores for the evaluation of the vegetative state

1. Alive with spontaneous respiration
2. Withdrawal response to pain
3. Spontaneous eye opening and closing
4. Spontaneous movement of extremities
5. Pursuit by eye movement
6. Emotional expression
7. Oral intake
8. Producing sounds
9. Obeying orders
10. Verbal response

Grading is made by adding each positive item. Range of scores: 0–10 points

amplitude of 0–10 volts into a 500 ohm resistive load, with a pulse rate of 1–120 pulses per s and a pulse width of 1.0–10 ms. This sequence of stimuli was applied daily during the daytime for 10–20 min every hour until there was recovery from the vegetative state.

The clinical evaluation of the physical condition of the prolonged coma cases included testing spontaneous respiration, recongnition of pain, spontaneous eye opening, spontaneous movement of extremities, pursuit by eye movement, emotional expression, oral intake of foods, producing sounds, obeying directions, and verbal response were checked and scored with each item receiving one point (Table 1). The vegetative state patients had 3–4 points. The evaluation was made weekly on the cases treated by chronic deep brain stimulation.

Results

Evaluation of Severity of the Brain Damage-Induced Prolonged Coma

In this study, 24 patients suffering from prolonged coma for more than 2 months were evaluated and classified by the results of EEG-recorded compressed spectral array, somatosensory evoked responses, and brain stem auditory responses (Table 2). The etiologies of the prolonged coma were cerebrovascular disease, head injury, brain tumor, brain abscess, and anoxia (Table 2). There were no differences in severity between the different causative conditions if no localized gross brain damage was present on CT.

There were two types of EEG changes on the compressed spectral array. One was a slow, monotonous spectrum which showed no fast wave but mainly delta band activity, and was recorded consistently in 8 cases. The other type showed a changeable spectrum which showed both slow and fast waves and a power spectrum that changed from time to time, and appeared in the other 16 cases (Fig. 3). All of the eight patients who showed the slow monotonous spectrum had needed

Table 2. Prolonged coma cases classified by EEG, SEP, and ABR and clinical effects on the vegetative state following chronic deep brain stimulation

Name	Age	Sex	Cause of coma	BSR	SEP·S-SEP	Respiration	Outcome without stimulation	Final outcome after stimulation neurological score
Group 1			Changeable spectrum (EEG)					
KS	32	F	Tumor	Normal	Normal	Spontaneous	Spontaneous recover	Excellent (4→9)
SH	49	F	Cerebral abscess	Normal	Normal	Spontaneous		
Group 2			Changeable spectrum					
RT	72	M	Vascular	Normal	Prolonged N_{20}	Spontaneous	Fell into vegetative state	
YY	86	M	Vascular	Prolonged V	Prolonged N_{20}	Spontaneous		
SA	68	M	Tumor	Prolonged V	Prolonged N_{20}	Spontaneous		
SO	64	M	Tumor	Prolonged V	Prolonged N_{20}	Spontaneous		
ES	65	F	Vascular	Prolonged V	Prolonged N_{20}	Spontaneous		
☆HS	43	M	Vascular	Prolonged V	Prolonged N_{20}	Spontaneous		Excellent (4→9)
☆SH	75	M	Trauma	Prolonged V	Prolonged N_{20}	Spontaneous		Excellent (4→9)
☆HY	24	M	Trauma	Prolonged V	Prolonged N_{20}	Spontaneous		Excellent (4→8)
☆TA	62	M	Vascular	Prolonged V	Prolonged N_{20}	Spontaneous		Good (4→7)
☆KS	22	M	Trauma	Prolonged V	Prolonged N_{20}	Spontaneous		Good (3→5)
Group 3			Changeable spectrum					
KK	83	M	Trauma	Prolonged V	No N_{20}	Spontaneous		
☆KM	74	F	Vascular	Prolonged V	No N_{20}	Spontaneous		Poor (2→3)
☆HO	48	M	Trauma	Prolonged V	No N_{20}	Spontaneous		Poor (2→3)
☆TS	41	F	Anoxia	Prolonged V	No N_{20}	Spontaneous		Poor (3→4)
			Slow monotonous spectrum (EEG)					
TJ	56	M	Vascular	No response	No N_{20}	Control		
YH	62	M	Vascular	No response	No N_{20}	Control		
YA	77	F	Vascular	No response	No N_{20}	Control		
HO	66	F	Vascular	Only I–III	No N_{20}	Assist	Death within 3–7 months	
HK	17	M	Vascular	Only I–II	No N_{20}	Control		
NA	78	F	Vascular	Prolonged V	No N_{20}	Assist		
KH	74	F	Tumor	Prolonged V	No N_{20}	Assist		
MF	52	M	Trauma	Prolonged V	No N_{20}	Assist		

☆ treated by chronic deep brain stimulation

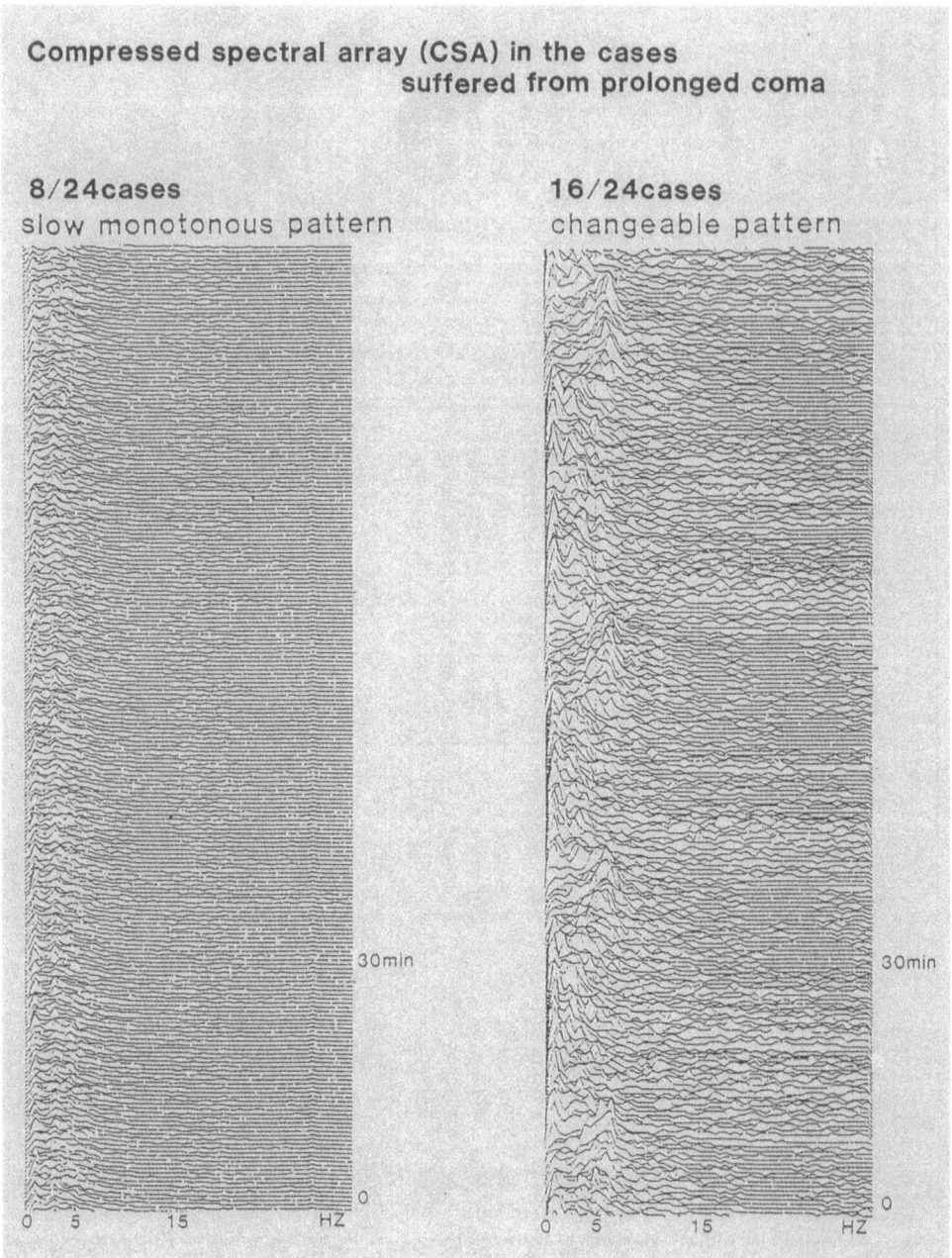

Fig. 3. Prolonged coma patients were classified into 2 groups by alteration of compressed spectral array

some assistance for respiration and died within 3–6 months after the onset of coma. The N_{20} wave on SER could not be recorded in any of these cases and wave V of the BSR could not be recorded in 5 cases. These 8 cases were not candidates for the chronic brain stimulation therapy.

The other 16 cases who showed a changeable spectrum were classified into three groups according to the changes in BSR and SER (Table 2). The first group (2 cases) who had normal SER and BSR were able to recover spontaneously to a conscious state by 3 months after the insult. These patients did not require deep brain stimulation. The other 14 patients were in the vegetative state. They were classified into two groups: ten cases (group 2) showed both wave V on the BSR and the N_{20} wave on SER, and their latency was prolonged more than 2 times standard deviation. The other 4 other cases showed wave V on BSR with long latency and low amplitude, but no recording of N_{20} on SER as seen in group 3 (Table 2). The second and third groups were thought to be suitable for treatment by advanced brain stimulation.

Effects of Chronic Deep Brain Stimulation

Eight of the 14 patients belonging to the second and third groups were treated by the chronic deep brain stimulation (Table 2). Thalamic stimulation was applied on 6 cases and 2 cases were stimulated at the midbrain reticular formation. Of the second group (long latency of wave V on BSR and N_{20} on SER), 6 cases were treated by deep brain stimulation. In two cases, the level of consciousness gradually improved during 2–3 months and verbal contact became possible 4 months after onset of the stimulation (Figs. 4, 5). At 5 months, the patients were able to sit and take food by themselves. They are evaluated as excellent examples of effect of the treatment. The other three cases in the second group showed very good behavioral and EEG arousal response and increases of both cerebral blood flow and glucose uptake in the cortex on both sides and in the thalamus during and after chronic stimulation. But these cases did not emerge from the vegetative state, although their neurological score came up to 6–7 points by chronic stimulation during a 3-month period. They are evaluated as examples of a good effect of the chronic deep brain stimulation.

On the other hand, three cases belonging to the third group, who did not show any N_{20} of SER, had good responses only during the stimulation of either the thalamic intralaminar nuclei or the midbrain reticular formation, but such arousal behavior did not continue after the stimulation. Even after 4 months' stimulation, they did not emerge from the vegetative state and there was no change of more than 1 or 2 points on the neurological score. These cases are evaluated as poor or having no effect.

Follow-up Result of Chronic Deep Brain Stimulation

Eight cases who were treated by chronic deep brain stimulation were followed up for more than 1 year. The neurological grade of these cases before stimulation was 2 points in 3 cases, 3 points in 1 case, and 4 points in 4 cases. Two

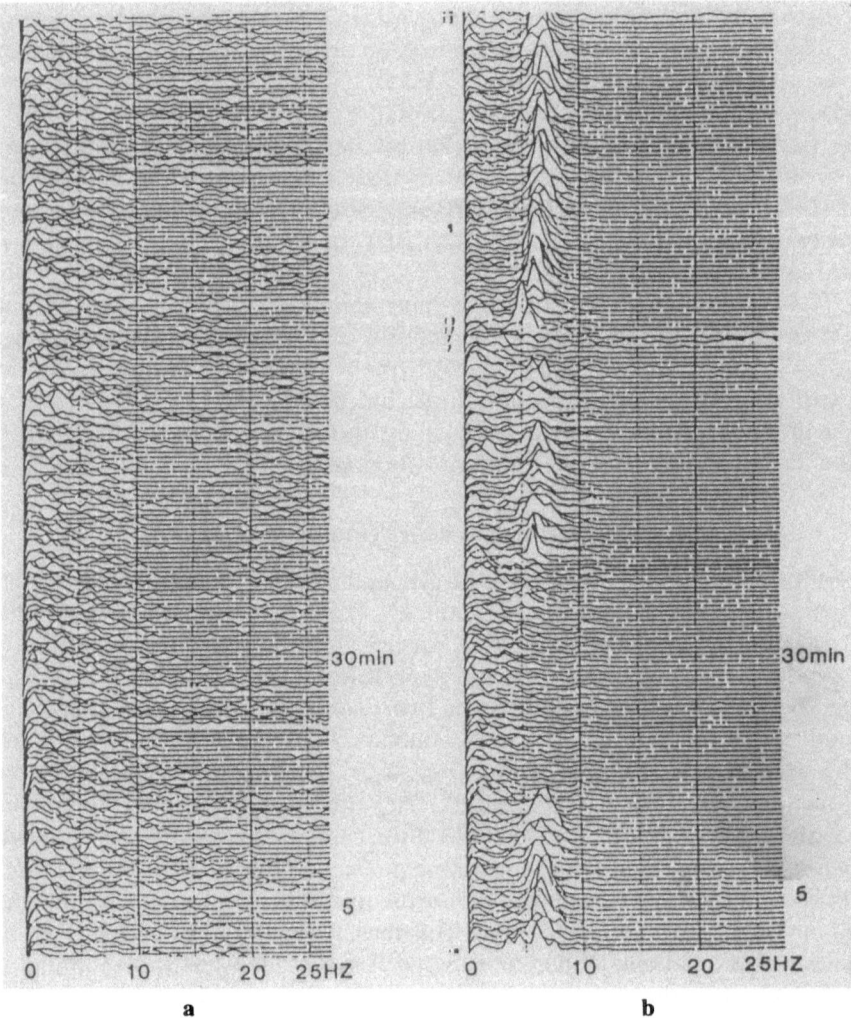

Fig. 4. The cases classified in the second group were able to improve from the vegetative state. Before (**a**) and 3 months after stimulation (**b**), the EEG changes markedly to a normal recording

months after beginning the chronic stimulation, it was very difficult to find any of the stimulation effects on the neurological score in all cases, but 4–5 months after beginning the chronic stimulation, all cases who eventually got excellent effects showed remarkable improvement on the neurological score as shown in Fig. 5. One year after beginning the chronic deep brain stimulation, 4 cases were able to emerge from the vegetative state (Fig. 5). The cases who had excellent scores from the chronic deep brain stimulation are in group 2 and their neurological scores before stimulation were 3-5 points (Table 2).

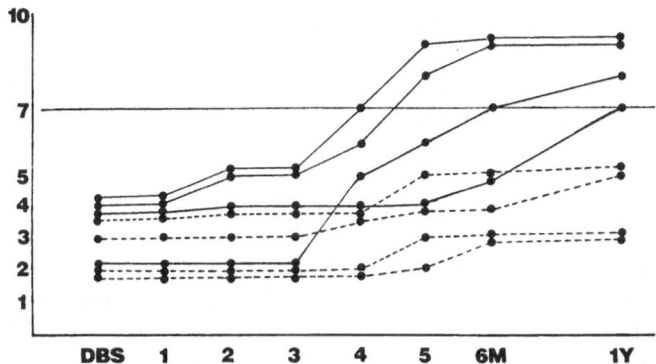

Fig. 5. Follow-up neurological score of deep brain stimulation on the vegetative cases

Discussion

Hassler et al. [2] treated a patient in a prolonged coma by repeated stimulation of the right pallidum and basal portion of the left lateropolor nucleus of the thalamus (VA) and Strum et al. [3] also reported a case in the vegetative state treated by chronic stimulation at the left reticular polaris nucleus. They recognized clear and immediate arousal reaction during the periods of stimulation. It had been shown in animal experiments that behavioral and EEG arousal pattern could be induced by acute stimulation at the brain stem reticular formation [5]. These arousal responses were induced by an increase of activity in the ascending reticular system [5,6].

Hassler et al. [2] and Strum et al. [3] did not explain the clinical effect which might emerge from the vegetative state because they did not use chronic brain stimulation for more than 1 month. They also did not report any criteria on patients in prolonged coma or in the vegetative state who received deep brain stimulation. As shown in this paper, deep brain stimulation not only induced the EEG and behavioral arousal responses during stimulation, but also induced remarkable increases of bilateral cortical regional blood flow and glucose uptake with a decrease of intracranial pressure. The arousal effect from prolonged coma appeared from 3–4 months after beginning the chronic stimulation. The cases who had excellent results by our chronic deep brain stimulation showed the neurophysiological state characterized by a changeable EEG pattern with a long latency, but not with the disappearance of both wave V on BSR and N_{20} on SEP.

According to the findings and clinical effects of chronic deep brain stimulation for relief from the vegetative state, its mechanisms are not only involved in the arousal response during stimulation of the ascending reticular activating system, but also have an important role in inducing neuronal plasticity when the cerebral blood flow and cerebral glucose uptake are associated with suppressed intracranial pressure caused by chronic deep brain stimulation.

Although the stimulating therapy for relief of the vegetative state is not a new one, for the first time chronic deep brain stimulation using telemetric high frequency transmitter receiver systems has been applied to the vegetative cases after checking electrophysiological activities for selection of the candidate.

Conclusion

From these clinical studies, the following might be concluded: (1) electrophysiological evaluation, including recording of EEG and multimodality evoked responses, were not only useful in predicting the prognosis of patients in prolonged coma but also in defining the candidates for chronic deep brain stimulation to obtain improvement from the vegetative state, (2) deep brain stimulation is an advanced treatment for relief from the vegetative state if the patients have a changeable spectrum on EEG with both a prolonged latency of wave V on BSR and N_{20} on SER (the second group by our classification) without any localized gross brain damage on CT, and (3) it takes 3–4 months to emerge from the vegetative state with chronic deep brain stimulation therapy. This indicates that the effects of chronic deep brain stimulation are not only related to the arousal response caused by stimulation of the ascending activation system, but that it also has an important role in increasing cerebral blood flow and glucose metabolism in both the cortex and the thalamus. This may induce neural plasticity in the central nervous system for maintaining consciousness in the vegetative cases.

Summary. In an attempt to facilitate recovery from the vegetative state, chronic deep brain stimulation with long-term implantation of a transmitter-receiver system has been applied to the thalamic intralaminar nuclei or the midbrain reticular formation in 8 cases. Three cases emerged from the vegetative state by this stimulation after 3–4 months. All of them had a changeable spectrum on EEG with a prolonged latency of wave V on BSR and N_{20} on SER and they had no localized gross brain damage on CT 2 months after the onset of unconsciousness.
    According to these clinical experiences, it is concluded that the chronic deep brain stimulation of the medial thalamic nucleus (CM•Pf) or the midbrain reticular formation (nuclei cuneiformis) is an advanced treatment for relief from the vegetative state if the patients do not have gross brain damage on CT but do have a changeable spectrum on EEG with both BSR and SER, even when those are of a longer latency.

References

1. McLardy T, Ervin F, Sweet W (1968) Attempted inset-electrodes arousal from traumatic coma: Neuropathological findings. Trans Am Neurol Ass 93:25–30
2. Hassler R, Dalle Ore G, Dieckmann G, Bricolo A, Dolce G (1969) Behavioral and EEG arousal induced by stimulation of unspecific projection systems in a patient with post-traumatic apallic syndrome. Electroencephalogr Clin Neurophysiol 27:306–310
3. Strum V, Kühner A, Schmitt HP, Assmus H, Stock G (1979) Chronic electrical

stimulation of the thalamic unspecific activating system in a patient with coma due to midbrain and upper brain stem infarction. Acta Neurochir (Wein) 47:235–244

4. Katayama Y, Miyazaki S, Yamamoto T, Maejima S, Tsubokawa T (1987) Chronic stimulation of the midbrain reticular formation in patients with traumatic prolonged coma. Oriental Medicine and the Pain Clinic 17:67–71

5. Moruzzi G, Magoun HW (1963) Brain stem reticular formation and activation of the EEG. Electroencephalogr Clin Neurophysiol 139:343–344

6. Dieckmann G (1968) Cortical synchronized and desynchronized responses evoked by stimulation of the putamen and pallidum in cats. J Neurol Sci 7:385–391

7. Tsubokawa T, Katayama Y (1985) Active neural process within the brain stem in production of coma. Neurol Med Chir (Tokyo) 25:503–514

8. Katayama Y, Tsubokawa T, Yamamoto T, Maejima S (1987) Chronic stimulation of the midbrain reticular formation and thalamus in patients suffering from prolonged coma caused by diffuse brain injury. Neurotraumatology 10:32

9. Yamamoto T, Katayama Y, Tsubokawa T, Maejima S, Yoshino A, Hiratama T (1988) Clinical indication of deep brain stimulation for comatose patients: Electrophysiological classification and long-term prognosis of persistent vegetative patients. Functional Neurosurgery 27:153–159

10. Tsubokawa T, Katayama Y, Miyasaki S (1989) Intracranial pressure changes in response to deep brain stimulation in traumatic prolonged coma patients. In: Hoff JH, Betz AL (eds) Intracranial Pressure VII, Springer-Verlag, Berlin Heidelberg, pp 703–705

11. Tsubokawa T, Katayama Y, Yamamoto T, Hirayama T (1985) Chronic stimulation of deep brain structures for treatment of chronic pain. In: Tasker RR (ed) Neurosurgery (State of arts review) vol 2, stereotaxic surgery. Henley and Belfus, Philadelphia, pp 235–255

24

Neurological Intensive Care: Past Lessons, Current Accomplishments, Future Directions

Allan H. Ropper[1]

Introduction

I wish to offer some perspective on new brain sparing and brain resuscitation therapies. In the end, we all must ask, "What does this mean for the patient"? Many experimental therapies undoubtedly protect the brain, but practical limitations in their administration, timing, and the need for careful patient selection will probably be more important determinants of clinical success than a drug's biological activity.

These clinical thoughts and observations result from over a decade's experience in dealing with therapies to spare and resuscitate the brain in a 14-bed neurological and neurosurgical intensive care unit at our hospital. I would like to discuss topics that have not been covered in great detail, such as the role of raised intracranial pressure in non-traumatic intracranial masses, the importance of the details of cardiopulmonary resuscitation (CPR) in brain resuscitation, and perhaps the most difficult practical clinical problem in neurological intensive care, the management of arterial hypertension in patients with intracranial catastrophes.

A Brief History of Neurological Intensive Care

Reviewing the development of intensive care facilities, and neurologic intensive care in particular, several historical perspectives emerge that may be helpful in guiding future work. In 1952, Scandinavia—especially Denmark—was struck by a poliomyelitis epidemic of unprecedented severity hospitalizing thousands of people. The hospital for communicable diseases in Copenhagen admitted almost 3,000 patients, over 10% of whom had severe respiratory muscle paralysis. Early in the epidemic all the patients were treated in the available tank and cuirasse

[1] St. Elizabeth's Hospital, Boston, MA 02135 USA
Formerly at the Massachusetts General Hospital

respirators. Of the first 31 patients with respiratory paralysis, 27 died. When the thirty-second patient, a 12-year-old girl, was nearing a terminal state of respiratory failure, the anesthetist Dr. Bjorn Ibsen was consulted. He inserted a tracheostomy tube and initiated positive pressure manual ventilation resulting in correction of respiratory acidosis. For the first time, the principles of the operating room had been brought to the intensive care unit. Over the next year almost all the practical features of respiratory intensive care and ventilator management were established [1]. Subsequent units, modeled after Ibsen's, were opened at 1958 in Oxford University, University Hospital of Baltimore, and Toronto General Hospital. Our own respiratory unit began in 1961. Through the 1960s and early 1970s, special care units proliferated in other fields, based mainly on this respiratory intensive care model.

In the 1930s and 1940s, Lundy, who introduced barbiturate anesthesia in the U.S. has been credited with pointing out that it was difficult to reduce intracranial pressure during neurosurgical procedures unless artificial ventilation was used [2]. It was not until 1972 that Adams and others [3] pointed out that hyperventilation could reduce ICP that was elevated by the use of volatile anesthetics. The 1960s and 1970s saw an explosion in the understanding of cerebral blood flow, vascular control, and intracranial pressure, initiated by Lundberg and colleagues [4] and resulting in the proliferation of ideas about secondary causes of brain damage that have been presented in this symposium. The aim of neurosurgical anesthesia is to apply the basic principles of cerebral physiology to intraoperative and postoperative neurosurgical patients.

These two very **practical responses** to patient problems, the respiratory unit and neurosurgical anesthesia, converged in the late 1970s to produce intergrated neurological and neurosurgical intensive care units. (Table 1) Each development was a response to an immediate patient care need, not the result of theoretical considerations. This practical orientation assured that treatment would be easy to apply and consistently very effective, therefore overcoming the variability inherent in each patient's individual situation that may have otherwise obscured a beneficial effect of therapy. We are now in a position to conceive of therapies, summarized in this symposium, that are based on a scientific understanding of brain pathophysiology. The old rules will probably apply; namely, any new brain resuscitation or brain sparing methods will still have to be easily administered in the emergency room or intensive care unit to be useful, no matter how strong their theoretical background. There are many examples in medicine and surgery of treatments that have an excellent scientific basis but do not improve patient outcome enough to justify general use, such as the EC/IC bypass.

The availability of ICUs has created an expectation that intensive care will be beneficial for all severely ill patients. However, studies to date demonstrate that intensive care improves outcome in only a selected population of patients, and that the costs per survivor are quite high. A study of the cost of providing intensive care to patients with nervous system diseases showed an average cost of approximately $20,000 per patient [5]. In addition, the risk of aggressive therapy in the ICU is usually underestimated when the prognosis without intervention is known to be poor. Because most damage to the brain is irreversible, extreme therapies seem justified in neurological intensive care. However, these more

Table 1. The development of intensive care units

Respiratory ICU	Neurosurgical anesthesia
Positive pressure with tank respirator (1949)	DeMartel (1913), Cushing (1917) use local anesthesia for neurosurgery
Severe European polio epidemic: anesthesia principles (tube, positive pressure, shock therapy) used in ventilatory failure (1952)	Thiopental anesthesia (1934), ICP monitoring and pathophysiology (1960's), Hyperventilation and barbiturate to reduce ICP (1972)
Respirator-polio-units (1955)	Neurosurgical postoperative wards (1970s)
Multidisciplinary ICUs (1958)	Head trauma research (1970s)
Proliferation of special care units (1965–1975)	Application of head trauma principles to cereberal hemorrhage

$$\searrow \qquad \swarrow$$

Neurological ICU

aggressive therapies usually carry a substantial risk, and it will be important to demonstrate that the advantages of new treatments exceed these risks. Moreover, technical progress in our field often has unfortunate or unwanted results, particularly brain death and the vegetative state. We should recognize that treatment with these new brain-sparing therapies will be expensive and utilize many resources. In order to be generally accepted these therapies will have to be more than marginally effective, and must not produce a new group of vegetative survivors.

It has taken approximately 15–20 years for clinical acceptance of each of the prior interventions in our field, such as barbiturate therapy, and intracranial pressure (ICP) monitoring. In the interval between introduction and acceptance into general practice, there are usually wide swings in opinion, first polarized on the extreme of uncritical acceptance, and later on the extreme of unreasonable rejection. Finally some sensible compromise is found based on a demonstration of efficacy under some circumstances. Both head injury and global ischemia are complex brain injuries with many factors influencing outcome. Therefore, it is unlikely that a single agent or a single approach will control secondary brain damage. Some combination of improved cardiopulmonary resuscitation (CPR), better brain substrate delivery, decreased metabolic demand, manipulation of cerebral vessels, and blockade of excitotoxins may be most likely to work. If the past is a guide, it may take another 15–20 years to reach a clinical consensus on the appropriate combination of therapies to minimize brain damage. During this time we should all continue to ask critically if these therapies helping the patient or are we simply trying to reproduce laboratory experiments in an ICU. On the other hand, we can afford to be patient, armed with the historical knowledge that it may take two decades to resolve these issues. If a therapy is very effective, it may be possible to change routine clinical practice in order to accommodate the new treatment.

Table 2. Massachusetts General Hospital Neuro-ICU admissions during a typical recent 10 month period (From [6])

Primary diagnosis	Number of cases
Postoperative tumor (all types)	131
Stroke or transient ishemic attack	90
Subarachnoid hemorrhage	87
Head trauma (nonoperable)	71
severe	48
mild	23
Cerebral hemorrhage (other than subarachnoid)	39
Subdural hematoma (acute and chronic)	31
Medical complication	29
Spinal cord trauma	20
Status epilepticus	18
Postoperative laminectomy	15
Postoperative AVM	12
Guillain-Barré syndrome	10
Encephalitis	8
Epidural hematoma	7
Giant carotid aneurysm	7
Myasthenia gravis	6
Cranial gunshot wound	5
Meningitis	4
Brain abscess	4
Acute global ischemia/carbon monoxide	3
Brain tumor, raised intracranial pressure	2
Other	11

Raised ICP in Non-Traumatic Mass Lesions

What precisely are the problems seen in the neurological intensive care unit? (Table 2). Catastrophic diseases, such as head injury, subarachnoid hemorrhage, and postoperative complications, are well represented in large ICUs such as ours. However, there is another group of illnesses, neuromedical or non-traumatic intracranial catastrophes, such as massive brain edema after stroke, cerebral hemorrhage, and encephalitis with brain swelling, in which much secondary damage is potentially caused by raised ICP. In some ways, this group allows a clearer study of the effects of ICP on outcome since systemic complications such as hypoxia, lung injury, and cardiac complications do not occur as often as in severe head injury. In addition, ICP in most of these instances rises rather slowly, allowing an opportunity for a more deliberate institution of therapy. It might be helpful to review the small amount of available clinical material concerning raised ICP in these non-traumatic lesions.

We have cared for 26 patients in the last several years with large middle cerebral artery infarcts and massive brain swelling. (Table 3 [7]). Of these, 6 had ICP contained by conventional medical means, in a few instances including barbiturates, and none became brain dead (although some had severe neurological def-

Table 3. Brain edema after large middle cerebral artery infaction (Massachussetts General Hospital 1980–1987)

	Survived	Brain dead	Medical death
ICP			
< 15 torr	5	0	1
> 15 torr	0	8	0
Not measured	4	5	3
Total	9	13	4

icits from the infarct). Of the eight whose ICP escaped therapy, all died—specifically, they were brain dead. It appears that raised ICP is closely associated with poor outcome in stroke edema, as in head injury.

Similarly, our group [8] and others, notably Papo et al., Janny et al., and more recently Waga and colleagues [9–11] as well as many other Japanese neurosurgeons, have been interested in prognostic factors, including ICP, in cerebral hemorrhage. Our experience (Fig. 1, Table 4) suggests, again, that if intracranial pressure is not controlled, patients ultimately become brain dead. All 4 of our comatose patients whose ICP could not be controlled with medical and surgical

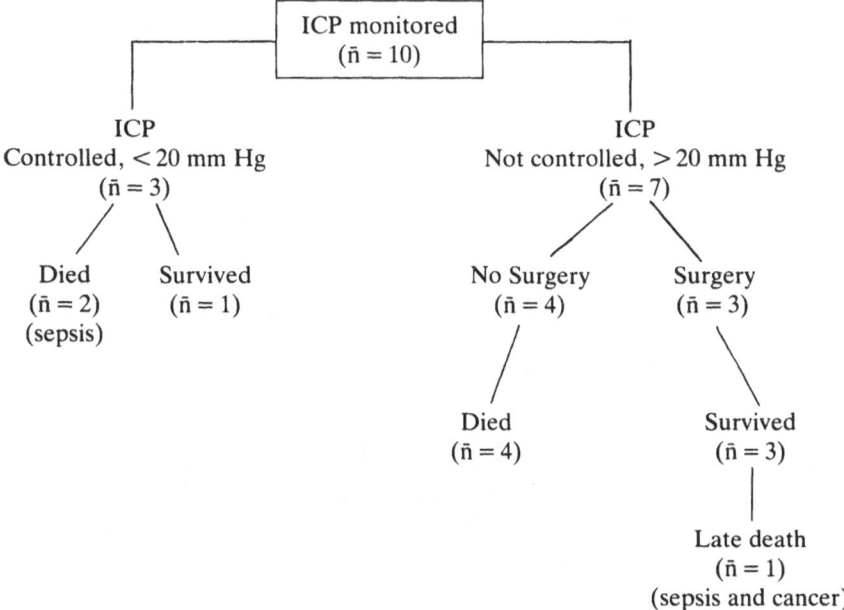

Fig. 1. Flow chart showing the relationship of daily mean intracranial pressure (ICP) to outcome in 10 patients with cerebral hemorrhage. (From [8]. Copyright 1984, American Medical Association)

Table 4. Characteristics of 10 consecutive patients with cerebral hemorrhage and intracranial pressure (ICP) monitoring

Patient Age, yr	Hematoma location	GCS*	ICP monitoring, hr/placement†	Initial ICP, mm Hg	ICP course‡	Surgery	Patient Outcome
1/62	L ganglionic	7	18/R	22	↑	Yes (ICP, 32 mm Hg)	R hemiparesis, ambulatory
2/54	R ganglionic	4	17/R	40	←	No	Died (ICP, >50 mm Hg)
3/55	L ganglionic	4	18/L	34	←	No	Died (ICP, >50 mm Hg)
4/53	R ganglionic	7	48/L	14	↑	Yes (ICP, 35 mm Hg)	Vegetative; died later of sepsis
5/57	R ganglionic	7	10/L	20	→	No	L hemiparesis; ambulatory
6/43	L ganglionic	11	48/R	6	↑	No	Died of sepsis
7/47	L thalamic	11	16/L	18	←	Yes (ICP, 30 mm Hg)	R hemiparesis; ambulatory
8/72	L ganglionic	10	20/L	9	↑	No	Died of cardiac arrest (ICP, <20 mm Hg)
9/69	R ganglioic	6	10/R	28	→	No	Died of sepsis
10/43	R frontotemporal	6	5 days/R	22	↑	No	Died (ICP, 47 mm Hg)

* Glasgow coma scale score at the time of the insertion of the ICP measuring device; †Time between hemorrhage and insertion of the ICP device and side on which device was placed; ‡Upward arrow inidicates progressively increasing ICP; *downward arrow*, declining ICP; and *horizontal arrow*, ICP initially less than 20 mm Hg that remained low. (From [8]. Copyright 1984, American Medical Association)

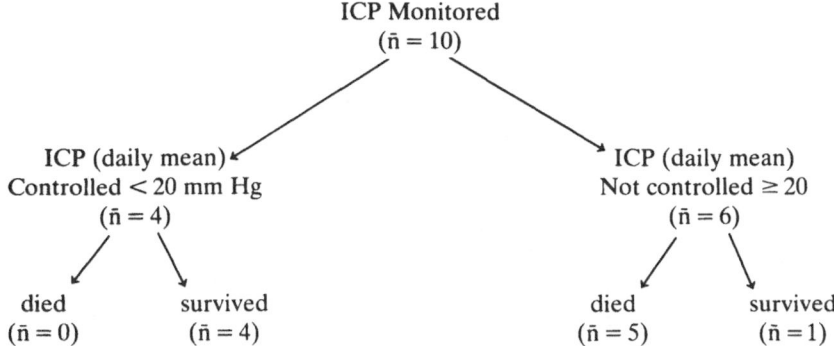

Fig. 2. Flow chart showing the relationship of daily mean intracranial pressure (ICP) to outcome in 10 patients with encephalitis. (From [12])

therapy became brain dead. Unfortunately, this is one area with great potential for leaving vegetative survivors or losing patients to medical complications, because of a lengthy period of coma.

Encephalitis in children, particularly those with Reyes syndrome, (we do not work with these cases with in our unit) seems to confirm the point that uncontrolled ICP is associated with a poor outcome. In adults, we have recently studied 10 consecutive patients with encephalitis [12] and found that uncontrolled ICP in six patients led to death in five (all brain dead) and survival in one (Fig. 2, Table 5). The four patients whose ICP could be controlled survived. Finally, in hepatic encephalopathy, where little work has been done on ICP, tentative evidence also suggests that intracranial hypertension is the most common cause of immediate death [13,14].

It is reasonable to conclude that survival in all of these non-traumatic intracranial catastrophes is predicated largely on control of ICP. While the final neurological deficit in survivors depends on the extent and location of the underlying lesion, the majority of deaths in these patients are due to brain death as a result of uncontrolled ICP. The remainder are medically related deaths, the incidence of which depends on the extent and detail of general care. This, in turn, is often determined by the perception of the patient's prognosis, often a poor one in these cases. An association between raised ICP and poor outcome does not necessarily imply a definite causal relationship between the two. A large study would be required to take the argument to its next logical step—that aggressive therapy to reduce ICP would improve outcome. It seems unlikely that short of an extensive multi-center trial, enough patients could be gathered in any of these categories to resolve this question in the near future. Barbiturates have been reasonably successful in treating ICP "failures" after conventional therapy in patients with non-traumatic lesions [15]. The use of hemicraniectomy, as primitive as it is, needs to be explored further in younger patients who are dying of raised ICP.

Allan H. Ropper

Table 5. Characteristics of 10 patients with adult encephalitis and ICP monitoring (From [12] with permission)

Case No.	Age (yrs)	Cause of encephalitis	GCS score*	1st Day of ICP monitoring†	Initial ICP (mm Hg)	Highest mean daily ICP	ICP course‡	Outcome
1	66	Herpes	4	9	10	16	↑↓	Poor: akinetic mute
2	57	Herpes	9	6	15	21	↑↑	Death: coma and pneumonia
3	62	Herpes	8	5	7	19	↑→	Fair: requires supervision
4	53	Herpes	4	13	5	20	↑↑	Death: come and pneumonia
5	29	Unknown	4	8	12	20	↑↑	Brain death
6	20	Herpes	3	9	6	22	↑↑	Excellent: returned to college
7	48	Acute hemorrhagic leukoencephalitis	3	3	22	26	↑	Brain death
8	63	Herpes	12	6	18	23	↑	Brain death
9	41	Herpes	14	3	9	10	↔	Good: returned to work
10	61	Herpes	7	7	11	15	↑↓	Vegetative

*Glasgow Coma Scale (GCS) score at time of insertion of intracranial pressure (ICP) measuring device; †Time between onset of neurological symptoms and insertion of the ICP device; ‡ ↑ = initial increase in ICP; ↓ = ICP controlled (<20 torr); ↔ = no change in initial ICP

Table 6. "New" CPR

Improved cardiac output and CBF	Other changes
Longer compression duration (50% of cycle)	Avoid excessive volume loading administration
Simultaneous ventilation—compression	Elimination of bicarbonate
Interposed abdominal compression	No calcium except in specific circumstances
Increased compression rate (100–120/min)	-Calcium channel blocker toxicity
	-Severe hypocalcemia
	-Severe hyperkalemia
	-Severe hypermagnesemia
	Increased use of epinephrine (α-adrenergic
	agents favored)

Newer Techniques in Cardiopulmonary Resuscitation

An important area in brain resuscitation that requires further expansion is the role of CPR. The first minutes after cardiac arrest are crucial in determining the extent of brain damage. More may be gained by refining CPR as a way of improving cerebral blood flow than almost any other intervention available to us at this time. In order to accomplish improved CPR, a new focus on the brain has begun in the field of resuscitation. Much of this work stems from the recognition that high intrathoracic pressures propel blood through the carotid arteries, and not direct cardiac compression. This is largely because venous valves prevent backflow.

Clinically validated methods that improve cardiac output during CPR have included (Table 6): a longer compression duration, compression for 50% of the cycle, simultaneous ventilation, and compression to produce high thoracic pressures [16], interposed abdominal compressions (controversial) [17], and, perhaps most importantly, an increased compression rate of a 100–120 per minute [18]. Volume loading, in an attempt to increase intrathoracic pressure has paradoxically decreased brain blood flow [19]. Some of these "new" CPR techniques have not yet been adopted into general practice but probably will eventually modify standard CPR. Apparently harmful features of previously practiced CPR are now being eliminated, including bicarbonate administration which shifts the hemoglobin saturation curve to the left, causes a paradoxical CNS acidosis, and may theoretically worsen neurological outcome [20], as well as depress myocardial function, and the administration of calcium which does not appear to improve outcome from CPR but it is theoretically implicated in post-anoxic tissue damage. It is therefore only used currently in patients with calcium channel blocker toxicity, severe hypocalcemia (e.g., for multiple blood transfusions), severe hyperkalemia, or hypermagnesemia. There has also been an increased emphasis on the use of α-adrenergic agents, but β-adrenergic stimulation does not appear to be useful and may be harmful. Therefore epinephrine has become the favored initial therapy, simultaneous with repeated attempts at cardioversion. Increased attention to these refinements as well as policies that promote rapid CPR availability may improve outcome from cardiac arrest as much as therapies directed at secondary damage. Nonetheless, there is probably a limit to even ideal CPR, and there will always be a need for secondary brain sparing therapies.

Blood Pressure Management

Finally, it is worth reviewing the current state of knowledge, as meager as it is, on the control of blood pressure, both after resuscitation and with raised ICP. Blood pressure management is perhaps the most difficult practical decision made in the neurologic intensive care unit. Unfortunately, almost all clinical decisions are made on the basis of theoretical considerations (Fig. 3). Normally, cerebral perfusion pressure is maintained in the 60–70 torr range unless blood pressure is

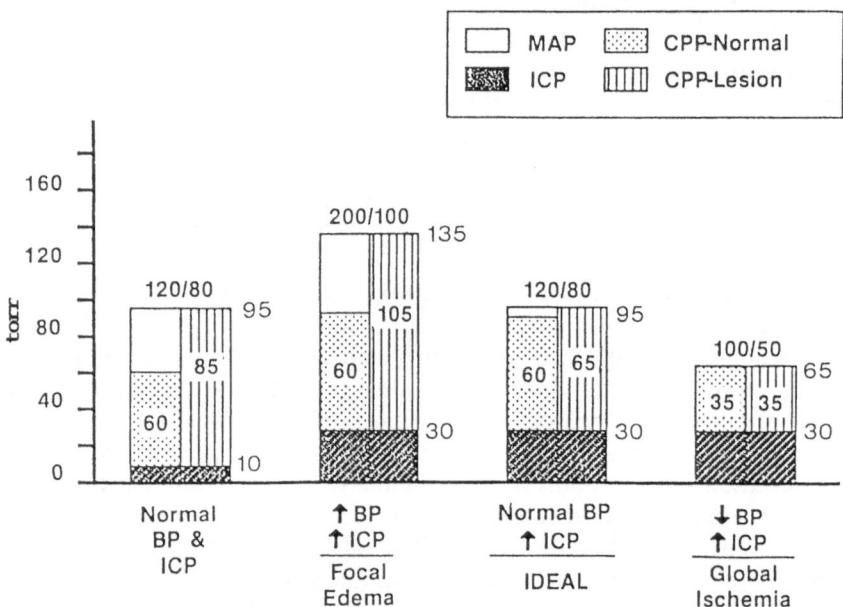

Fig. 3. The relation between mean blood pressure (MAP), ICP, and cerebral perfusion pressure (CPP) in adults is shown in four clinical situations. In each bar the *left vertical panel* shows CPP in normal regions of brain, and the *right vertical panel* shows CPP in damaged regions where loss of autoregulation allows passive transmission of pressure and edema formation. Normal and desirable CPP is approximately 60 torr. From left to right: (first bar) when MAP and ICP are normal, CPP in damaged regions, for example, in regions of infarction, is elevated. Treatment is aimed toward preventing severe uses in blood pressure that would potentially cause edema. (second bar) Hypertension produces very high CPP in damaged areas, resulting in edema. (third bar) The result of reducing MAP to normal levels, "ideal therapy" in the presence of a lesion that raises ICP, is to bring CPP closer to normal levels and reduce edema formation. (fourth bar) Hypotension with raised ICP compromises perfusion in normal and damaged regions. (From [6])

Table 7. Antihypertensive agents in head injury and raised ICP

Nitroglycerin and nitroprusside (variably raise ICP)
Propranolol* vs hydralazine [22]
Verapamil (raises ICP) [23]
Nifedipine (clinical decompensation with brain metastasis) [24]
Nicardipine (increases CSF pressure) [26]
Nimodipine (experimental—lowers ICP more than BP) [27]
Nifedipine vs chlorpromazine vs reserpine vs furosemide vs thiopental* [28]

*Opinion given of superior drug

extremely deranged. When there is an intracranial lesion with disrupted vasculature and blood pressure is simultaneously elevated, some areas of the brain are exposed to much higher perfusion pressures, potentially increasing regional edema. The precise level of blood pressure at which edema is worsened by hypertension in patients is not known. As a general rule, we have taken a mean blood pressure of 125–135 torr as our threshold for instituting antihypertensive therapy. On the other hand, there is an obvious need to avoid lowering perfusion pressure in patients with raised ICP, since this potentially leads to ischemia earlier than it would in a patient without raised ICP.

Most clinical work in this field has been concerned with the choice of antihypertensive agent, not the precise limits for blood pressure treatment. After head injury, most patients have a hyperdynamic state characterized by increased cardiac output, mild to moderate hypertension, tachycardia, decreased or normal systemic and pulmonary vascular resistance, increased pulmonary shunting, and raised circulating catecholamines [21]. Robertson and colleagues [22] compared propranolol and hydralazine after head injury and found that the β-blocker decreases arterial catecholamines and stabilizes hemodynamic abnormalities better. (Table 7). Another area of interest has been the use of calcium channel blockers. The potential stabilization of cerebral vascular tone, ease of use, and minimal side effects suggest that these may be excellent drugs in an intensive care unit setting. Nifedipine is probably the agent of choice in patients with accelerated or malignant hypertensive emergencies but not with intracranial masses (labetalol may be an alternative). Bedford et al. [23] and Catchings et al. [24] reported cases in which patients decompensated from an intracranial mass after the administration of verapamil or nifedipine, respectively. Experimental work in cats has suggested that nifedipine-induced hypotension can produce an increase in ICP, resulting in a large decrease in cerebral perfusion pressure [25]. Nicardipine has been shown by Nishikawa and colleagues [26] to increase spinal fluid pressure in awake patients without intracranial lesions. The only suggestions that calcium channel blockers used as antihypertensive agents may not be potentially harmful for ICP comes from Hadley's group, who showed that nimodipine slightly reduced ICP out of proportion to a reduction in arterial blood pressure, thus improving perfusion pressure [27].

Perhaps the most useful study has been performed by Hayashi and colleagues [28], who compared nifedipine, chlorpromazine, reserpine, furosemide, and thiopental in 38 patients with cerebral hemorrhage. The results can be summarized as follows: the first three drugs generally raised intracranial pressure, furosemide had an insignificant effect on arterial pressure but slightly reduced ICP, and thiopentol reduced both mean blood pressure and ICP (but ICP more), resulting in a net improvement in cerebral perfusion pressure. This represents a revival of barbiturates as antihypertensive agents. We have been using β-blockers or barbiturates almost exclusively in our unit for blood pressure control in patients with raised ICP, and only rarely do we see a drop in blood pressure that exceeds the reduction in intracranial pressure. Quite often, however, there is a parallel reduction in both pressures. My impression is that ICP reduction after several doses of barbiturates is often simply a passive result of blood pressure reduction. However, with the first few doses of barbiturates, perfusion pressure usually improves unless profound hypotension occurs.

A current recommendation, based on anecdotal experience, is to treat patients with an intracranial mass and raised ICP with intermittent doses of barbiturates to keep mean blood pressure below 125–135 torr (systolic below approximately 180 torr). Propranolol,—better yet labetolol—can be used as adjuncts. Some may prefer to add furosemide. Midazolam is a rapidly acting water soluble benzodiazepine that has little cardiovascular effect but may reduce ICP in some patients [29]. It has a substanial risk of respiratory arrest in unventilated patients. Its use is still being investigated; limited experience in our unit suggests it may lower ICP slightly and keep blood pressure stable because it is a sedative, but is less effective than barbiturates in the presence of an intracranial mass.

Perhaps one way to study the effects of changing blood pressure on perfusion pressure in patients is the use of transcranial Doppler, pioneered by Aaslid and used for this purpose by Hassler et al. [30]. We have seen interesting changes in the Doppler pattern as blood pressure is reduced or ICP rises, on the way to brain death [31].

The Future of Neurology ICUs

These points show how much more clinical work is necessary to make our knowledge of cerebral physiology useful. It is a lesson of which we are constantly reminded in the ICU environment. It is also difficult to predict the future course of neurological critical care, particularly since the clinical success of brain sparing may change its future role. It is clear that several social and philosophical issues may impede progress in this field. Perhaps neurologists, neurosurgeons, and intensivists can prevent this by taking the lead in proposing criteria for withdrawal of support from severely brain–damaged patients. That would be preferrable to having society impose. restrictions upon us. In formulating these guidelines, we must be flexible enough to accommodate many different clinical situations. Is it possible that the Neuro ICU of the future will be a technological marvel of brain sparing and resuscitation? Will it be for our patients what the polio-respiratory units of the early 1950s were for patients with respiratory failure? It is worth reminding ourselves that respiratory units were virtually put out of business by the polio vaccine. Major advances in prevention usually have a more dramatic impact than attempts at treating the damage once it has already occurred. The substantial scientific and clinical gains which have been discussed in this symposium continue to add to our body of knowledge about sparing the damaged brain.

Summary. Neurological intensive care has evolved from respiratory units which were established during the polio epidemic of the 1950s, as well as from clinical work in brain resuscitation and intracranial pressure (ICP) management. Few of the modern techniques used in Neuro-ICUs have been confirmed to be effective, but a few sound principles have emerged.

Several studies have suggested a relationship between raised ICP and outcome in nontraumatic brain masses, such as edema following stroke, spontaneous hemorrhage, and encephalitis. The appropriate methods of treatment for raised ICP are currently being delineated. Investigators are paying particular attention to the role of hypertension in exaggerating raised ICP, and to the deleterious effects of drugs, especially antihypertensive medications, in these circumstances.

The relationship between various methods of cardiac resuscitation, cerebral blood flow, and clinical outcome are just beginning to be explored, with only preliminary insights currently available.

References

1. Pontoppidan H, Wilson RS, Rie MA, Schneider RC (1977) Respiratory intensive care.
2. Lundy JS (1935) Intravenous anesthesia: Preliminary report of the use of two new thiobarbiturates. Mayo Clinic Proc 10:536
3. Adams RW, Grouvert GA, Sundt TM (1972) Halothane, hypocapnia, and cerebrospinal fluid pressure in neurosurgery. Anesthesia 37:510
4. Lundberg N (1960) Continuous recording and control of ventricular fluid pressure in neurosurgical practice. Acta Psychiatr Neurol (Scand) 149 (Suppl):1
5. Butler PW, Bone RC, Field T (1985) Technology under Medicare diagnosis related groups. Chest 87:229
6. Ropper AH (1988) Neurological and neurosurgical intensive care. Aspen, Maryland
7. Ropper AH, Shafran B (1984) Brain edema after stroke. Arch Neurol 41:26
8. Ropper AH, King RB (1984) Intracranial pressure monitoring in comatose patients with cerebral hemorrhage. Arch Neurol 41:725–728
9. Papo I, Janny P, Caruselli G, Colnet G, Luongo A (1979) Intracranial pressure time course in primary intracerebral hematoma. Neurosurg 4:504
10. Janny P, Colnet G, Georget A, Chazal J (1978) Intracranial pressure with intracerebral hemorrhage. Surg Neurol 10:37
11. Waga S, Miyazaki M, Okada M, Tochio H, Matsushima S, Tanaka Y (1986) Hypertensive putaminal hemorrhage: analysis of 182 patients. Surg Neurol 26:159
12. Barnett GH, Ropper AH, Romeo J (1988) Intracranial pressure and outcome in adult encephalitis. J Neurosurg 68:585
13. Hanid MA, Davies M, Mellon PJ, Silk DB, Strunin L, McCabe JJ, Williams R (1980)

Clinical monitoring of intracranial pressure in fluminant hepatic failure. Gut 21: 866,1980.

14. Gazzard BD, Portmann B, Murray-Lyen JM (1975) Causes of death in fulminant hepatic failure and relationship to quantitative histological assessment of parenchymal damage. Q J Med 44:615

15. Woodcock J, Ropper AH, Kennedy SK (1982) High dose barbiturates in nontraumatic brain swelling: ICP reduction and effect on outcome. Stroke 13:785

16. Chandra N, Weisfeldt M, Tsitlik J, Vaghaiwalla F, Snyder LD, Hoffecker M, Rudikoff MT (1981) Augmentation of carotid flow during CPR by ventilation at high airway pressure with simultaneous chest compression. Am J Cardiol 48:1053

17. Voorhees WD, Ralston SH, Babbs CF (1984) Regional blood flow during CPR with abdominal counterpulsation in dogs. Am J Emerg Med 2:123

18. Paraskos JA: External compression without adjuncts. Circulation 74:IV33–6, 1986 (Review)

19. Ditchey RV, Lindenfeld J (1984) Potential adverse effect of volume on perfusion of vital organs during CPR. Circulation 69:181

20. Berenji KJ, Wolk, Killip T (1975) Cerebrospinal fluid acidosis complicating therapy of experimental CPR. Circulation 52:319

21. Clifton GL, Robertson CS, Kyper K, Taylor AA, Dhenke RD, Grossman RG (1983) Cardiovascular response to severe head injury. J Neurosurg 59:447

22. Robertson CS, Clifton GL, Taylor AA, Grossman GG (1983) Treatment of hypertension associated with head injury. J Neurosurg 59:455

23. Bedford RF, Dacey R, Winn HR, Lynch C (1983) Adverse impact of a calcium entry-blocker (verapamil) on intracranial pressure in patients with brain tumor. J Neurosurg 59:800

24. Catchings TT, Prough DS, Kelly DL, Higgins AC (1985) Symptoms of clinically silent intracranial mass lesions precipitated by treatment with nifedipine. Surg Neurol 24:151

25. Griffin JP, Cottrell JE, Hartung J, Shwiry B (1983) Intracranial pressure during nifedipine-induced hypotension. Anesth Analg 62:1078

26. Nishikawa T, Omote K, Namiki A, Takahashi T (1986) The effects of nicardipine on cerebrospinal fluid pressure in humans. Anesth Analg 65:507

27. Hadley MN, Spetzler RF, Fifield MS, Bichard WD, Hodak JA (1987) The effect of nimodipine on intracranial pressure. J Neurosurg 66:387

28. Hayashi M, Kobayashi H, Kawano H, Handa Y, Hirose S (1988) Treatment of systemic hypertension and intracranial hypertension in cases of brain hemorrhage. Stroke 19:314

29. Griffin JP, Cottrell JE, Shwiry B, Hartung J, Epstein J, Lim K (1984) Intracranial pressure, mean arterial pressure, and heart rate following midazolam or thiopental in humans with brain tumors. Anesthesiology 60:491

30. Hassler W, Steinmetz H, Gawlowski J (1988) Transcranial Doppler ultrasonography in rasied ICP and intracranial circulatory arrest. J Neurosurg 68:754

31. Ropper AH, Kenhe Sm, Wechsler LR (1987) Transcranial Doppler in brain death. Neurology 37:1733

25

Brain Resuscitation: a Clinical Perspective

J. Douglas Miller[1]

Introduction

The special demands and problems of brain resuscitation have become recognised within the field of critical care medicine. This knowledge has largely been gained from experimental studies carried out in many laboratories which have defined the thresholds of brain ischemia and hypoxia, and the precise mechanisms responsible for the cascade of events that lead from hypoxia to brain cell damage and death.

Two of the most frequent clinical problems managed in the neurological critical care unit are head injury and cerebrovascular disease, in the form of hemorrhagic (subarachnoid hemorrhage and intracerebral hematoma) and vascular occlusive disorders (stroke). Common to these problems, and to many of the other acute brain disorders, is the production of brain ischemia. The lessons learned from laboratory studies of focal and global ischemia, complete and partial, and with and without reperfusion, can now be applied to clinical solutions. In this way, clinical management can be based on firm pathophysiological principles, and there is an exciting prospect of novel treatments based upon our emerging understanding of the molecular basis in the development of ischemic brain damage.

Neuro-intensve care is emerging as a distinct subspecialty of critical care medicine. The pivotal feature that distinguishes it from the general field is the practice of continuous monitoring of intracranial pressure (ICP). Due largely to the pioneering studies of Pierre Janny [1], Nils Lundberg [2] and the subsequent series of International Intracranial Pressure Symposia started by Mario Brock [3–9], there is widespread recognition that intracranial pressure is pathologically raised in most of the brain disorders that result in impairment of consciousness.

[1] Department of Clinical Neurosciences, University of Edinburgh, Western General Hospital, Edinburgh EH4 2XU, Scotland, UK

These include head injury, subarachnoid hemorrhage, major cerebral infarction, bacterial and viral infections of the brain, brain tumors, disorders of the circulation of the CSF and metabolic encephalopathies due to energy lack or toxins.

In the clinical section of this book/two aspects of neuro-intensive care receive emphasis: physiological monitoring and structural imaging methods specific to the brain, and management and therapy of acute brain disorders. In considering brain-specific therapies, an important question remains. Should we begin immediately to use new brain protective agents on a trial basis or should we be making better use of treatments that we already know?

Neuromonitoring and Investigation

Intracranial Pressure

A number of methods and devices are now available to enable accurate long-term recordings of ICP. While intraventricular pressure monitoring remains the standard against which all other methods must be judged, there is increasing interest in the use of transducers implanted into the ventricular catheter, the subdural space, or even directly into brain parenchyma. The threshold level of raised ICP at which therapeutic measures should begin remains a matter of clinical opinion. Some believe that any elevation over 15 mmHg should signal therapeutic measures, most adopt a level of 20 or 25 mmHg, while others believe that ICP can be allowed to rise as high as 30 mmHg before taking steps to lower it if more than 48 h have elapsed since the injury or ictus and arterial pressure is maintained at a satisfactory level. It is mandatory to monitor arterial pressure continuously throughout the period of ICP monitoring and to repeatedly or continuously estimate the cerebral perfusion pressure from the difference between arterial and intracranial pressure.

The presence of spontaneous waves of increased ICP, particularly of the plateau wave variety, is an indication of reduced intracranial compliance. This can be measured directly by the instillation of bolus injections into the lateral ventricle and measurement of the resultant rise in ICP. The results may be expressed in terms of the pressure volume index (the notional volume of CSF required to produce a tenfold rise in ICP) or as the volume pressure response (the change in ICP per change in CSF volume). Such measurements must be made on a number of occasions and can be made only during relatively stable periods of ICP recording. They are more time consuming than they might first appear and it is essential to be scrupulous about avoiding the added risk of infection during these volume addition maneuvers.

In the last few years interest has been growing in computer assisted analysis of the ICP waveform to determine whether it can yield information about intracranial compliance and the likelihood that ICP may become elevated at a future point in time. Bray and his associates [10] have carried out Fast-Fourier transform of the ICP waveform and have stated that an increase in its high frequency components is an indication of reduced intracranial compliance, that it correlates with reduction in the pressure volume index (PVI), and can be used as an noninvasive method of calculating PVI. Takizawa and his colleagues showed in experimental models that different mechanisms of elevating ICP produced divergent amplitude changes in the different harmonics of the ICP waveform and in the amplitude transfer function, relating the amplitude of ICP harmonics to the corresponding arterial pressure wave harmonic components [11–13]. Of note was an increase in PCO_2 specifically increased low frequency harmonics. Piper and his colleagues have also found that an increase in the low frequency harmonic amplitudes is related to rises in ICP due to vascular causes, while increases in the high frequency harmonics are associated with non-vascular causes such as brain edems and other factors that may alter brain compliance [14].

It appears likely that ICP monitoring will not only become useful for identifying the level of pressures and whether treatment is needed, but may also give an indication of the cause of raised intracranial pressure, vascular or non-vascular in origin, and indicate the best form of treatment to be applied.

Cerebral Blood Flow and Brain Metabolism

The ideal measurement would provide multiregional assessment of cerebral blood flow and tissue energy metabolism, be obtainable at bedside, and be accomplished in a short time, like under a minute. Despite the enormous advances that have been made in the techniques of measuring aspects of cerebral tissue blood flow and energy metabolism, this ideal system has yet to be developed and, at the present time, all measurements of CBF represent something of a compromise.

The Kety Schmidt technique, based upon tracer measurements from arterial and jugular bulb blood, provides estimates of global cerebral blood flow and the uptake or production of such metabolites as can be measured from those blood samples. The measurements take some time to carry out, during which steady state conditions must apply in the brain. The method can, however, be applied at bedside. Recent studies of blood flow and energy metabolism in the severely damaged brain have been reported by Robertson and colleagues of Baylor University [15]. Simple measurement of the arterial to jugular bulb difference in oxygen content is also useful although it must be appreciated that abnormal findings will be obtained only when the major part of the cerebral circulation is affected by either hyperemia or ischemia. These methods are unsuitable for detection of focal disturbances of CBF.

The use of radioactive isotopes of Xenon, Technicium, and other compounds has enabled two-dimensional or tomographic brain maps to be obtained showing the distribution of multifocal CBF or cerebral blood transit times. These techniques can also be applied at bedside and are useful for identifying focal derangements of blood flow. Inevitably, the simpler and shorter such measurements are, the greater is their imprecision. The most accurate assessments of blood flow and metabolism on a regional tomographic basis can be obtained from positron emission tomography but this inevitably requires transport of the patient from the critical care unit.

At the other end of the spectrum of flow measurements is the transcranial Doppler technique that allows segments of intracranial arteries to be insonated through the thinner parts of the skull to permit a continuous display of the systolic, diastolic, and mean velocity profile of blood flow in these larger arteries. When blood flow velocity is in the normal range of around 55 cms/sec, then increases in velocity can be interpreted as increases in blood flow. Under conditions of cerebral vasospasm, however, abnormally high velocity measurements are obtained in the range of 150–250 cms/sec, possibly related to narrowing of the artery. Under such pathological circumstances, a successful measure that causes the spastic vessel to relax and bloodflow through it to improve is seen as a reduction in flow velocity.

Brain Electrical Activity

In any monitoring system in which protection of the function of the brain is a goal, measurement of the electrical activity of the brain is clearly of fundamental importance. Measurements of ICP, blood pressure, cerebral pefusion pressure, and cerebral blood flow can only be indications of the potential limitation of energy supply to the brain and a warning of physiologically unsafe conditions. Modern electronics technology has largely overcome the problems of recording the minute electrical potentials from cerebral tissue within the electrically hostile environment of the intensive care unit. A number of techniques are available to compress or condense the information obtained from brain electrical activity recording to make it an acceptable continuous monitoring tecthique to be used over long periods of time.

The cerebral function monitor and cerebral function analysing monitor (CFAM) both produce a condensed record indicating the average amplitude of the EEG recording from two electrodes, one for each hemisphere. The CFAM provides, in addition, a continuous indication of the relative amounts of EEG activity in different frequency bands. The compressed spectral array also provides this information allowing easy identification of a shift of the predominant EEG activity into the lower frequency zones. Such methods are valuable in the detection of failing brain electrical activity, with loss of amplitude or slowing of brain electrical activity, and in the detection of seizure activity in patients who may be artifically ventilated and paralysed.

The measurement of evoked potentials permits assessment of specific neural pathways in the central nervous system. If patients are subjected to multimodelity measurements—auditory, visual, somatasensory and motor evoked potentials—then a considerable amount of information can be obtained concerning the location and extent of areas of brain damage and dysfunction. Such multimodality measurements are time-consuming and there are many potential sources of artifacts when measurements are made in the intensive care unit on patients who may have difficulty with regulation of blood pressure and ICP, body temperature, and other variables. Some believe that it may be more valuable to repeat a single type of measurement more frequently and under more limited conditions. An example would be measurement of the central conduction time of the somatosensory evoked potential as reported by Symon and his colleagues in the neurosurgical operating theatre and intensive care unit [16].

Computer Tomographic and Magnetic Resonance Imaging

Although these methods of evaluating brain injury require transport of the patient from the critical care unit and are directed towards identifying disorders of structure rather than function in the brain, nevertheless, most comatose patients in the neurointensive care unit will have had one or an other of these studies carried out. In addition to the information concerning the abnormal brain morphology, computed tomography is also a useful predictor of raised ICP. Absence of the third ventricle and the perimesencephalic CSF cisterns, and a mid-line shift away from a mass lesion with expansion of the contralateral ventricle are both reliable indicators hypertension.

Magnetic resonance (MR) imaging permits clear identification, more accurately than computerized tomography (CT), of brain areas in which the tissue water content is increased. In other words, magnetic resonance imaging will accurately identify areas of brain edema. This has been verified by comparisons between MR measurements of the T_1 and T_2 relaxation times and direct measurements of tissue water content from both experimental and human subjects. MR has also been used to explore mechanisms of effect of drugs used in the treatment of brain edema. Bell and his colleagues studied patients with peritumoral brain edema treated with steroids and with intravenous mannitol [17]. They found that steroids caused no change in tissue water content despite an improvement in the neurological status of all patients, whereas mannitol caused a significant decrease in tissue water content estimated from T_1 times in areas of tumour and peritumoural edema, but not in normal brain tissue.

Management and Therapy

In managing patients with acute brain disorders in intensive therapy unit (ITU) the goal is to maintain all variables as nearly as possible within the normal range. The practice of deliberately dehydrating patients with acute brain damage in order to prevent brain edema has fallen into disfavor because of the risk that such conditions may lead to hypovolemia. The application of sedative drugs, head-up tilt, and positive end expiratory pressure may in these circumstances result in a critical fall in arterial blood pressure. The importance of avoiding elevations of body temperature is well appreciated. Hyperthermia causes increased ICP, accelerates the rate of formation of brain edema, and increases cerebral metabolic demand at a time when the energy supply may be in jeopardy. There is also growing awareness of the importance of avoiding even moderate degrees of hyponatremia. Reduction in the serum sodium concentration, particularly in women, encourages the formation of osmotic brain edema, raised ICP, and brain herniation.

Meticulous attention to the positioning of the patient, ensuring a clear airway, avoidance of respiratory efforts against the ventilator, and maintenance of normal body temperature, arterial and central venous pressures, blood volume electrolytes, gases and pH should reduce the need for specific therapy for raised intracranial pressure. This is important because there is no universally successful treatment for raised ICP and some forms of therapy carry significant risks for the patient.

Existing Therapies for Intracranial Hypertension

Hyperventilation with reduction of arterial PCO_2 between 25 and 30 mmHg will cause cerebral vasoconstriction in most patients, reduce cerebral blood volume and the ICP. The penalty is that cerebral blood flow also falls; where the flow has already been restricted due to extravascular causes, the superimposition of hypocapnia may reduce blood flow in parts of the brain to ischemic levels. In most cases the consequent accumulation of acid metabolites will reverse the vasoconstriction so that this is a self-limiting process. Nevertheless, in certain critically ill patients the use of hyperventilation may take the cerebral arterio-venous oxygen content difference above 9 ml/dl into what is generally considered to be the ischemic range.

Hyperventilation has a limited duration of effect, related to accumulating lactacidosis in the extracellular fluid and CSF. In experimental studies, constant hyperventilation applied for 6 h or more caused cerebral blood flow to fall at first but then it had a tendency to return to baseline levels despite the continuing hypocapnia. It is likely, therefore, that the ICP-reducing effect of hyperventilation is similarly limited in its duration.

In cases where ICP is being monitored from an intraventricular catheter, controlled drainage of CSF against a positive pressure can be usefully employed to prevent rises in ICP. This therapeutic avenue is open only in cases where the ventricles are of normal or increased size. Sadly, in many cases of head injury, metabolic encephalopathy and other acute brain disorders, the ventricular system is already reduced in volume, the ventricles hard to locate, and fluid difficult to drain without blockage of the ventricular catheter. Although the ICP may be held down by this method, brain shift due to a mass lesion will not be ameliorated and the clinical benefit of the method may have limitations other than in cases of hydrocephalus.

Intravenous mannitol is for many the mainstay of control of intracranial hypertension in the ITU. The agent should be given in the lowest effective volume and the baseline osmolality levels checked at intervals. No further effect of mannitol upon ICP will be obtained when baseline osmolality exceeds 330 mOsm/1; indeed administration of mannitol at this stage may precipitate renal failure. There is dispute about the mechanism of effect of mannitol and almost certainly multiple mechanisms, both vascular and non-vascular, are involved. Barbiturate and other sedative drugs have been widely used for the control of raised intracranial pressure in acute brain damage, as well as in a protective capacity when temporary occlusion of cerebral vessels is required during neurosurgical procedures. The mechanism of such agents is to reduce the energy demand of the tissue, consequently its blood flow and blood volume, and hence to reduce ICP. One problem common to all sedative drugs is that the metabolic depressant effects are systemic as well as cerebral and there is always a risk of reduction of blood pressure and cardiac output. In the presence of any pre-existing hypovolemia the reductions in blood pressure may be profound and may themselves impair CBF and brain function. Despite earlier optimistic reports, in three randomised clinical trials of the use of barbiturate therapy in head-injured patients no overall beneficial effect was identified, possibly because the beneficial effects were counter-balanced by the side-effects of therapy, which in one series consisted of arterial hypotension in 50% of the treated patients [18–20].

These risks have become well recognised and barbiturates and other sedative durgs are now used with more circumspection and probably with greater success in a more limited role in controlling raised intracranial pressure in specific patients. The issue of which patients are likely to benefit most from barbiturates is addressed elsewhere in this volume.

Steroid therapy is undeniably beneficial in patients who are suffering from neurological dysfunction and intracranial hypertension related to chronic perifocal edema, most commonly around brain tumors. The very success of steroid therapy under these limited conditions has led to its more widespread application in acute brain disorders. Thus, steroid therapy has been evaluated in acute stroke and acute head injury. Despite a number of trials, no convincing evidence of benefit has been demonstrated [21–23]. A multi-center study of ultra-high dose steroid therapy given very early after severe head injury is still in progress in West Germany.

There is little doubt when raised ICP and brain shift are due to a mass lesion that can be removed, that this is the best treatment. There is, however, much more controversy concerning the role of bony decompression to afford relief for the diffusely swollen brain. The proponents of this procedure argue that the additional space provided gives satisfactory reduction of ICP and diminution of the risks of brain herniation [24]. The counter argument is that the decompressed brain often swells into the defect, obstructs its venous supply and infarcts, producing major neurological deficit [25]. The patients that appear to benefit from surgical decompression are almost always those who would have been satisfactorily managed by other methods. At the present time there is no conclusive evidence for surgical decompression in diffuse brain injury.

Newer Approaches to Control of Raised Intracranial Pressure and Amelioration or Prevention of Ischemic Brain Damage

A number of new approaches are now being evaluated or considered for the prevention or limitation of secondary brain damage in patients. These are directed towards the reduction of intracranial hypertension, improvement of cerebral blood flow to alleviate brain ischemia, or to inhibition or blockade at the molecular and membrane receptor level of processes that lead to neuronal membrane dysfunction and intracellular damage.

New sedative drugs are being evaluated, with the purpose of identifying agents that have less cardio-depressant qualities but retain the capacity to reduce $CMRO_2$, CBF, CBV and ICP. Propofol, diazepam, and chlormethiazole are examples. The problem that is most frequently encountered is that agents with little cardio–depressent effect have correspondingly little effect on ICP. Common to all agents is the fundamental fact that in order for sedative drugs to work, brain electrical activity must be retained. If the EEG is already flat then no form of sedative drug will have an effect upon the ICP.

Cerebral tissue acidosis has been proposed as an important mechanism by which brain ischemia can lead to cell damage. In this volume, the role of bicarbonates in reversing brain acidosis is explored. A trial of the buffering agent, tromethamine (THAM), is in progress in patients with severe head injury in two centers in the United States [26]. This clinical study follows laboratory experiments in which THAM was shown to reduce the mortality that followed experimental fluid percussion brain injury.

The no-reflow phenomenon has been proposed as a mechanism by which temporary reduction in CBF that results from surges in ICP can proceed to cause permanent brain damage because flow does not resume uniformly throughout the brain after the ICP falls and cerebral perfusion pressure is restored. A combination of heparin and urokinase has been proposed in an experimental study reported in this volume by Hashimoto.

Entry of calcium into brain neurons is held by many to represent a final common pathway to the production of disruption of the cell membrane and intracellular structures [27]. At the present time a trial on patients with severe head injury of the calcium channel blocking agent, Nimodipine, is in progress in a number of centers in Europe. A particular concern is that calcium channel blockers may produce cerebral vasodilation, and in the circumstances of reduced brain compliance that may be present in head injured patients, this could, in theory, lead to severe rises in ICP. As of now, this problem does not appear to have arisen.

More recently the excitotoxic hypothesis of brain damage caused by calcium entry via receptor-gated channels has received much attention [27]. Trials of scopolamine and NMDA receptor antagonists have been carried out in experimental models of brain damage. Consideration is presently being given to clinical trials of agents suitable for use in head injured patients.

The liberation of singlet oxygen and other free radical compounds during the arachidonic acid cascade is believed to cause structural damage to cell membranes in the brain [28]. The demonstration that free radical quenching agents can ameliorate at least some of the sequelae of experimental brain injury has led to proposals to mount a clinical trial of superoxide dismutase in certain types of human head injury. Against such a proposal is the observation that both mannitol and barbiturates have some free radical scavenging properties and are already in widespread use in head injury. Furthermore, alcohol is known to increase the liberation of free radical compounds and is commonly present in patients with head injury. Despite some data from Albin that alcohol increases the severity of experiental focal brain contusion, there is no date from human head injury to support the notion that alcohol significantly worsens traumatic brain damage [29].

These examples serve to show that while experimental studies may point the way towards the use of certain physiological approaches and newer drugs to obtain improvement in outcome from head injury in man, considerable caution must be exercised before applying the results of laboratory experiments to therapeutic efforts in the neuro-intensive care unit.

Summary. Brain resuscitation is now recognised as a separate area within the field of critical care, based on an understanding of pathophysiology that derives largely from experimental studies of cerebral ischemia. This has led to the development of certain forms of monitoring that are specific to neurological critical care. These include recording intracranial pressure and measurement of craniospinal compliance, measuring cerebral blood flow and metabolism, and monitoring brain electrical activity. At the same time, computed tomography and magnetic resonance imaging provide detailed information on the structural changes in the brain that correspond to physiological disturbances.

Management of patients with brain disorders consists of three components: provision of a normal physiological milieu, reduction of raised intracranial pressure and, most recently, the prospect of specific drug therapy directed to reversal, at a molecular level, of the processes by which brain ischemia leads to neuronal dysfunction and cell death.

References

1. Guillaume J, Janny P (1951) Manometrie intracranienne continue: Interêt de la méthode et premiers resultats. Rev Neurol 84:131–142
2. Lundberg N (1960) Continuous recording and control of ventricular fluid pressure in neurosurgical practice. Acta Psychiatr Neurol Scand 36 (Suppl 149):1–193
3. Brock M, Dietz H (1972) Intracranial pressure. Springer, Berlin
4. Lundberg N, Ponten U, Brock M (1975) Intracranial pressure II. Springer, Berlin
5. Beks JWF, Bosch DA, Brock M (1976) Intracranial pressure III. Springer, Berlin
6. Shulman K, Marmarou A, Miller JD, Becker DP, Hochwald GM, Brock M (1980) Intracranial pressure IV. Springer, Berlin
7. Ishii S, Nagai H, Brock M (1983) Intracranial pressure V. Springer, Berlin
8. Miller JD, Teasdale GM, Rowan JO, Galbraith SL, Mendelow AD (1986) Intracranial pressure VI. Springer, Berlin
9. Hoff JT, Betz AL (1989) Intracranial pressure VII. Springer, Berlin
10. Bray RS, Sherwood AM, Halter JA, Robertson C, Grossman RG (1986) Development of a clinical monitoring system by means of ICP waveform analysis. In: Miller JD, Teasdale GM, Rowan JO, Galbraith SL, Mendelow AD (eds) Intracranial pressure VI. Springer, Berlin, pp 260–270
11. Takizawa H, Gabra-Sanders T, Miller JD (1986) Spectral analysis of the CSF pulse wave at different locations in the craniospinal axis. J Neurol Neurosurg Psychiatry 49:1135–1141
12. Takizawa H, Gabra-Sanders T, Miller JD (1986) Change of frequency spectrum of the CSF pulse wave caused by supratentorial epidural brain compression. J Neurol

Neurosurg Psychiatry 49:1367–1373
13. Takizawa H, Gabra-Sanders T, Miller JD (1987) Changes in the CSF pulse wave spectrum associated with raised intracranial pressure. Neurosurgery 20:355–361
14. Piper IR, Dearden NM, Miller JD (1989) Can waveform analysis of ICP separate vascular from non-vascular causes of intracranial hypertension? In: Hoff JT, Betz AL (eds) Intracranial pressure VII. Springer, Berlin, pp 157–163
15. Robertson CS, Narayan RK, Gokaslan ZL, Pahwa R, Grossman RG, Caram P, Allen E (1989) Cerebral arteriovenous oxygen difference as an estimate of cerebral blood flow in comatose patients. J Neurosurg 70:222–230
16. Branston NM, Ladds A, Symon L, Wang AD (1984) Comparison of the effects of ischaemia on the early components of the somatosensory evoked potential in the brain stem, thalamus, and cerebral cortex. J Cereb Blood Flow Metab 4:68–81
17. Bell BA, Smith MA, Kean DM, McGhee CNJ, MacDonald HL, Miller JD,Barnett GH, Tocher JL, Douglas RHB, Best JJK (1987) Brain water measured by magnetic resonance imaging: Correlation with direct estimation and changes following mannitol and dexamethasone. Lancet I:66–69
18. Schwartz ML, Tator CH, Rowed DW, Reid SR, Megura K, Andrews DF (1984) The University of Toronto head injury treatment study: A prospective, randomised comparison of pentobarbital and mannitol. Can J Neurol Sci 11:434–440
19. Ward JD, Becker DP, Miller JD, Choi SC, Marmarou A, Wood C, Newlon P, Keenan R (1985) Failure of prophylactic barbiturate coma in the treatment of severe head injury. J Neurosurg 62:383–88
20. Eisenberg HM, Frankowski RF, Contant CF, Marshall LF, Walker MD (1988) High dose barbiturate control of elevated intracranial pressure in patients with severe head injury. J Neurosurg 69:15–23
21. Cooper PR, Moody S, Clark WK, Kirkpatrick J, Maravilla K, Gould AL, Drane W (1979) Dexamethasone and severe head injury: A prospective double blind trial. J Neurosurg 51:307–316
22. Braakman R, Schouten HJA, vanDishoeck BM, Minderhoud JM (1983) Megadose steroids in severe head injury: Results of a prospective double blind clinical trial. J Neurosurg 58:326–330
23. Dearden NM, Gibson JS, McDowall DG, Gibson RM, Cameron MM (1986) Effect of high dose dexamethasone on outcome from severe head injury. J Neurosurg 64:81–88
24. Ransohoff J, Vallo B, Gage EJ, Epstein F (1971) Hemicraniectomy in the management of acute subdural hematomas. J Neurosurg 34:70–76
25. Cooper PR, Rovit RL, Ransohoff J (1976) Hemicraniectomy in the management of acute subdural hematoma: A reappraisal. Surg Neurol 5:25–29
26. Rosner MJ, Elias JG, Coley I (1989) Prospective randomised trial of THAM therapy in severe brain injury: Preliminary results. In: Hoff JT, Betz AL (eds) Intracranial pressure VII. Springer, Berlin, pp 611–615
27. Meldrum B (1990) Protection against ischaemic neuronal damage by drugs acting on excitatory neurotransmission. Cereb Brain Metab Rev 2:27–57
28. Siesjo BK, Agardh DH, Bengtsson F (1989) Free radicals and brain damage. Cereb Brain Metab Rev 1:165–211
29. Albin MS, Bunegin L (1986) An experimental study of craniocerebral trauma during ethanol intoxication. Crit Care Med 14:841–846

Index